U.S CLIMATE CHANGE
TECHNOLOGY PROGRAM

U.S. Department of Energy (Lead-Agency)

U.S. Department of Agriculture

U.S. Department of Commerce, including

National Institute of Standards and Technology

U.S. Department of Defense

U.S. Department of Health and Human Services, including

National Institutes of Health

U.S. Department of Interior

U.S. Department of State, including

U.S. Agency for International Development

U.S. Department of Transportation

U.S. Environmental Protection Agency

National Aeronautics and Space Administration

National Science Foundation

Other Participating Research and Development Agencies

Executive Office of the President, including

Council on Environmental Quality

Office of Science and Technology Policy

Office of Management and Budget

U.S. Climate Change Technology Program
1000 Independence Avenue, SW
Washington, DC 20585

202-586-0070

http://www.climatetechnology.gov

U.S. Climate Change Technology Program

STRATEGIC PLAN

September 2006

To the Reader:

We are pleased to be able to present this *Strategic Plan* for the Climate Change Technology Program (CCTP). The technology strategy detailed in this *Plan* is an essential element of a comprehensive climate change strategy that includes undertaking short-term actions to reduce greenhouse gas emissions intensity, advancing climate science, and promoting international cooperation.

CCTP was created by the President in 2002—and subsequently authorized in the Energy Policy Act of 2005—to coordinate and prioritize the Federal Government's portfolio of investments in climate-related technology research, development, demonstration, and deployment (RDD&D). The portfolio totaled about $3 billion in Fiscal Year 2006.

The *Plan* expands on the themes presented in CCTP's *Vision and Framework for Strategy and Planning*. It provides the underpinnings for a robust RDD&D effort that can make advanced technologies available sooner and at a lower cost. It takes a century-long look at the nature of the climate change challenge and the potential for technological solutions across a range of uncertainties. Most anthropogenic greenhouse gases emitted over the course of the 21st century will come from equipment and infrastructure not yet built, a circumstance that poses significant opportunities to reduce or eliminate these emissions. The technologies outlined in this *Plan*—hydrogen, biorefining, clean coal, carbon sequestration, nuclear fission and fusion, advanced concepts in buildings, industry, transportation and electric energy storage and distribution, and others—have the potential to transform our economy in fundamental ways that can address not just climate change, but energy security, air quality, and other pressing needs.

The *Plan* articulates a vision of the role for advanced technologies, defines a supporting mission for the CCTP, establishes guiding principles for Federal R&D agencies to use in formulating R&D portfolios, outlines approaches to attain CCTP's strategic goals, and identifies a series of next steps toward implementation. We believe this *Plan* will strengthen the U.S. research enterprise and stimulate U.S. innovation and advance technology development in myriad ways. It is our hope that this *Plan* will inspire similar initiatives in other nations and enhance international collaboration on development and deployment of these technologies.

This document is the outcome of a long process involving governmental working groups, expert review, and a public comment period that stimulated thoughtful and energetic dialogue. It is our hope that with publication of the *Plan*, this dialogue will continue to inform and improve the Program.

The United States is working to ensure a bright and secure energy and economic future for our Nation and a healthy planet for future generations. Through a combination of near-term actions, enhanced scientific understanding of climate change, advanced technology development, and international cooperation, this future can become a reality.

Carlos M. Gutierrez
Secretary of Commerce

Chair, Committee on
Climate Change
Science and Technology Integration

Samuel W. Bodman
Secretary of Energy

Vice-Chair, Committee on
Climate Change
Science and Technology Integration

John H. Marburger III, Ph.D.
Director, Office of
Science and Technology Policy
Executive Director, Committee on
Climate Change
Science and Technology Integration

Foreword

I n February 2002, President George W. Bush reorganized the overarching management structure that coordinates and directs U.S. climate change research and development activities. Under this new structure, climate change science and climate-related technology research programs are integrated to an extent not seen previously. The Climate Change Science Program (CCSP), led by the Department of Commerce, was established to reduce the uncertainties in climate science and develop science-based resources to support decision makers. The Climate Change Technology Program (CCTP), the counterpart organization to CCSP, led by the Department of Energy, was formed to coordinate the Federal Government's portfolio of climate-related technology research and development activities, including technology deployment and adoption activities, which were supported by nearly $3 billion in Fiscal Year 2006, and to focus efforts on the subset of priority activities that are part of the President's National Climate Change Technology Initiative.

The CCTP's *Research and Current Activities* and *Technology Options for the Near and Long Term* reports provided detailed looks at, and introduced the public to, an array of technologies with the potential to reduce greenhouse gas emissions. Our goal with this *Strategic Plan* is to provide a long-term planning context, taking into account the many uncertainties, in which the nature of both the challenges and the opportunities for advanced technologies are illuminated and balanced. Along with its short companion document, CCTP's *Vision and Framework for Strategy and Planning*, the *Plan* provides an inspiring vision of what may be possible and a basis for setting priorities for research through its technology strategies and investment criteria. It also highlights those opportunities that are ripe for advancement.

The *Plan* was guided by the leadership of the Cabinet-level Committee on Climate Change Science and Technology Integration and its Interagency Working Group of agency deputies. It was prepared by an interagency team of six working groups, under the direction of CCTP Deputy Director Dr. Robert C. Marlay. Experts from many different disciplines have made significant contributions to the *Plan*. Without their efforts, this document would not have been possible.

Further, the *Plan* has benefited greatly from hundreds of comments received during the public comment period following release of the draft *Plan* in September 2005. We have been gratified by the quantity and quality of the comments we received, which reflect the importance of the Program. We were able to accommodate most comments, but even those we did not accept challenged our thinking and made the *Plan* stronger. We thank all of those individuals and groups who took time to make comments. As the *Plan* is forward-looking, we expect that public input will be an aspect of future updates to the *Plan*.

Stephen D. Eule
Director
U.S. Climate Change Technology Program

September 2006

U.S. Climate Change Technology Program

Strategic Plan

TABLE OF CONTENTS

FIGURES

TABLES

BOXES

ACRONYMS AND ABBREVIATIONS

ABR	Advanced Burner Reactor	CCCSTI	Committee on Climate Change Science and Technology Integration
AFCI	Advanced Fuel Cycle Initiative		
AFV	Alternative Fuel Vehicles	CCP	Carbon Capture Project
AGAGE	Advanced Global Atmospheric Gases Experiment	CCS	Carbon Capture and Sequestration
		CCSP	U.S. Climate Change Science Program
ANL	Argonne National Laboratory	CCTP	U.S. Climate Change Technology Program
APS	Aerosol Polarimetery Sensor		
AUV	Autonomous Underwater Vehicles	CDIAC	Carbon Dioxide Information Analysis Centre
BC	Black Carbon	CEM	Continuous Emissions Monitor
BES	Office of Basic Energy Sciences, U.S. Department of Energy	CETC	Natural Resources Canada CANMET Energy Technology Center
BESAC	Basic Energy Sciences Advisory Committee	CEQ	Council on Environmental Quality
		CFC	Chlorofluorocarbon
BP	British Petroleum	CH_4	Methane
Btu	British Thermal Unit	CHP	Combined Heat and Power (system)
		CMM	Coal Mine Methane

CMOP	Coalbed Methane Outreach Program	**GtC-eq.**	Gigatonnes (10^9 tonnes or metric tons) of Carbon Equivalent (emissions)
CO$_2$	Carbon Dioxide		
COL	Construction and Operating License	**GWP**	Global Warming Potential
CSLF	Carbon Sequestration Leadership Forum		
CSP	Competitive Solicitation Program	**H$_2$**	Molecular Hydrogen
CSRP	Carbon Sequestration Regional Partnerships	**H$_2$S**	Hydrogen Sulfide
		HAP	Hazardous Air Pollutants
CT	Computed Tomography	**HCFC**	Hydroclorofluorocarbon (refrigerant)
CVD	Chemical Vapor Deposition	**HFC**	Hydrofluorocarbon
		HHS	U.S. Department of Health and Human Services
DAI	Dangerous Anthropogenic Interference		
DG	Distributed Generation	**HNLC**	High Nutrient, Low Chlorophyll
DOC	U.S. Department of Commerce	**HSHL**	High Spectral Resolution LIDAR
DoD	U.S. Department of Defense	**HTS**	High-Temperature Superconductivity (e.g. wire)
DOE	U.S. Department of Energy		
DOI	U.S. Department of the Interior	**HVACR**	Heating, Ventilation, Air Conditioning, and Refrigeration
DOS	U.S. Department of State		
DOT	U.S. Department of Transportation	**HVDC**	High Voltage Direct Current
DPC	Domestic Policy Council		
		IAEA	International Atomic Energy Agency
ECBM	Enhanced Coal-Bed Methane	**ICF**	Inertial Confinement Fusion
EIA	Energy Information Administration	**IEA**	International Energy Agency
EJ	Exajoule (10^{18} Joules)	**IEOS**	Integrated Earth Observation System
EMF	Energy Modeling Forum, Stanford University	**IFE**	Inertial Fusion Energy
		IGCC	Integrated Gasification Combined Cycle
EOR	Enhanced Oil Recovery	**IMSS**	Image Multi-Spectral Sensor
EPA	U.S. Environmental Protection Agency	**IPCC**	Intergovernmental Panel on Climate Change
ESP	Early Site Permit		
Euratom	European Atomic Energy Community	**IPHE**	International Partnership for the Hydrogen Economy
EU	European Union		
FACE	Free-Air CO$_2$ Enrichment	**ITER**	International Thermonuclear Experimental Reactor (Latin for "the way")
FACTS	Flexible Automated Control Transmission Systems		
		ITS	Intelligent Transportation Systems
FC	Fuel Cell	**IWG**	Interagency Working Group
FCT	Fuel Cell Turbine		
FES	Fusion Energy Sciences, U.S. Department of Energy, Office of Science	**kg**	Kilogram
		kW	Kilowatt
FFRDC	Federally Funded Research and Development Center	**kWe**	Kilowatt (electric)
		kWh	Kilowatt-hour
FHA	Federal Highway Administration		
FTC	Federal Trade Commission	**LANL**	Los Alamos National Laboratory
FTIR	Fourier Transform Infrared Spectroscopy	**LCCP**	Life-Cycle Climate Performance
FY	Fiscal Year	**LED**	Light-Emitting Diode
GDP	Gross Domestic Product	**LFG**	Landfill Gas
Gen IV	Generation IV	**LH$_2$**	Liquefied Hydrogen
GEO	Group on Earth Observations	**LIBS**	Laser Induced Breakdown Spectroscopy
GEO-SEQ	Geological Sequestration (project)	**LIDAR**	Light Detection and Ranging
		LNLC	Low Nutrient, Low Chlorophyll
GEOSS	Global Earth Observation System of Systems		
		MEA	Monoethanolamine
GHG	Greenhouse Gas	**MFE**	Magnetic Fusion Energy
GIF	Generation IV International Forum (nuclear power)	**MiniCAM**	Mini Climate Assessment Model (Pacific Northwest National Laboratory)
GOSAT	Greenhouse Gas Observing SATellite	**MM**	Measuring and Monitoring
Gt	Gigatonnes (10^9 tonnes or metric tons)	**MOF**	Microporous Metal Organic Frameworks
		mpg	miles per gallon
GtC	Gigatonnes (10^9 tonnes or metric tons) of Carbon	**mph**	miles per hour

MSPI	Multi-angle SpectroPolarimetric Imager
MtC	Megatonnes Carbon
MWe	Megawatt electric
N_2O	Nitrous Oxide
NACP	North American Carbon Program
NAE	National Academy of Engineering
NAS	National Academy of Sciences
NASA	National Aeronautics and Space Administration
NEC	National Economic Council
NEPO	Nuclear Energy Plant Optimization (Program)
NCCTI	National Climate Change Technology Initiative
NERAC	Nuclear Energy Research Advisory Committee
NETL	National Energy Technology Laboratory
NH_3	Ammonia
NIF	National Ignition Facility
NNSA	National Nuclear Security Administration, U.S Department of Energy
NO_X	Nitrogen Oxides
NOAA	National Oceanic and Atmospheric Administration
NRC	National Research Council or Nuclear Regulatory Commission
NRCan	Natural Resources Canada
NREL	National Renewable Energy Laboratory
NSC	National Security Council
NSCR	Non-Selective Catalytic Reduction
NSF	National Science Foundation
NSTX	National Spherical Torus Experiment
NVFEL	National Vehicle and Fuels Emission Laboratory
OC	Organic Carbon
OCO	Orbiting Carbon Observatory
ODS	Ozone-Depleting Substance
OMB	Office of Management and Budget
ONR	Office of Naval Research
ORNL	Oak Ridge National Laboratory
OSTP	Office of Science and Technology Policy
PEM	Polymer Electrolyte Membrane
PFC	Perfluorocarbons
PM	Particulate Matter
PNNL	Pacific Northwest National Laboratory
POU	Point Of Use
PPPL	Princeton Plasma Physics Laboratory
PV	Present Value

Quad	Quadrillion Btus (10^{15} Btus)
R&D	Research and Development. Also used generically to mean RD&D and RDD&D
RD&D	Research, Development, and Demonstration
RDD&D	Research, Development, Demonstration, & Deployment
RFI	Request for Information
SCR	Selective Catalytic Reduction
SF_6	Sulfur Hexafluoride
SNAP	Significant New Alternatives Program
SOFeX	Southern Ocean Iron Fertilization Experiment
SOIREE	Southern Ocean Iron Enrichment Experiment
SO_X	Sulfur Oxides
SQUIDS	Superconducting Quantum Interference Devices
SRES	Special Report on Emissions Scenarios (of the IPCC)
T&D	Transmission and Distribution
TgC	Teragrams of Carbon
Tg CO_2	Teragrams Carbon Dioxide
Tg CO_2-eq.	Teragrams Carbon Dioxide Equivalent (emissions)
UN	United Nations
UNDP	United Nations Development Program
UNEP	United Nations Environmental Program
UNFCCC	United Nations Framework Convention on Climate Change
USAID	U.S. Agency for International Development
USDA	U.S. Department of Agriculture
USGEO	United States Group on Earth Observation
VAM	Ventilation Air Methane
VOC	Volatile Organic Compounds
W/m^2	Watts per Square Meter
WCRP	World Climate Research Program
WG	Working Group
WMO	World Meteorological Organization
WOCE	World Ocean Circulation Experiment
WRE	T. Wigley, R. Richels, and J. Edmonds

Introduction

The 21st century will see substantial changes in economic and social development around the world, with accompanying transformations in the way the world uses energy and its natural resources. The 20th century witnessed revolutionary innovations in technologies used to produce goods and services, power homes and buildings, and transport people and goods. These innovations have contributed significantly to the prosperity that the United States and many other countries currently enjoy. Continued innovations will be just as important in making possible a prosperous future for countries around the world as they are for enabling sound stewardship of the environment, including the Earth's climate system (Figure 1-1).

As a party to the United Nations Framework Convention on Climate Change (UNFCCC),[1] the United States shares with many other countries the UNFCCC's ultimate objective, that is, the "...stabilization of greenhouse gas[2] concentrations in Earth's atmosphere at a level that would prevent dangerous anthropogenic interference with the climate system . . . within a time-frame sufficient to allow ecosystems to adapt naturally to climate change, to ensure that food production is not threatened, and to enable economic development to proceed in a sustainable manner." Meeting this objective will require a sustained, long-term commitment by all nations over many generations.

Figure 1-1. Courtesy: NASA, Hasler Laboratory for Atmospheres Goddard Space Flight Center, *Credit: Nelson Stockli*

I've asked my advisors to consider approaches to reduce greenhouse gas emissions, including those that tap the power of markets, help realize the promise of technology and ensure the widest-possible global participation….Our actions should be measured as we learn more from science and build on it. Our approach must be flexible to adjust to new information and take advantage of new technology. We must always act to ensure continued economic growth and prosperity for our citizens and for citizens throughout the world.

PRESIDENT BUSH (6/11/01)

---

[1] The UNFCCC was adopted by 157 countries in 1992; as of May 24, 2004, 189 Parties, including the European Economic Community, had ratified the UNFCCC.

[2] Greenhouse gases (GHGs) are those gaseous constituents of the atmosphere, both natural and anthropogenic, that absorb and emit radiation at specific wavelengths within the spectrum of infrared radiation emitted by the Earth's surface, the atmosphere and clouds. This property causes the greenhouse effect. Water vapor, carbon dioxide (CO_2), nitrous oxide (N_2O) methane (CH_4), and ozone (O_3) are the primary GHGs in the Earth's atmosphere. Moreover, there are a number of entirely human-made GHGs in the atmosphere, such as the halocarbons and other chlorine- and bromine-containing substances, dealt with under the Montreal Protocol. Besides CO_2, N_2O, and CH_4, the Kyoto Protocol deals with the GHGs sulphur hexafluoride (SF_6), hydrofluorocarbons (HFCs), and perfluorocarbons (PFCs). Gases dealt with under the Montreal Protocol are excluded from the Climate Change Technology Program (CCTP) purview.

Although scientific understanding of climate change continues to evolve, the potential ramifications of increasing accumulations of carbon dioxide (CO_2) and other greenhouse gases (GHGs) in the Earth's atmosphere have heightened attention on anthropogenic sources of GHG emissions and various means to mitigate them. Most long-term, prospective analyses project significant increases of anthropogenic GHG emissions over the next century, stemming primarily from global population growth, economic expansion, and a continuation of existing patterns and trends in energy use (combustion of fossil fuels), land use, and industrial and agricultural production. Energy is the biggest source of emissions. Over 80 percent of current anthropogenic GHG emissions are energy related, and although projections vary considerably, a tripling of global energy demand by 2100 is not unimaginable. The International Energy Agency (2004) estimates that 1.6 billion people lack access to electricity. Governments around the world are working to ensure that their people have access to energy to power economic development.

A realistic climate change policy, therefore, must embrace these and other legitimate concerns. Indeed, the most effective way to meet this challenge is not to focus solely on GHG emissions, but on a broader agenda that promotes economic growth, provides energy security, reduces pollution, and mitigates GHG emissions. It is within this context that U.S climate change policy has been developed. Accordingly, the United States places special emphasis on the fundamental importance of science and technology as a means of achieving climate goals in ways that support these other societal goals. More than $25 billion has been so invested since 2001. Ultimately, meeting these complementary goals may entail fundamental changes in the way the world produces and consumes energy, operates industrial enterprises, grows food and fiber, and manages and uses its land.

The United States has established and implemented a robust and flexible climate change policy that harnesses the power of markets and technological innovation, uses the best available science, maintains economic growth, and encourages global participation. Major elements of this approach include implementing policies and measures to slow the growth in GHG emissions, advancing climate change science, accelerating technology development, and promoting international collaboration.

For the near term, the President has set a national goal to reduce the GHG emissions intensity of the U.S. economy by 18 percent between 2002 and 2012.[3] To this end, the Administration has developed an array of policy measures, including financial incentives, voluntary programs, and other Federal efforts. These include the Climate VISION,[4] Climate Leaders,[5] ENERGY STAR®,[6] and SmartWay Transport Partnership[7] programs, all of which work with industry to voluntarily reduce emissions. The Department of Energy encourages entity-wide emissions reductions through its Voluntary Reporting of Greenhouse Gases program, which was authorized under section 1605 of the Energy Policy Act of 1992. The Department of Agriculture's conservation programs provide incentives for actions that increase carbon sequestration[8] in trees and soils. Energy efficiency, alternative fuels, renewable and nuclear energy, methane capture, and other GHG reduction programs and financial incentives are also underway.

The Energy Policy Act of 2005, which the President signed into law in August 2005, also promotes clean energy technologies. The Act authorizes approximately $11 billion (net) in tax credits over 10 years for a broad range of clean technologies, including those associated with energy efficiency and conservation, renewable energy, clean vehicles and alternative fuels, clean coal, nuclear power, and other technologies, all of which can potentially contribute to reducing GHG emissions. It also: authorizes loan guarantee programs that may help to accelerate the commercialization of advanced energy technologies with potential to reduce GHG emissions in the future; mandates the use of renewable fuels (such as ethanol and biodiesel) in gasoline, increasing their use from 4.0 billion gallons in 2006 to 7.5 billion gallons in 2012; authorizes standby support coverage for certain regulatory delays for up to six new nuclear plants; and authorizes the setting of efficiency standards for more than a dozen additional energy-using products.

---

3 Intensity means emissions per unit of economic output. See White House Fact Sheet on Climate Change, www.whitehouse.gov/news/releases/2003/09/20030930-11.html.

4 See http://www.climatevision.gov.

5 See http://www.epa.gov/climateleaders.

6 See http://www.energystar.gov.

7 See http://www.epa.gov/smartway.

8 See http://www.usda.gov/news/releases/2003/06/fs-0194.htm.

Internationally, the Administration believes that well-designed multilateral collaborations focused on achieving practical results can accelerate development and commercialization of new technologies, and the United States has brought together key nations to jointly tackle some tough energy challenges. These multilateral collaborations in hydrogen, carbon sequestration, nuclear power, methane recovery and use, and fusion energy mirror the main strategic thrusts of our domestic technology research programs. They address a number of complementary energy concerns, such as energy security, climate change, and environmental stewardship. These programs are discussed in greater detail in Chapter 2.

The technology-focused approach that puts climate change in the context of broader development goals is gaining adherents in many parts of the world. In July 2005, the Group of Eight Leaders meeting at Gleneagles, Scotland agreed to a Plan of Action on Climate Change, Clean Energy, and Sustainable Development[9] that is based on over fifty specific, practical activities, mostly focused on technology development and deployment. Later that same month, the United States, Australia, China, India, Japan, and South Korea announced they were joining together to accelerate clean development under the new Asia-Pacific Partnership on Clean Development and Climate (APP).[10] The focus of APP will be on helping each country meet nationally-designed strategies for improving energy security, reducing harmful pollution, promoting economic development, and addressing the long-term challenge of climate change.

This integrated approach, which supports the UNFCCC objective, forms the long-term planning environment in which this *Plan* was developed. Significant progress toward meeting the climate change goals can be facilitated over the course of the 21st century by new and revolutionary technologies that can reduce, avoid, capture, or sequester GHG emissions, while also continuing to provide the energy-related and other services needed to sustain economic growth. The U.S. strategy for developing these technologies, both in the near and long term, is outlined in this *Plan*, which builds on America's strengths in innovation and technology.[11] The United

States is committed to leading the development of these new technologies.

The *Plan* takes a century-long look at the nature of this challenge, across a range of planning uncertainties, and explores an array of opportunities for technological solutions.[12] The *Plan* articulates a vision for new and advanced technology in addressing climate change concerns, defines a supporting planning and coordination mission, and provides strategic direction to the Federal agencies in formulating a comprehensive portfolio of related technology research, development, demonstration, and deployment (R&D).[13] The *Plan* establishes six strategic goals and seven approaches to be pursued toward their attainment and identifies a series of next steps toward implementation.

1.1 U.S. Leadership and Presidential Commitment

Soon after assuming office, the President initiated a cabinet-level climate change policy review and directed that innovative approaches for addressing climate change concerns be developed in accordance with a number of basic principles. Specifically, the approaches should: (1) be consistent with the long-term goal of stabilizing GHG concentrations in the atmosphere; (2) be measured as we learn more from science and build on it; (3) be flexible to adjust to new information and take advantage of new technology; (4) ensure continued economic growth and prosperity; (5) pursue market-based incentives and spur technological innovation; and (6) be based on global participation, including developing countries.

In June 2001, the Administration released an interim report of the Cabinet-level climate change working group, and President Bush unveiled, among other initiatives, the National Climate Change Technology Initiative (NCCTI).[14] Backed by significant levels of Federal investment in climate change R&D and

---

9 See http://www.whitehouse.gov/news/releases/2005/07/20050708-2.html.

10 See http://www.whitehouse.gov/news/releases/2005/07/20050727-9.html.

11 See technology roadmaps as described in Chapters 4, 5, 6, 7, 8, and 10.

12 To achieve GHG stabilization, emission reductions must continue beyond the 100-year CCTP planning horizon.

13 Throughout this report, the use of the term "R&D" is meant generally to include research, development, demonstration, and technology adoption programs. However, where relevant, the report distinguishes research and development from demonstration and deployment, as each activity has different rationales, different appropriate roles for the private sector, and different associated policy instruments.

Cabinet-Level Committee on Climate Change Science and Technology Integration

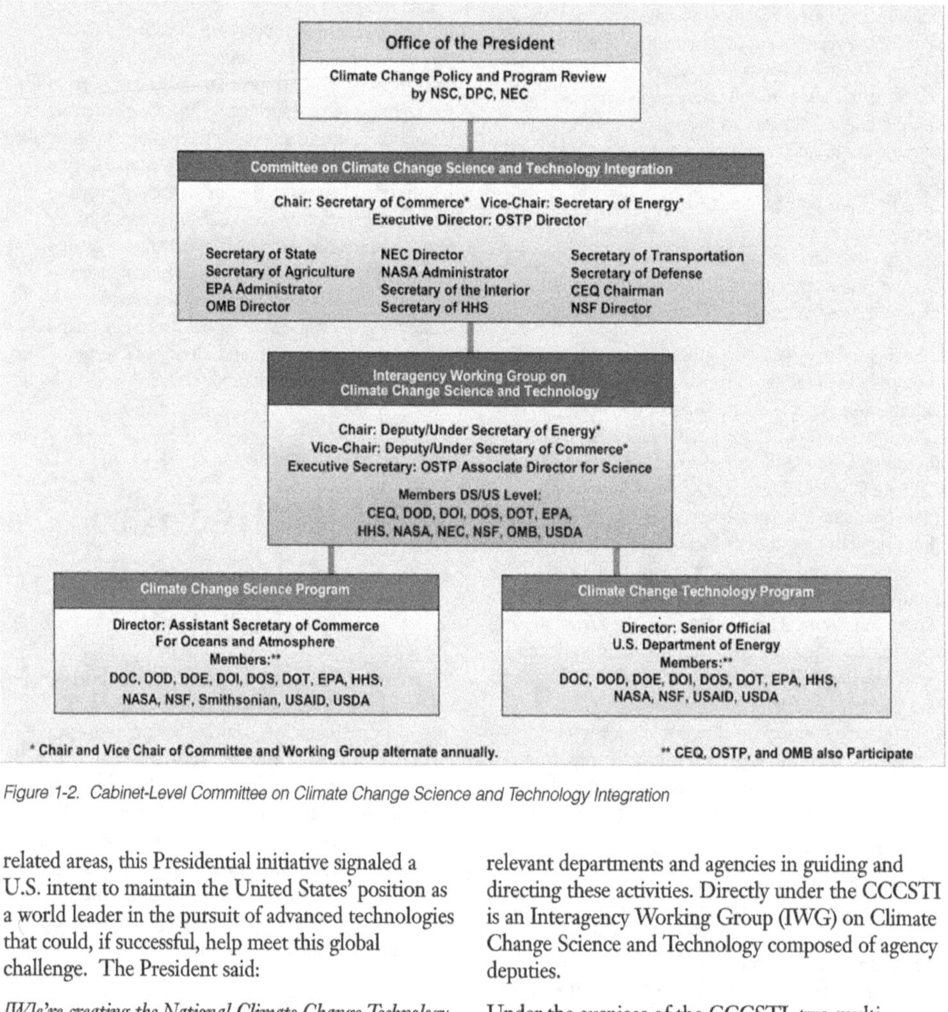

Figure 1-2. Cabinet-Level Committee on Climate Change Science and Technology Integration

related areas, this Presidential initiative signaled a U.S. intent to maintain the United States' position as a world leader in the pursuit of advanced technologies that could, if successful, help meet this global challenge. The President said:

[W]e're creating the National Climate Change Technology Initiative to strengthen research at universities and national labs, to enhance partnerships in applied research, to develop improved technology for measuring and monitoring gross and net greenhouse gas emissions, and to fund demonstration projects for cutting-edge technologies.

In February 2002, the President reorganized Federal oversight, management, and administrative control of climate-change-related activities. He established a Cabinet-level Committee on Climate Change Science and Technology Integration (CCCSTI) and charged it with coordinating and advancing, in an integrated fashion, climate change science and technology research. This action directly engaged the heads of all

relevant departments and agencies in guiding and directing these activities. Directly under the CCCSTI is an Interagency Working Group (IWG) on Climate Change Science and Technology composed of agency deputies.

Under the auspices of the CCCSTI, two multi-agency programs were established to coordinate Federal activities in climate change scientific research and advance the President's vision under his Climate Change Research Initiative and National Climate Change Technology Initiative. These are known, respectively, as the U.S. Climate Change Science Program (CCSP), led by the Department of Commerce, and the U.S. Climate Change Technology Program (CCTP), led by the Department of Energy (Figure 1-2).

---

[14] White House Rose Garden speech: www.whitehouse.gov/news/releases/2001/06/20010611-2.html.

1.2 U.S. Climate Change Science Program

CCSP is an interagency research planning and coordinating entity responsible for facilitating the development of a strategic approach to Federally supported research, integrated across the participating agencies. Collectively, the activities under CCSP constitute a comprehensive research program charged with investigating natural and human-induced changes in the Earth's global environmental system, monitoring important climate parameters, predicting global change, and providing a sound scientific basis for national and international decision-making. Its principal aim is to improve understanding of climate change and its potential consequences. CCSP operates under the direction of the Assistant Secretary of Commerce for Oceans and Atmosphere and reports through the IWG to the CCCSTI (Figure 1-2).

Regarding climate change science, on May 11, 2001, the President asked the National Academies National Research Council (NRC) to examine the state of knowledge and understanding of climate change. The resulting NRC report concluded that "the changes observed over the last several decades are most likely due to human activities, but we cannot rule out that some significant part of these changes is also a reflection of natural variability." The report also noted that there are still major gaps in our ability to measure the impacts of GHGs on the climate system. Major advances in understanding and modeling of the climate system, including its response to natural and human-induced forcing, and modeling of the factors that influence atmospheric concentrations of GHGs and aerosols, as well as the feedbacks that govern climate sensitivity, are needed to predict future climate change with greater confidence.

In July 2003, CCSP released its strategic plan[15] for guiding climate research. The plan is organized around five goals: (1) improving the knowledge of climate history and variability; (2) improving the ability to quantify factors that affect climate; (3) reducing uncertainty in climate projections; (4)

improving the understanding of the sensitivity and adaptability of ecosystems and human systems to climate change; and (5) exploring options to manage risks. In Fiscal Year 2005, the Federal Government spent about $2 billion on research related to advancing climate change science.[16]

A subsequent NRC review[17] of the CCSP strategic plan concluded that the Administration is on the right track, stating that the plan "articulates a guiding vision, is appropriately ambitious, and is broad in scope." The NRC's report also identified the need for a broad global observation system to support measurements of climate variables.

In June 2003, the United States hosted more than 30 nations at the inaugural Earth Observation Summit, which resulted in a commitment to establish an intergovernmental, comprehensive, coordinated, and sustained Earth observation system.[18] The data collected by the system will be used for multiple societal benefit areas, including better climate models, improved knowledge of the behavior of CO_2 and aerosols in the atmosphere, and the development of strategies for carbon sequestration.

Since that initial meeting, two additional ministerial summits have been held, and the intergovernmental partnership has grown to nearly 60 nations. At the most recent meeting, Earth Observation Summit III

---

15 See http://www.climatescience.gov/Library/stratplan2003/final/default.htm.

16 See Appendix A and http://www.usgcrp.gov/usgcrp/Library/ocp2004-5/default.htm.

17 See http://books.nap.edu/catalog/10139.html.

18 See http://www.earthobservationsummit.gov.

in Brussels, a Ten-Year Implementation Plan for the Global Earth Observation System of Systems (GEOSS) was adopted, and the intergovernmental Group on Earth Observations was established to begin implementation of the 2-, 6-, and 10-year targets identified in the plan. The U.S. contribution to GEOSS is the Integrated Earth Observation System (IEOS). In April 2005, the USG Committee on Environment and Natural Resources (CENR) released the *Strategic Plan for the U.S. Integrated Earth Observation System*[19] that addresses the policy, technical, fiscal, and societal benefit components of this integrated system, and established the U.S. Group on Earth Observation (USGEO).

1.3 U.S. Climate Change Technology Program

CCTP is the technology counterpart to CCSP. It is a multi-agency planning and coordinating entity, led by the Department of Energy, aimed at accelerating the development of new and advanced technologies to address climate change. It works with participating agencies (Table 1-1), provides strategic direction for the CCTP-related elements of the overall Federal R&D portfolio, and facilitates the coordinated

Federal Agencies Participating in the U.S. Climate Change Technology Program and Examples of Related Activities

AGENCY*	SELECTED EXAMPLES OF CLIMATE CHANGE-RELATED TECHNOLOGY R&D ACTIVITIES
DOC	Instrumentation, standards, ocean sequestration, decision support tools
DoD	Aircraft, engines, fuels, trucks, equipment, power, fuel cells, lasers, energy management, basic research
DOE	Energy efficiency, renewable energy, nuclear fission and fusion, fossil fuels and power, carbon sequestration, basic energy sciences, hydrogen, electric grid and infrastructure
DOI	Land, forest, and prairie management, mining, sequestration, geothermal, terrestrial sequestration technology development
DOS	International science and technology cooperation, oceans, environment
DOT	Aviation, highways, rail, freight, maritime, urban mass transit, transportation systems, efficiency and safety
EPA	Mitigation of CO_2 and non-CO_2 GHG emissions through voluntary partnership programs, including ENERGY STAR®, Climate Leaders, Green Power, combined heat and power, state and local clean energy, methane and high-GWP gases, and transportation; GHG emissions inventory
HHS	Environmental sciences, biotechnology, genome sequencing, health effects
NASA	Earth observations, measuring, monitoring, aviation equipment, operations and infrastructure efficiency
NSF	Geosciences, oceans, nanoscale science and engineering, computational sciences
USAID	International assistance, technology deployment, land use, human impacts
USDA	Carbon fluxes in soils, forests and other vegetation, carbon sequestration, nutrient management, cropping systems, forest and forest products management, livestock and waste management, biomass energy and bio-based products development

Agency titles for the acronyms above are shown in the list of Abbreviations and Acronyms

Table 1-1. Federal Agencies Participating in the U.S. Climate Change Technology Program and Examples of Related Activities.

[19] See http://iwgeo.ssc.nasa.gov.

planning, programming, budgeting, and implementation of the technology development and deployment aspects of U.S. climate change strategy, including advancing the President's NCCTI. The CCTP operates under the direction of a senior-level official at the Department of Energy and reports through the IWG to the CCCSTI.

The Potential Role of Technology

Analyses documented in the literature (see Chapter 3) show that accelerated advances in technology have the potential to facilitate progress towards meeting climate change goals and, under certain assumptions, to significantly reduce the cost of such progress over the course of the 21st century, compared to what otherwise would be the case without accelerated advances in technology.[20] Further, it is expected that the new technologies would create substantial opportunities for economic growth.

CCTP aims to achieve a balanced and diversified portfolio, including a broad range of deployment activities focusing on: energy-efficiency enhancements; low-GHG-emission energy supply technologies; carbon capture, storage, and sequestration methods; and technologies to reduce emissions of non-CO_2 gases. Conducting this R&D will help reduce technology risk and improve the prospects that such advanced technologies can be adapted to market realities, better positioning them for eventual commercialization.

Together, CCSP and CCTP will help lay the foundation for future progress. Advances in climate change science under CCSP can be expected to improve the knowledge of climate change and its potential impacts. As a result, uncertainties about the causes and effects of climate change and increasing concentrations of GHGs will be better understood, as will the potential benefits and risks of various courses of action.

Similarly, advances in climate change technology under the CCTP can be expected to bring forth an expanded array of advanced technology options, at reasonable costs, that can meet a range of societal needs, including reducing GHG emissions. The pace and scope of needed technology change will be driven partially by future trends in GHG emissions that are uncertain. The complex relationships among

population growth; economic development; energy demand, mix, and intensity; resource availability; technology; and other variables make it difficult to predict with confidence future GHG emissions on a 100-year time scale. Progress in the CCSP will provide much of the information needed to guide and pace future decisions about climate change mitigation. CCTP will provide the means for enabling and facilitating that progress.

Three publications issued by the CCTP provide more information about CCTP and related technologies in the CCTP R&D portfolio (see Appendix A). The *Vision and Framework for Strategy and Planning* provides strategic direction and guidance to the Federal agencies developing new and advanced global climate change technologies. The *Research and Current Activities* report provides an overview of the science, technology, and policy initiatives that make up the Administration's climate change technology strategy. And the CCTP report, *Technology Options for the Near and Long Term*, provides details on the 85 technologies in the R&D portfolio.[21] (Figure 2-1)

1.4 Continuing Process

The United States, in partnership with others, is now embarked on a near- and long-term global challenge, guided by science and facilitated by advanced technology, to address concerns about climate change and increasing concentrations of GHGs. This CCTP *Strategic Plan* is a first step toward guiding Federal investments in R&D to accelerate technologies that will address these concerns. It is hoped that this *Plan* will form the basis for continuing dialogue with the public and interested partners. The *Plan* will be updated periodically, as needed. As noted earlier, the *Plan* is but one component of a comprehensive approach to climate change, which includes policy measures, financial incentives, and voluntary and other Federal programs aimed at slowing the growth of U.S. GHG emissions and reducing GHG intensity. To the extent that technology development needs to be complemented by additional and supporting policies and measures to spur adoption, CCTP intends to evaluate a number of options along these lines (see "Next Steps" in Chapter 10).

---

[20] For example, see Battelle (2000) and IPCC (2000).

[21] All three documents are available at www.climatetechnology.gov. The internet-based version of the report on Technology Options is updated periodically.

1.5 References

Battelle Memorial Institute. 2000. *Global energy technology strategy: addressing climate change.* Washington, DC: Battelle Memorial Institute. http://www.pnl.gov/gtsp/docs/infind/cover.pdf

Intergovernmental Panel on Climate Change (IPCC). 2000. *Special report on emissions scenarios.* Cambridge, UK: University Press. http://www.grida.no/climate/ipcc/emission/index.htm.

Intergovernmental Panel on Climate Change (IPCC). 2001a. *Climate change 2001: mitigation.* Working Group III to the Third Assessment Report. Cambridge, UK: Cambridge University Press. http://www.grida.no/climate/ipcc_tar/wg3/index.htm.

Intergovernmental Panel on Climate Change (IPCC). 2001b. *Climate change 2001: the scientific basis.* Working Group I to the Third Assessment Report. Cambridge, UK: Cambridge University Press. http://www.grida.no/climate/ipcc_tar/wg1/index.htm.

International Energy Agency (IEA). 2004. *World energy outlook* 2004. Paris, France: Organization for Economic Cooperation and Development.

National Research Council. 2004. *Implementing climate and global change research: a review of the final U.S. Climate Change Science Program Strategic Plan.* Washington, DC: National Academies Press.

National Research Council. 2001. *Climate change science: an analysis of some key questions.* Committee on the Science of Climate Change. Washington, DC: National Academy Press.

The White House. June 11, 2001. *Climate change policy review – initial report.* www.whitehouse.gov/news/releases/2001/06/climatechange.pdf.

Vision, Mission, Goals, and Approaches

The Climate Change Technology Program (CCTP) is a multi-agency planning and coordination activity, led by the Department of Energy, which organizes and supports an associated portfolio of Federal R&D.[1] It was established under the Committee on Climate Change Science and Technology Integration in 2002 and subsequently authorized in the Energy Policy Act of 2005.

CCTP constitutes the technology component of a comprehensive U.S. approach to climate change that includes undertaking short-term actions to reduce greenhouse gas (GHG) emissions intensity, advancing climate change science, and promoting international cooperation. CCTP's purpose is to accelerate the development, and reduce the cost of, new and advanced technologies, as well as promote the deployment of advanced technologies and best practices that could avoid, reduce, or capture and store GHG emissions. CCTP was established by President Bush to implement his National Climate Change Technology Initiative (NCCTI) and coordinate existing efforts. This initiative brings to bear America's strengths in innovation and technology to address climate change concerns.

Figure 2-1. CCTP has a number of publications that are available at http://www.climatetechnology.gov.

CCTP provides strategic direction and leadership through interagency coordination of R&D planning, programming, and budgeting. To support these functions, CCTP conducts analyses, technology assessments, and progress reviews. CCTP's interagency working groups have representatives from all participating agencies, and this *Plan* is an important product of this interagency effort.

The focus of this *Plan* is technology research, development, demonstration, and deployment, and it provides a broad roadmap to the future. It does not, however, provide detailed roadmaps for specific technologies. Such roadmaps are referenced in the appropriate sections throughout the document. Moreover, the *Plan* does not, nor is it intended to, provide a comprehensive mitigation strategy. It is not a policy document. It does not, for example, address technology incentives (e.g., tax credits) or regulation (e.g., energy efficiency standards), but it does set a course for evaluating such policies. Further, the *Plan*

makes no judgments as to what constitutes a dangerous level of GHGs in the atmosphere. For century-long planning, the technology portfolio must be designed to be robust in the face of a range of plausible concentration target scenarios. Finally, the *Plan* is not a budget document, but provides a framework to set budget priorities.

The *Plan* was prepared by an interagency team and includes the contributions of a broad range of government and non-government experts from many different disciplines. It is a "first-of-its-kind" document that provides a comprehensive, long-term look at the nature of the climate change challenge and defines clear and promising roles for advanced technologies. CCTP has also published other reports (Figure 2-1).

This chapter outlines the vision, mission, strategic goals, core approaches, R&D prioritization process, and management of the program.

---

[1] Throughout this report the use of the term "R&D" is meant generally to include research, development, demonstration, and technology adoption programs. See also Footnote 14, Chapter 1.

2.1 Vision and Mission

CCTP seeks to attain, in partnership with others, a technological capability that could provide on a global scale abundant, clean, secure, and affordable energy and other services needed to power economic growth, while simultaneously achieving substantial reductions in emissions of GHGs (CCTP Vision). With leadership in R&D and progress in technology development, CCTP aims to inspire broad interest inside and outside of government and internationally, in an expanded global effort to develop and commercialize advanced technologies.

THE CCTP VISION

is to attain on a global scale, in partnership with others, a technological capability that can provide abundant, clean, secure, and affordable energy and other services needed to encourage and sustain economic growth, while simultaneously achieving substantial reductions in emissions of GHGs and mitigating the potential risks of climate change and increasing GHG concentrations.

THE CCTP MISSION

is to stimulate and strengthen the scientific and technological enterprise of the United States, through improved coordination and prioritization of multi-agency Federal climate change technology R&D programs and investments, and to provide global leadership, in partnership with others, aimed at accelerating development and facilitating adoption of technologies that can attain the CCTP vision.

CCTP will strive to stimulate and strengthen the scientific and technological enterprise of the United States, through improved coordination and prioritization of multi-agency Federal climate change technology R&D programs and investments. CCTP will provide support to decision-makers so that they can address issues, make informed decisions, weigh priorities on related science and technology matters, and provide strategic direction. It will do this through recommendations based on multi-agency planning, portfolio reviews, interagency coordination, technical assessments, and other analyses. CCTP will also continue to work with and support the participating agencies in developing plans and carrying out activities needed to achieve the CCTP's vision and strategic goals (CCTP Mission).

2.2 Strategic Goals

The ultimate objective of the U.N. Framework Convention on Climate Change—stabilizing GHG concentrations at a level that would prevent dangerous anthropogenic interference—provides an important (though not the only) planning context for CCTP's *Strategic Plan*. Two considerations that arise from this are relevant to long-term technology R&D planning and guidance. First, the level of GHGs in the Earth's atmosphere implied by the UNFCCC is not known and is likely to remain a key planning uncertainty for some time.[2] Accordingly, CCTP's strategic goals are not based on any hypothesized level of stabilized GHG concentrations, but rather encompass a range of levels. Second, stabilizing the atmospheric GHG concentration at any level implies that global additions and withdrawals of GHGs to and from the atmosphere must achieve a net balance. This means that the growth in net GHG emissions would need to slow, eventually stop, and then reverse and approach levels that are low or near zero. The technological challenge is to develop new systems that could help achieve this goal affordably.

Set against this backdrop is the growing global appetite for energy to power economic growth and development. Energy-related GHG emissions account for over four-fifths of total anthropogenic emissions, and the technologies employed to produce and use energy will have a direct bearing on future emissions. Different countries will approach climate objectives in different ways and in the context of other urgent needs. For most developing countries, the

---

2   Additional scientific research is required to determine the level of GHG concentrations that would prevent dangerous anthropogenic interference with the climate system. The CCSP's principal aim is to improve understanding of climate change and its potential impacts, which will inform CCTP.

overriding goal will continue to be economic development to reduce poverty and advance human well-being. Increased global energy use is needed to help lift out of poverty the nearly two billion people who lack even the most basic access to modern energy services. Addressing this "energy poverty" is one of the world's key development objectives, as lack of energy services is associated with high rates of disease and child mortality.

All countries can be expected to seek to ensure that energy sources are secure, affordable, and reliable, and to seek approaches that address other environmental concerns, in addition to climate change, such as air pollution and conservation. Opportunities for new and advanced technologies that can address multiple societal objectives, including GHG reduction, present themselves in a number of areas.

These opportunities form the basis for CCTP's six strategic goals as follows:

1. Reduce emissions from energy end use and infrastructure.

2. Reduce emissions from energy supply.

3. Capture and sequester carbon dioxide.

4. Reduce emissions of non-carbon dioxide GHGs.

5. Improve capabilities to measure and monitor GHG emissions.

6. Bolster basic science contributions to technology development.

To the extent that agency missions and other priorities allow, each participating CCTP agency will align the relevant components of its R&D portfolio in ways that are consistent with and supportive of one or more of these goals.

These six CCTP strategic goals focus primarily on mitigating GHG emissions to make progress toward stabilizing atmospheric GHG concentrations. They are not intended to encompass the broad array of technical challenges and opportunities that may arise from climate change. These may include such research areas as: mitigating vulnerabilities and adaptation of natural and human systems to climate change; addressing effects of acidification of the oceans; geoengineering to reduce radiative forcing through modification of the Earth's surface albedo or stratospheric sunlight scattering; and others. Such topics are important, but they are beyond the scope of this *Plan*.

Figure 2-2. A major and growing source of GHG emissions is closely tied to transportation.

Credit: iStockphoto

■ CCTP Goal 1
Reduce Emissions from Energy End Use and Infrastructure

Major sources of anthropogenic carbon dioxide (CO_2) emissions are closely tied to the use of energy in transportation (Figure 2-2), residential and commercial buildings, and industrial processes. Improving energy efficiency and reducing GHG-emissions intensity in these economic sectors through a variety of technical advances and process changes present large opportunities to decrease overall GHG emissions.

In addition, application of advanced technology to the electricity transmission and distribution (T&D) infrastructure (the "grid") can have dual effects on reducing GHG emissions. First, there is a direct contribution to energy and CO_2 reductions resulting from increased efficiency in the T&D system itself. Second, there can be an indirect contribution by enabling, through modernized systems, expanded use of low-emission electricity and distributed generating technologies (such as wind, cogeneration of heat and power, geothermal, and solar power), and by improved management of system-wide energy supply and demand. Emissions reduction from energy efficiency gains and reduced energy use could be among the most important contributors to strategies aimed at overall CO_2 emissions reduction.

Types of technological advancements and deployment activities applicable to this goal include, but are not limited to:

◆ **Efficiency, Infrastructure, and Equipment.** Development and increased use of highly efficient motor vehicles and transportation systems, buildings equipment and envelopes, industrial combustion and process technology, and components of the electricity grid can significantly reduce CO_2 emissions, avoid other kinds of environmental impacts, and reduce the life-cycle costs of delivering the desired products and services. Process technology includes non-energy sources of CO_2, such as the calcination of carbonates to produce cement and lime.

◆ **Transition Technologies.** So-called "transition" technologies, such as high-efficiency natural gas-fired power plants, are not completely free of GHG emissions, but are capable of achieving significant reductions of GHG emissions in the near and mid terms by significantly improving or displacing higher GHG-emitting technologies in use today. Ideally, transition technologies would also be compatible with more advanced GHG-free technologies that would follow in the future.

◆ **Enabling Technologies.** Enabling technologies contribute indirectly to the reduction of GHG emissions by making possible the development and use of other important technologies. The example of a modernized electricity grid, mentioned above, is seen as an essential step for enabling the deployment of more advanced end-use technologies and distributed energy resources that can reduce GHG emissions. An intelligent electricity grid integrated with smart end-use equipment would further raise system performance. Another example is storage technologies for electricity or other energy carriers.

◆ **Alternatives to Industrial Processes, Feedstocks, and Materials.** Manufacturing, mining, agriculture, construction, services, and other commercial and industrial activities will require feedstocks and other material inputs to production.[3] In addition to the energy efficiency improvements discussed above, opportunities for lowering CO_2 and other GHG emissions from industrial and commercial activities include: replacing current feedstocks with those produced through processes (or complete resource cycles) that have low or net-zero GHG emissions (e.g., bio-based feedstocks); reducing the average energy intensity of material inputs; and developing alternatives to current industrial processes and products.

◆ **Improved Land Use and Management Practices.** Improved land use and management practices can reduce emissions of CO_2 in agriculture through energy-efficient and conservation practices. Additionally, long-term local and regional planning can help integrate and optimize residential, business, and transportation systems.

CCTP Goal 2
Reduce Emissions from Energy Supply

Current global energy supplies are dominated by fossil fuels—coal, petroleum products, and natural gas—that emit CO_2 when burned. A transition to a low-carbon future would likely require the availability of multiple energy supply technology options characterized by low or net-zero CO_2 emissions. Many such energy supply technologies are available today or are under development. When combined with improved energy carriers (e.g., electricity, hydrogen), they offer prospects for both reducing

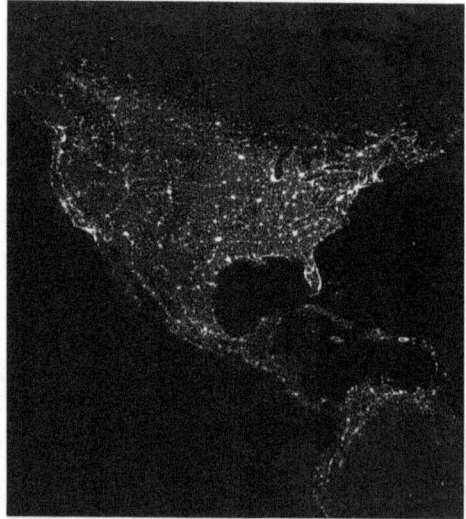

Figure 2-3. *Earth's city lights suggest where the world's energy use is concentrated. Advanced technologies in various forms of energy supply, including electricity generation, could significantly reduce or avoid future GHG emissions.*

Credit: Craig Mayhew and Robert Simmons, NASA GSFC

---

3 Producing feedstocks and materials can and does result in net emissions of GHGs.

GHG emissions and improving overall economic efficiency. Examples include the following:

◆ **Electricity.** Electricity will remain an important energy carrier in the global economy in the future (Figure 2-3). While substantial improvements in efficiency can reduce the growth of electricity consumption, the prospects of increased electrification and growing demand, especially in the developing regions of the world, still imply significant increases in electricity supply. Reducing GHG emissions from electricity supply could be achieved through further improvements in the efficiency of fossil-based electricity generation technologies, deployment of renewable technologies, increased use of fossil-fuel-based power systems in conjunction with CO_2 capture and sequestration (see Goal 3), increased use of nuclear energy, and the development of fusion energy or other novel power sources.

◆ **Hydrogen, Bio-Based, and Low-Carbon Fuels.** The world economy will have a continuing need for portable, storable energy carriers for heat, power, and transportation. A promising energy carrier is hydrogen, which can be produced in a variety of ways, including carbon-free or low-carbon methods using nuclear, wind, hydroelectric, solar energy, biomass, or fossil fuels combined with carbon capture and sequestration. Hydrogen and other carriers, such as methanol, ethanol, and other biofuels, could serve both as a means for energy storage and as energy carriers in transportation and other applications, if such carriers were to be produced by carbon-free or low-carbon methods.

CCTP Goal 3
Capture and Sequester Carbon Dioxide

Fossil fuels will likely remain a mainstay of global energy production well into the 21st century. Transforming fossil-fuel-based combustion systems into low-carbon or carbon-free energy processes would enable the continued use of the world's plentiful coal and other fossil energy resources. Such a transformation would require further development and application of technologies to capture CO_2 and store it using safe and acceptable means, removing it from the atmosphere for the long term. In addition, large amounts of CO_2 could be removed from the atmosphere and sequestered on land or in oceans

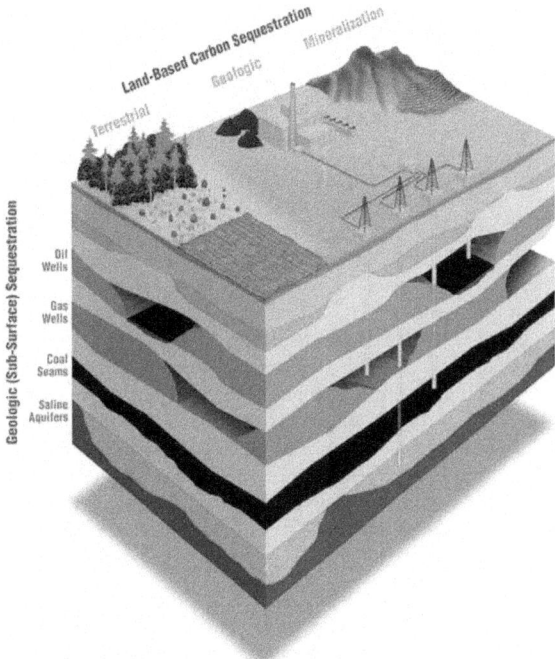

Figure 2-4. *Capture, storage, and other forms of CO_2 sequestration could significantly reduce emissions to the atmosphere and slow the growth of GHG concentrations.*

Credit: DOE/NETL

through improved land, forest, and agricultural management practices, changes in products and materials, and other means. Focus areas include:

◆ **Carbon Capture and Storage.** Advanced techniques are under development that could capture CO_2 from such sources as coal-burning power plants, oil refineries, hydrogen production facilities, and various high-emitting industrial processes. Carbon capture would be linked to geologic storage — long-term storage in geologic formations, such as depleted oil and gas reservoirs, deep coal seams, saline aquifers, other deep injection reservoirs, and chemical or other forms of storage (Figure 2-4).

◆ **CO_2 Sequestration.** Land-based, biologically-assisted means for removing CO_2 from the atmosphere and sequestering it in trees, soils, or other organic materials have proven to be relatively low-cost means for long-term carbon storage. Enhancing the ocean's biological CO_2 sink could also play a role. An understanding of the efficacy and environmental effects of the preceding approaches is needed.

CCTP Goal 4
Reduce Emissions of Non-CO$_2$ GHGs

GHGs other than carbon dioxide, including methane (CH$_4$), nitrous oxide (N$_2$O), sulfur hexafluoride (SF$_6$) and others, are more potent per unit weight as radiant energy absorbers than CO$_2$.[4] In addition, the atmospheric concentration of troposphere ozone (O$_3$), another GHG, is increasing due to human activities. The Intergovernmental Panel on Climate Change (IPCC) estimated that the cumulative effects of such gases since pre-industrial times account for about 40 percent of the anthropogenic radiative forcing[5] from GHGs. Reducing emissions of these other GHGs is an important climate change goal and key component of a comprehensive climate change technology strategy. Many categories of technologies are relevant to the attainment of this CCTP goal. Highlights include:

◆ **Methane Collection and Utilization.** Improvements in methods and technologies to collect methane and detect leaks from various

Figure 2-5. *Emissions of gases other than CO$_2$, such methane from landfills, coal mining operations, and natural gas pipelines, add to greenhouse concentrations in the atmosphere. Their capture and return to marketable uses can reduce their contributions to global warming.*

Credit: EPA

sources, such as landfills (Figure 2-5), coal mines, natural gas pipelines, and oil and gas exploration operations, can prevent this GHG from escaping to the atmosphere. These methods are often cost-effective, because the collected methane is a fuel that can be used directly or sold at natural gas market prices.

◆ **Reducing N$_2$O and Methane Emissions from Agriculture.** Improved agricultural management practices and technologies, including altering application practices in the use of fertilizers for crop production, dealing with livestock waste, and improved management practices in rice production, are key components of the strategy to reduce other GHGs.

◆ **Reducing Use of High Global-Warming-Potential (GWP) Gases.** Hydrofluorocarbons and perfluorocarbons have substituted for ozone-depleting chlorofluorocarbons in a number of industries, including refrigeration, air conditioning, foam blowing, solvent cleaning, fire suppression, and aerosol propellants. These and other high-GWP synthetic gases are generally used in applications where they are important to complex manufacturing processes or provide safety and system reliability, such as in semiconductor manufacturing, electric power transmission and distribution, and magnesium production and casting. Because they have high GWPs, methods to reduce leakage and use of these chemicals can contribute to UNFCCC goal attainment and include the development of lower-GWP alternatives to achieve the same purposes.

◆ **Black Carbon Aerosols.** Programs aimed at reducing airborne particulate matter have led to significant advances in fuel combustion and emission control technologies in both transportation and power generation sectors. Further advances can continue to reduce future black carbon aerosol emissions. Reduced emissions of black carbon, soot, and other chemical aerosols can have multiple benefits. Apart from improving public health and air quality, they can reduce radiative forcing in the atmosphere.

---

4 These calculations are made on the basis of instantaneous radiative forcing, not time-integrated radiative forcing (i.e., not accounting for different atmospheric lifetimes).

5 Radiative forcing is a measure of the overall energy balance in the Earth's atmosphere. It is zero when all energy flows in and out of the atmosphere are in balance, or equal. If there is a change in forcing, either positive or negative, the change is usually expressed in terms of watts per square meter (W/m²), averaged over the surface of the Earth. When it is positive, there is a net "force" toward warming, even if the warming itself may be slowed or delayed by other factors, such as the heat-absorbing capacity of the oceans or the energy absorption needed for the melting of natural ice sheets.

CCTP Goal 5
Improve Capabilities to Measure and Monitor GHG Emissions

Improved technologies for measuring, estimating, and monitoring GHG emissions and the flows of GHGs across various media and boundaries will help characterize emission levels and mark progress in reducing emissions. With enhanced means for GHG measuring and monitoring, future strategies to reduce, avoid, capture, or sequester CO_2 and other GHG emissions can be better supported, enabled, and evaluated (Figure 2-6). Key areas of technology R&D related to this goal may be grouped into four areas:

- **Anthropogenic Emissions.** Measurement and monitoring technologies can enhance and provide direct and indirect emissions measurements for various types of emissions sources using data transmission and archiving, along with inventory-based reporting systems and local-scale atmospheric measurements or indicators.

- **Carbon Capture, Storage, and Sequestration.** Advances in measurement and monitoring technologies for geologic storage can assess the integrity of subsurface reservoirs, transportation and pipeline systems, and potential leakage from geologic storage. Measurement and monitoring systems for terrestrial sequestration are also needed to integrate carbon sequestration measurements of different components of the landscape (e.g., soils versus vegetation) across a range of spatial scales.

- **Non-CO_2 GHGs.** Monitoring the emissions of methane, nitrous oxide, black carbon aerosols, hydrofluorocarbons, perfluorocarbons, and sulfur hexafluoride is important because of their high global warming potential (GWP) and, for some, their long atmospheric lifetimes. Advanced technologies can make an important contribution to direct and indirect measurement and monitoring approaches for both point and diffuse sources of these emissions.

- **Integrated Measuring and Monitoring System Architecture.** An effective measurement and monitoring capability is one that can collect, analyze, and integrate data across spatial and temporal scales, and at many different levels of

Figure 2-6. Integrated systems of instruments for measuring and monitoring GHG emissions and inventories can help to inform strategies for climate change technology development and associated emissions mitigating options.

Credit: NASA

resolution. This may require technologies such as sensors and continuous emission monitors, protocols for data gathering and analysis, development of emissions accounting methods, and coordination of related basic science and research in collaboration with the Climate Change Science Program and the U.S. Integrated Earth Observation System.[6]

CCTP Goal 4
Bolster Basic Science Contributions to Technology Development

Advances arising from basic scientific research are fundamental to future progress in applied technology research and development. The dual challenges—addressing global climate change and providing the energy supply needed to meet future demand and sustain economic growth—will likely require discoveries and innovations well beyond what today's science and technology can offer (Figure 2-7). Science must not just inform decisions, but provide the underlying knowledge foundation upon which new technologies can be built. The CCTP

---

6 See http://ostp.gov/html/EOCStrategic_Plan.pdf.

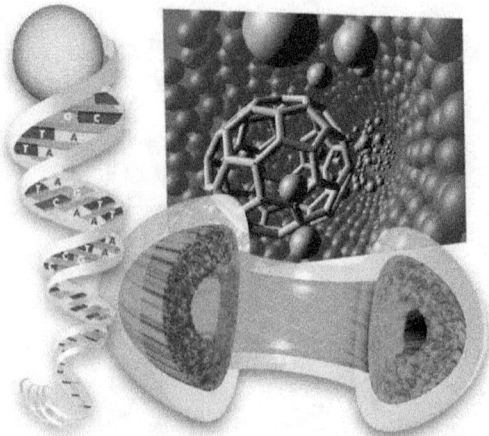

Figure 2-7. *Fundamental discoveries from basic research can help overcome technical barriers to progress in applied climate change technology research programs and illuminate entirely new pathways to innovative solutions.*

Courtesy: DOE, Office of Science

framework aims to strengthen the basic research enterprise so that it will be better prepared to find solutions and create new opportunities, including fostering new ideas and approaches that may be outside current R&D thrusts. CCTP will focus on several ways to meet this goal:

◆ **Fundamental Research.** Fundamental research provides the underlying foundation of scientific knowledge necessary for carrying out applied activities of research and problem solving. It is the systematic study of properties and natural behavior that can lead to greater knowledge and understanding of the fundamental aspects of phenomena and observable facts, but without prior specification toward applications, processes, or products. It includes scientific study and experimentation in the physical, biological, and environmental sciences, as well as interdisciplinary areas, such as computational sciences. Fundamental research is the source of much of underlying knowledge that will enable future progress in CCTP.

◆ **Strategic Research.** Strategic research is basic research that is inspired by technical challenges in the applied research and development programs. This is research that could lead to fundamental discoveries (e.g., new properties, phenomena, or materials) or scientific understanding that could be applied to solving specific problems or technical barriers impeding progress in advancing technologies in energy supply and end use; carbon capture, storage, and sequestration; other GHGs; and monitoring and measurement.

◆ **Exploratory Research.** Innovative concepts are often too risky or multi-disciplinary for one

program mission to support. Sometimes they do not fit neatly within the constructs of other mission-specific program goals. Therefore, not all of the research on innovative concepts for climate-related technology is, or should be, aligned directly to one of the existing Federal R&D mission-related programs. The climate change challenge calls for new breakthroughs in technology that could dramatically change the way energy is produced, transformed, and used in the global economy. Basic, exploratory research of innovative and novel concepts, not elsewhere covered, is one way to uncover such "breakthrough technology" and strengthen and broaden the R&D portfolio.

◆ **Integrated Planning.** Effective integration of fundamental research, strategic research, exploratory research, and applied technology development presents challenges to, and opportunities for, both the basic research and applied research communities. These challenges and opportunities can be effectively addressed through innovative and integrative planning processes that place emphasis on communication, cooperation, and collaboration among the many associated communities, and on workforce development, to meet the long-term challenges. CCTP seeks to encourage broadened application of successful models and best practices in this area.

2.3 Core Approaches

Consistent with the principles established by the President, CCTP will employ seven core approaches to stimulate participation by others and ensure progress towards attaining CCTP's strategic goals:

1. Strengthen climate change technology R&D.

2. Strengthen basic research contributions.

3. Enhance opportunities for partnerships.

4. Increase international cooperation.

5. Support cutting-edge technology demonstrations.

6. Ensure a viable technology workforce of the future.

7. Provide supporting technology policy.

Chapter 10 outlines next steps for CCTP for each of these core approaches.

APPROACH 1:
Strengthen Climate Change Technology R&D

The Federal Government is engaged in many research and technology development activities that contribute to meeting the President's climate change goals, investing about $3 billion in Fiscal Year 2006 in related technology R&D, including demonstration and deployment activities (Appendix A). Strengthening these activities, however, does not necessarily mean spending more money—it can also mean spending available resources more wisely by appropriately prioritizing activities and reallocating resources, or by leveraging them with the work of others.

To strengthen the current state of the U.S. climate change technology R&D, the CCTP has made, and will continue to make, recommendations (Figure 2-8) to the Cabinet-level Committee on Climate Change Science and Integration (CCCSTI) to sharpen the focus of, and provide support for, climate change technology R&D in a manner consistent with the mix and level of R&D investment required by the nature of the technical challenge.

APPROACH 2:
Strengthen Basic Research Contributions

A base of supporting fundamental research is essential to the applied research and development for technology development. The CCTP approach includes strengthening basic research in Federal research facilities and academia by focusing efforts on key areas needed to develop insights or breakthroughs relevant to climate-related technology R&D. A strong and creative science program is necessary to support and enable technical progress in CCTP's portfolio of applied research and development programs, explore novel approaches to new challenges, and bolster the underlying knowledge base for new discoveries.

Fundamental discoveries can reveal new properties and phenomena that can be applied to development of new energy technologies and other important systems. These can include breakthroughs in our understanding of biological functions, properties and phenomena of nano-materials and structures, computing architectures and methods, plasma science,

Figure 2-8. CCTP is charged with reviewing the Federal portfolio of related R&D and making recommendations to strengthen it. For recent results of a series of technical workshops that provided input to this process, see http://www.climatetechnology.gov.

Courtesy: Energetics, Inc., 2006

environmental sciences, and other emerging areas of scientific discovery.

APPROACH 3:
Enhance Opportunities for Partnerships

Federal research is but one element of the overall strategy for development and adoption of advanced climate change technologies. Engagement in this process by private entities, including business, industry, agriculture, construction, and other sectors of the U.S. economy, as well as by non-Federal governmental entities, such as the States and non-governmental organizations, is essential to make R&D investments wisely and to expedite innovative and cost-effective approaches for reducing GHG emissions.

Public-private partnerships can facilitate the transfer of technologies from Federal and national laboratories into commercial application. Partnering can also advise and improve the productivity of Federal research, and can promote the adoption of new technologies and best management practices. Private partners also benefit because those who are engaged in Federal R&D gain rights to intellectual property and gain access to world-class scientists, engineers, and laboratory facilities. This can help motivate further investment in the commercialization of technology.

Today, partnering is a common mode of operation in most Federal R&D programs, but it can be improved. Opportunities exist for private participation in virtually every aspect of Federal R&D. With respect to climate change technology R&D, the CCTP seeks to expand these opportunities in R&D planning, program execution, and technology demonstrations, leading ultimately to more efficient and timely commercial deployment.

Another important aspect of this approach is non-research and development partnerships. Such partnerships arise when there are mutual benefits and aligned interests between governmental and non-governmental entities in promoting progress toward CCTP strategic goals. Examples include the Climate VISION,[7] Climate Leaders,[8] ENERGY STAR®,[9] and SmartWay Transport Partnership[10] programs, all of which encourage industry to undertake activities that can lead to reduced GHG emissions.

APPROACH 4:
Increase International Cooperation

Given the global nature of climate change concerns, and in recognition of the contributions being made abroad, the CCTP seeks to engage other nations—government to government—in large-scale cooperative technology research initiatives. Such cooperation can prove beneficial to the success of U.S. technology development initiatives, through leveraging of resources, partitioning of research activities addressing large-scale and multi-faceted complex problems, and sharing of results and knowledge created. Indeed, in certain areas of climate change technology research and development, such as advanced wind turbine design and nuclear fission and fusion energy research, U.S. technical expertise can be augmented by many advanced technical capabilities that reside outside the United States. The U.S. Government has initiated or joined several multilateral cooperative agreements, such as the International Partnership for the Hydrogen Economy, Carbon Sequestration Leadership Forum, Generation IV International Forum, international Methane-to-Markets Partnership, and the ITER project to develop fusion as a commercially viable power source.

Since June 2001, the United States has launched bilateral partnerships with Australia, Brazil, Canada,

China, Central America, Germany, the European Union, India, Italy, Japan, Mexico, New Zealand, Republic of Korea, the Russian Federation, and South Africa on issues such as climate change science, energy and sequestration technologies, and climate change policy approaches. The countries covered by these bilateral partnerships account for about three quarters of global GHG emissions. In addition, the United States is a leader in the 59-member country Global Earth Observations System of Systems.

In related developments in July 2005, President Bush and the G-8 Leaders agreed on a far-reaching Plan of Action to speed the development and deployment of clean energy technologies to achieve the combined goals of addressing climate change, reducing harmful air pollution, and improving energy security in the United States and throughout the world. The G-8 will work globally to advance climate change policies that spur and sustain economic growth, and improve the environment.

Also in July 2005, the United States joined with Australia, China, India, Japan, and South Korea to accelerate clean development under a new Asia-Pacific Partnership on Clean Development and Climate. These countries together account for roughly half of the world's population, economic growth, energy use, and GHG emissions. The focus of this Partnership is on voluntary practical measures taken by these six countries in the Asia-Pacific region to create new investment opportunities, build local capacity, and remove barriers to the introduction of clean, more efficient technologies. APP countries will work in partnership to meet nationally-designed strategies for improving energy security, reducing air pollution, and addressing the long-term challenge of climate change.

CCTP seeks to expand on these and other international opportunities to stimulate international participation in the development of new and advanced climate change technologies, foster capacity building in developing countries, encourage cooperative planning and joint ventures, and enable more rapid development, transfer, and deployment of advanced climate change technology.

---

7 See http://www.climatevision.gov.

8 See http://www.epa.gov/climateleaders.

9 See http://www.energystar.gov.

10 See http://www.epa.gov/smartway.

APPROACH 5:
Support Cutting-Edge Technology Demonstrations

While the government role is targeted primarily toward long-term, high-risk research and development activities, demonstrations of cutting-edge climate change technologies also constitute an important aspect of the goal to advance climate change technologies. Demonstrations help advance a cutting-edge technology from the laboratory to the commercial market. After a concept has been proven in principle, pilot or full-scale demonstrations are needed to identify real-world performance issues. Subsequently, demonstrations can prove the viability of the technology for deployment.

Technology demonstrations afford unique opportunities to reduce investment risks. They clarify the parameters affecting a technology's cost and operational performance. They identify areas needing further improvement or cost reduction. Federal leadership through technology demonstrations can strongly influence decisions of private-sector investors and other non-government parties.

Because technology demonstrations can be costly, which can reduce funds available for related research, it is often advantageous for demonstrations to be cost-shared with industry to the extent practical. Demonstrations are most beneficial when the technology is ready for demonstration and when the results are shared with the public. Open and competitive processes will help to ensure fairness of opportunity among all interested parties.

APPROACH 6:
Ensure a Viable Technology Workforce of the Future

The development and deployment, on a global scale, of new and advanced climate change technologies will require a skilled workforce and an abundance of intellectual talent, well versed in associated concepts and disciplines of science and engineering.

Workforce development and education are an added benefit of Federal research funding, since portions of research grants often support graduate and undergraduate research stipends.

CCTP-funded research provides an opportunity to promote education, and experience for students in science, math, and engineering education, and to encourage talented individuals to focus their careers on this global endeavor.[11] Such efforts could be coordinated with other countries, and particularly in emerging economies of the developing world, where much of 21st century emissions are expected to occur.

APPROACH 7:
Provide Supporting Technology Policy

While some advanced climate change technologies may be sufficiently attractive to penetrate the marketplace on a large scale without supporting policy or incentives, others may not. For example, technologies that capture or sequester CO_2 or others that afford certain climate change-related advantages are expected to remain more expensive than competing technologies, even considering further technical progress. Widespread adoption of these technologies would likely require support from appropriate government policies, potentially including market-based incentives.

As Federal efforts to advance technology go forward, broadened participation by the private sector in these efforts is important both to the acceleration of innovation and to the adoption of technologies. Extending such participation beyond R&D partnering and demonstrations (Approaches 3 and 5 above) can be encouraged by appropriate and supporting technology policy. This is evidenced today, in part, by a number of market-based incentives already in place, by others proposed by the Administration[12], and by still others soon to be implemented in accord with the provisions of Title XIII of the Energy Policy Act of 2005.[13]

Finally, Title XVI of the Energy Policy Act of 2005 sets the stage for future development of policies that would facilitate new technology adoption. This Title requires that the Administration identify barriers to

---

11 The CCTP approach is linked with the American Competitiveness Initiative (ACI), which aims to build the technologically skilled workforce of the future. ACI will double the Federal commitment to the most critical research programs emphasizing the physical sciences, modernize and make permanent the research and development tax credit, reform the job training system, and strengthen math and science education. See http://www.whitehouse.gov/stateoftheunion/2006/aci/.

12 Federal Climate Change Expenditures Report to Congress, March 2005. http://www.whitehouse.gov/omb/legislative/fy06_climate_change_rpt.pdf.

13 Financial incentives in Title XIII for technologies related to climate change goals are scored at more than $11 billion over 10 years.

the commercialization and deployment of GHG-intensity-reducing technologies, and develop recommendations for the removal of these barriers to facilitate the deployment of these technologies and practices.[14] This effort will be led by the Secretary of Energy.

2.4 Prioritization Process

An important role of the CCTP is to provide strategic direction for, and to strengthen, the Federal portfolio of investments in climate change technology R&D. The CCTP continues to prioritize the portfolio of Federally-funded climate change technology R&D consistent with the President's NCCTI. The CCTP has identified within its portfolio a subset of NCCTI priority activities for Fiscal Year 2006 and Fiscal Year 2007 (listed in Appendix B), defined as discrete research and development activities that address technological challenges, which, if successful, could advance technologies with the potential to dramatically reduce, avoid, or sequester GHG emissions.

Prioritization of Federal technology R&D activities related to climate change is a dynamic process that has evolved over time in response to emerging knowledge. This evolution is expected to continue.

Through coordinated interagency planning, the CCTP priorities will be reviewed periodically in conjunction with the Federal budget process, and recommendations will be made through the Interagency Working Group (IWG) to the CCCSTI.

This CCTP *Strategic Plan* provides a government-wide basis for guiding the formulation of the comprehensive Federal climate change technology R&D portfolio, identifying high priority investments, gaps, and emerging opportunities, and organizing future CCTP-related research. The CCTP planning activities are informed by results of studies, inputs from many and diverse sources, technical workshops, assessments of technology potentials, analyses regarding long-term energy and emissions outlooks, and modeling by a number of groups of a range of technology scenarios over the next 100 years (see Chapter 3). Additionally, these planning activities are guided by several important portfolio planning principles and investment criteria.[15]

Portfolio Planning Principles

The CCTP adheres to three broad principles. The first principle, given the many attendant uncertainties about the future, is that the whole of the individual R&D investments should constitute a balanced and diversified portfolio. Considerations include the realizations that: (1) no single technology will likely meet the challenge alone; (2) investing in R&D in advanced technologies involves risk since the results of these investments are not known in advance (some may fail, and others may not prove as successful as hoped); and (3) a diverse array of technology options can hedge against risk and provide flexibility in the future, which may be needed to respond to new information that could change strategy. The CCTP portfolio also strives to balance short- and long-term technology objectives.

A balanced and diversified portfolio must hedge risks, for example, by investing in projects that will pay off under different states of the future world. For this purpose, it is helpful to understand the major sources of uncertainty, such as the level of future GHG emissions, energy prices, technology costs and performance, and other factors. CCTP's tools in this regard are partially addressed in Chapter 3, but further work in terms of portfolio analysis and expected benefits and costs will be required.

The second principle is to ensure that factors affecting market acceptance are addressed. To enable widespread deployment of advanced technologies, each technology must be integrated within a larger technical system and infrastructure, not just as a component. Market acceptance of technologies is influenced by myriad social and economic factors. The CCTP's portfolio planning process must be informed by, and benefit from, private sector and other non-Federal inputs, examine the lessons of historical analogues for technology acceptance, and apply them as a means to anticipate issues and inform R&D planning.

Third and perhaps most importantly, the anticipated timing regarding the commercial readiness of the advanced technology options is an important CCTP planning consideration. Energy infrastructure has a long lifetime, and change in the capital stock occurs slowly. Once new technologies are available, their adoption takes time. Some technologies with low or near-net-zero GHG emissions may need to be available and moving into the marketplace decades before their maximum market penetration is achieved.

---

[14] See text of EPACT Title XVI at http://www.climatetechnology.gov.

[15] These criteria are consistent with the Better R&D Investment Criteria outlined in The President's Management Agenda. See http://www.whitehouse.gov/OMB/budget/fy2002/mgmt.pdf.

Portfolio Planning and Investment Criteria

Within the planning framework of vision, mission, goals, approaches, and portfolio investment principles, the CCTP's prioritization process applies four criteria (Box 2-1). CCTP will evaluate the merits of competing investments based on maximum expected benefits versus costs (Criterion #1), subject to consideration of the distinct roles of the public and private sectors in R&D (Criterion #2).[16] In addition, because of the risk of spreading resources across too many areas, CCTP focuses on technologies with the potential for large-scale application (Criterion #3). Nevertheless, technologies expected to have limited

---

BOX 2-1

CCTP PORTFOLIO PLANNING AND INVESTMENT CRITERIA

1. **Maximizing Expected Return on Investment.**
R&D investments that have the prospect to generate maximum expected benefits per dollar of investment receive priority in investment planning. Benefits are defined with respect to expected contributions to the attainment of CCTP goals, particularly GHG reductions, but also include other considerations, such as cost-effectiveness, improved productivity, and reduction of other pollutants. Climate change benefits are long-term public goods. Discount rates must be appropriate to the context, particularly when applied to very long-term impacts. This criterion includes considerations of development and deployment risks, and the hedging of risks across multiple projects. Projects with high risk, but low emissions-reduction potential should be removed from the CCTP R&D portfolio.

2. **Acknowledging the Proper and Distinct Roles for the Public and Private Sectors.** The CCTP portfolio recognizes that some R&D is the proper purview of the private sector; other R&D may be best performed jointly through public-private partnerships; and still other R&D may be best performed by the Federal sector alone. In cases where public support of R&D is warranted, technology development and adoption require cooperation and engagement with the private sector. History demonstrates that early involvement in technology R&D by the business community increases the probability of commercialization. A key consideration in the investment process is the means for engaging the talents of the private sector using innovative and effective approaches.

3. **Focusing on Technology with Large-Scale Potential.** The scope, scale, and magnitude of the climate change challenge suggest that relatively small, incremental improvements in existing technologies will not enable full achievement of CCTP goals. Every technology option has limits of various kinds. Such limits need to be identified, explored, and understood early in the planning process. Technology options should be adaptable on a global scale and have a clear path to commercialization. High-priority investments, including exploratory research, will focus on technology options that could, if successful, result in large mitigation contributions, accumulated over the span of the 21st century. For technologies on the lower end of this criterion, benefits should be deliverable earlier in the century and/or be particularly compelling from a marginal benefit/cost perspective.

4. **Sequencing R&D Investments in a Logical, Developmental Order.** Investments must be logically sequenced over time. Supporting a robust and diversified portfolio does not mean that all technology options must be supported simultaneously, or that all must proceed at an accelerated pace. Logical sequencing of R&D investments takes into account (i) the expected times when different technologies may need to be made available and cost-effective, (ii) the need for early resolution of critical uncertainties, and (iii) the need to demonstrate early success or feasibility of technologies upon which other technology advancements may be based.

---

[16] In a market economy, commercially-orientated R&D is primarily a private sector matter. Government support of basic and applied R&D is warranted when the social benefits of such R&D outweigh the benefits that can be captured by innovating businesses and their customers, leading to inadequate investment in such R&D by the private sector.

<table>
<tr><td>

BOX 2-2

CCTP WORKING GROUPS

Energy End Use – Led by DOE
- Hydrogen End Use
- Transportation
- Buildings
- Industry
- Electric Grid and Infrastructure

Energy Supply – Led by DOE
- Hydrogen Production
- Renewable and Low-Carbon Fuels
- Renewable Power
- Nuclear Fission Power
- Fusion Energy
- Low Emissions Fossil-Based Power

CO_2 Sequestration – Led by USDA
- Carbon Capture
- Geologic Storage
- Terrestrial Sequestration
- Ocean Storage
- Products and Materials

Other (Non-CO_2) Gases – Led by EPA
- Energy & Waste – Methane
- Agricultural Methane and Other Gases
- High Global-Warming-Potential Gases
- Nitrous Oxide
- Ozone Precursors and Black Carbon

Measuring and Monitoring – Led by NASA
- Application Areas
- Integrated Systems

Basic Research – Led by DOE
- Fundamental Research
- Strategic Research
- Exploratory Research
- Integrative R&D Planning

</td></tr>
</table>

impact on overall GHG emissions may still be given priority if they can deliver reductions earlier in the century and/or are particularly cost-compelling. Finally, the timing of investments is an important consideration in the decision process. The CCTP planning process gives weight to the logical sequencing of research (Criterion #4), where the value in knowing whether a technological advance is successful can have a cascading effect on the sequencing of later investments.

Application of Criteria

The CCTP's review, planning, and prioritization process will rely on ongoing reviews of strategies for technology development, buttressed by analysis, and of the overall R&D portfolio's adequacy to make progress toward attaining each CCTP strategic goal. There will be an emphasis on identifying gaps and key opportunities for new initiatives that will be accompanied by periodic realignments. The process is not easily reduced to quantitative analysis due, in part, to the large number of variables and uncertainties associated with the nature of the climate change technology challenge and, in part, to the CCTP's century-long planning horizon. Nevertheless, the prioritization criteria discussed above will be applied by the participating agencies to the maximum extent practicable and augmented by inputs from various sources.

The first step in the prioritization process is to establish a baseline, or inventory, of the existing portfolio of R&D activities across the participating agencies. The criteria used to compile this portfolio baseline, which closely track CCTP strategic goals, are listed in Appendix A. The resulting multi-agency baseline inventory accounted for more than $3 billion in R&D activities in Fiscal Year 2006. This inventory will be periodically updated.

The second step in the process is to use the insights gained from the scenario modeling[17] and other analyses to identify the more important elements of a diversified strategy and to assess the potential contributions of each activity within the CCTP portfolio. This assessment may affirm some elements of the portfolio, challenge others, and identify gaps and promising opportunities. Once a full set of candidate investments is identified, the prioritization criteria can be applied to each proposed investment activity. This step will require continuing

---

[17] See Chapter 3 for a discussion of the scenario analyses.

development of analytical tools and methods, including assessments of various technologies and their limitations.

The results of this process for Fiscal Year 2007, as evidenced in the Administration's budget request, are shown in Appendix A. Within this overall CCTP portfolio, NCCTI priorities are identified in Appendix B. Finally, key initiatives that advance multiple policy goals, while also addressing major thrusts of CCTP strategic goals, are highlighted throughout the *Plan* in their respective technology areas.[18] A current list of the key initiatives, with links to current programmatic information, may be found at the CCTP website.[19]

The current CCTP portfolio reflects a "snapshot" in time of an ongoing review and realignment that takes into account new and changing emphases among competing national priorities. In the years ahead, CCTP's portfolio and planning emphasis is expected to evolve as more studies and analyses are conducted, technology assessments are completed, additional gaps and opportunities are identified, and new developments and scientific knowledge emerge.

2.5 Management

The CCTP is a multi-agency R&D planning and coordination activity. It accomplishes its work by engaging and assisting the Federal R&D agencies in their respective efforts to plan, prioritize, and coordinate research activities to meet CCTP goals. CCTP also works with the Administration to formulate overall budget guidance and recommend adjustments, where appropriate, to the Federal R&D portfolio to align it more closely with CCTP goals. CCTP also has a role in coordinating non-R&D-related activities, monitoring progress, and accounting for related investments. As discussed below, the CCTP's management functions include executive direction, interagency planning and integration, agency implementation, external interactions, and program support.

Executive Direction

The CCTP exercises executive direction through the CCCSTI, and its associated IWG on Climate Change Science and Technology. The IWG is comprised of agency deputies who can adopt and implement plans and actions coordinated by the Group. The IWG also provides guidance on strategy and reviews, and approves CCTP strategic planning documents.

A CCTP Steering Group, composed of senior-level representatives from each participating Federal agency, provides a venue for agencies to raise and resolve issues regarding CCTP and its functions as a facilitating and coordinating body. It also assists the CCTP Director in accessing needed information and resources within each agency. The Steering Group assists in developing agency budget crosscuts and proposals, conveying information and actions back to the agencies, and supporting the CCTP mission. In addition, it ensures that consistent guidance and direction is given to the CCTP Working Groups, and it helps formulate recommendations and advice to the CCCSTI through the IWG.

Interagency Planning and Integration

Six multi-agency CCTP Working Groups (WGs), aligned with the six CCTP strategic goals (Box 2-2), are primarily responsible for carrying out the mission and staff functions of CCTP in a coordinated manner. The WGs are assisted by subgroups, as appropriate, and by technical staff drawn from participating agencies, affiliated laboratories and facilities, and other available consulting staff. The WGs are expected to:

◆ Serve as the principal means for interagency deliberation and development of CCTP plans and priorities, and for the formulation of guidance for supporting analyses in their respective areas.

◆ Provide a forum for exchange of inputs and information relevant to planning processes, including workshops and other meetings.

◆ Engage, cooperate with, and coordinate inputs from relevant agencies.

◆ Identify ongoing R&D activities and gaps, needs, and opportunities for the near and long term.

◆ Support relevant interaction with CCSP science studies and analyses.

◆ Formulate advice and recommendations to present to the IWG and CCCSTI.

◆ Assist in preparing periodic reports to Cabinet members and the President.

---

[18] See also CCTP *Vision and Framework for Strategy and Planning* (2005).

[19] See http://www.climatetechnology.gov.

Agency Implementation

CCTP relies on participating Federal agencies and their respective R&D portfolios to contribute to achieving CCTP goals. At the same time, the CCTP recognizes that agencies must balance these priorities with other mission requirements. The Program depends on these agencies to place appropriate priority on implementing CCTP programs. Each participating agency has representation in the groups that facilitate the setting of CCTP priorities and follow-through on attaining these priorities. Agency heads and deputies serve on the CCCSTI and IWG, and top agency officials make up the CCTP Steering Group. Agency executives and senior-level managers also serve as chairs and members of the CCTP Working Groups. Once CCTP plans, programs, and priorities are set and approved, the agencies are expected to contribute to their execution and completion.

External Interactions

The CCTP taps expert opinion and technical input from various external parties through advisory groups, program peer review, conferences, international partnerships, and other activities. In addition, CCTP staff convenes technical workshops and meetings with experts both inside and outside the Federal Government. CCTP activities are of interest to a number of external parties, including State and local governments, regional planning organizations, academic institutions, national laboratories, and non-governmental organizations. They are of interest, as well, to foreign governments and international organizations, such as the Organization for Economic Cooperation and Development, the International Energy Agency, various global and regional compacts, and the IPCC. CCTP needs to communicate its activities to such entities and provide coordinated support, through the relevant agency programs, for enhanced external and international cooperation by engaging with and supporting activities of mutual interest.

Program Support

CCTP staff will provide technical and administrative support and day-to-day coordination of program integration, strategic planning, product development, communication, and representation. CCTP staff will: (1) provide support for the Working Groups and the Steering Group; (2) foster integration of activities to support CCTP goals; (3) conduct and support strategic planning activities that facilitate the prioritization of R&D activities and decision-making on the composition of the CCTP R&D portfolio, including conducting analytical exercises that support planning (such as technology assessments and scenario analysis); (4) develop improved methods, tools, and decision-making processes for climate technology planning and management, R&D planning, and assessment; (5) develop products that communicate CCTP's plans as well as the progress of the Program toward meeting its goals; (6) coordinate interagency budget planning and reporting; (7) assist and support the Administration in representing U.S. interests in the proceedings of the United Nations' IPCC Fourth Assessment Report; and (8) coordinate agency support of international cooperative agreements.

2.6 Strategic Plan Outline

The chapters that follow provide a century-long planning context, goal-oriented strategies for technology development, and a summary of conclusions and next steps. Chapter 3 provides a synthesis assessment on energy-economic modeling and forecasting of future global GHG emissions, based on a number of representative works in the literature. Chapter 3 also includes a number of insights regarding opportunities for advanced technologies drawn from scenarios analyses. Chapters 4 through 9 focus in depth on each of CCTP's six strategic goals. Each chapter outlines elements of a technology development strategy, highlights ongoing work, and suggests promising areas for future research. Chapter 10 provides a summary of conclusions regarding CCTP and its strategic goals, and identifies a series of next steps within the context of each of CCTP's seven approaches.

Synthesis Assessment of Long-Term Climate Change Technology Scenarios

I f the Climate Change Technology Program (CCTP) is to develop plans, carry out activities, and help shape an R&D portfolio that will advance its vision and mission and make progress toward achieving its strategic goals, the Program needs a long-term planning context, one that is informed by analyses from many sources using a variety of models and other decision support tools. Useful information to help shape the portfolio can come from assessments of the potential contributions that successful advancement of technologies could make to achieving CCTP's strategic goals. Analysis can also help determine the technology performance requirements needed to achieve greenhouse gas (GHG) emission reduction goals and the costs.

Such assessments are complex and must consider many uncertainties, and hence they must include a range of assumptions about the future. Specifically, a technology strategy aimed at influencing global GHG emissions over the course of the 21st century would need to consider population change, varying rates of regional economic development, differing regional technological needs and interests, and availability of natural resources. In addition, the long-term costs of GHG emission reductions will depend in part on future technological innovations, many of which are unknown, and on other factors that could either promote or discourage the use of various technologies in the future. Finally, the uncertainties inherent in climate science and the fact that value judgments are involved make it difficult to determine a level at which atmospheric GHG concentrations in the Earth's atmosphere would meet the stabilization objective of the United Nations Framework Convention on Climate Change noted in Chapter 1.

Scenario analysis using various types of models is one valuable tool that can be used to assist in planning under uncertainty. Scenarios can present alternative views about the rate of future GHG emissions growth to help gauge the scope of the potential challenge by methodically and consistently accounting for the complex interactions among economic and demographic factors, energy supply and demand, the advance of technology, and GHG emissions. Scenarios can also investigate various proposed pathways toward achieving different levels of GHG

The synthesis assessment of global climate-change technology scenarios examined anaylses based on a variety of models and tools.
Courtesy: DOE/PNNL

emissions reductions. The results can provide relative indications of the potential emissions reductions and economic benefits of particular classes of technology under a range of different assumptions, and a better understanding of the factors and constraints that might affect their market penetration.

Many research organizations, university-based teams, government agencies, and other groups have engaged in scenario analyses to explore these topics. This chapter reviews and synthesizes the results of such efforts to gain insights, under a range of uncertainties, about the scale of the technological challenge, the potential contributions of various advanced technologies to GHG emissions reductions or avoidances, the timing of technology deployment, and associated economic benefits. These insights will be used to guide CCTP in developing an effective climate change technology strategy and associated portfolio of technology R&D.

3.1 The Greenhouse Gases

GHGs are gases that absorb infrared radiation. In the Earth's atmosphere, the GHGs cause what is commonly known as the "greenhouse effect." As shown in Figure 3-1, the GHGs include[1] carbon dioxide (CO_2), methane (CH_4), nitrous oxide (N_2O), and substances with very high global warming potential (GWP),[2] such as the halocarbons and other chlorine and bromine containing substances.[3] CO_2 emissions from the burning of fossil fuels, other industrial activity, and land-use changes and forestry, account for the majority of GHG emissions. In 2000, the combined emissions of methane, nitrous oxide, and high-GWP gases accounted for about one-quarter of all GHG emissions (after converting the non-CO_2 gases to a CO_2-equivalency basis, in terms of gigatons carbon equivalent, or GtC-eq.).

As a GHG resulting from human activities, methane's contribution is second only to CO_2. Methane, on a kilogram-for-kilogram basis, is 23 times more effective than CO_2 at trapping radiation in the atmosphere over a 100-year time period (although it has a shorter lifetime in the atmosphere). Methane is emitted from various energy-related activities (e.g., natural gas, oil and coal exploration, and coal mining), as well as from agricultural sources (e.g., emissions from cattle digestion and rice cultivation; and waste disposal facilities, landfills, and wastewater treatment plants). Methane emissions have declined in the United States since the 1990s, due to voluntary

programs to reduce emissions and regulation requiring the largest landfills to collect and combust their landfill gas.[4]

Another important gas is nitrous oxide (N_2O), which is emitted primarily by the agricultural sector through direct emissions from agricultural soils and indirect emissions from nitrogen fertilizers used in agriculture, as well as from fossil fuel combustion, especially from fuel used in motor vehicles; adipic (nylon) and nitric acid production; wastewater treatment and waste combustion; and biomass burning (EPA 2005).

Global Emissions of GHGs in 2000
(% of total GtC-eq. using GWPs)

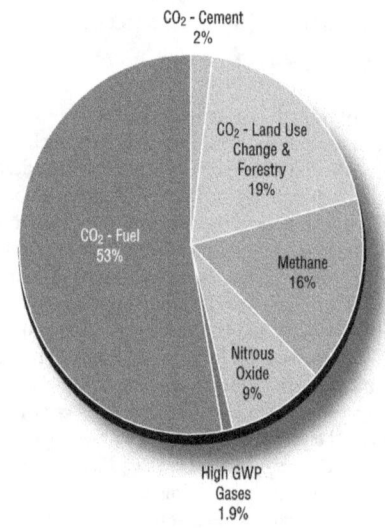

2000 Global Emissions
(Carbon Equivalents)

CO_2 - Cement 2%

CO_2 - Land Use Change & Forestry 19%

CO_2 - Fuel 53%

Methane 16%

Nitrous Oxide 9%

High GWP Gases 1.9%

Figure 3-1. Global Emissions of GHGs in 2000 (percentage of total GtC-eq. using GWPs) (Sources: http://www.epa.gov/methanetomarkets/docs/methanemarkets-factsheet.pdf and http://www.mnp.nl/edgar/model/)

---

[1] Water vapor and ozone are also GHGs.

[2] Global warming potentials (GWPs) are used to compare the abilities of different GHGs to trap infared radiation in the atmosphere. GWPs are based on the radiative efficiency (radiation-absorbing ability) of each gas relative to that of carbon dioxide (CO_2), as well as the decay rate of each gas (the amount removed from the atmosphere over a given number of years) relative to that of CO_2. The GWP provides a simplified construct for converting emissions of various gases into a common measure. However, it is important to note that GWPs are only an approximate metric for considering the relative impacts of different gases on the radiative balance of the Earth's atmosphere, because atmospheric lifetimes differ dramatically among gases. For example, methane's lifetime in the atmosphere is shorter than CO_2, so the radiative impact of a ton of methane emitted today will attenuate more quickly over time with respect to the radiative impact of a ton of CO_2 emitted today. Hence, the timeframe over which the GWP is developed critically affects the relative magnitude of the contributions of various gases (typically a 100-year timeframe is used), and GWPs serve as an indication of the relative importance of different gases, not as a precise measurement of the relative long-term impacts on the Earth's radiative balance.

[3] The ozone-depleting halocarbons and other chlorine- and bromine-containing substances are addressed by the Montreal Protocol and are not directly addressed by this Plan. Besides CO_2, N_2O, and CH_4, the Intergovernmental Panel on Climate Change (IPCC) definitions of GHGs include sulfur hexafluoride (SF_6), hydrofluorocarbons (HFCs), and perfluorocarbons (PFCs), often collectively called the "F-gases."

[4] See http://www.epa.gov/methane/voluntary.html.

Other non-CO_2 GHGs, including certain fluorine-containing halogenated substances such as sulfur hexafluoride (SF_6), hydrofluorocarbons (HFCs), and perfluorocarbons (PFCs), often collectively called the "F-gases," are used or produced by a variety of industrial processes. In most cases, emissions of these F-gases were relatively low in 1990 but have since grown rapidly. The sources of these non-CO_2 GHG emissions are discussed in more detail in Chapter 7.

The radiation-trapping capacities of GHGs vary considerably. GHGs also have different lifetimes in the atmosphere. In addition, some anthropogenic emissions (such as aerosols) can have cooling effects. Combining these effects, the Intergovernmental Panel on Climate Change (IPCC) estimated the key anthropogenic and natural factors causing changes in warming and cooling from year 1750 to year 2000,[5] as shown in Figure 3-2. The figure shows warming and cooling in terms of radiative forcing—the change in the balance of infrared radiation coming into and going out of the atmosphere. Positive radiative forcing leads to warming, and negative radiative forcing leads to cooling.

The differences in the characteristics of GHGs and other radiatively important substances, as well as the potential differences in rates of the growth of their emissions over time, influence the formulation of strategies to stabilize overall GHG concentrations through emissions mitigation. In addition to the climate implications of increasing atmospheric CO_2, there are also concerns about chemical and biological impacts to the ocean. As discussed in Chapter 6, an increase in ocean CO_2 concentrations and the potential effect on ocean chemistry and biology provide additional motivation, apart from climate change and considerations of global temperature, to mitigate emissions and stabilize atmospheric CO_2 concentrations.

 **Factors Affecting Future GHG Emissions**

Most analyses of future GHG emissions indicate that, in the absence of actions taken to address climate change, increases will occur in both emissions of GHGs and their atmospheric concentrations. The

Global Mean Radiative Forcing of the Climate System for the Year 2000, Relative to 1750

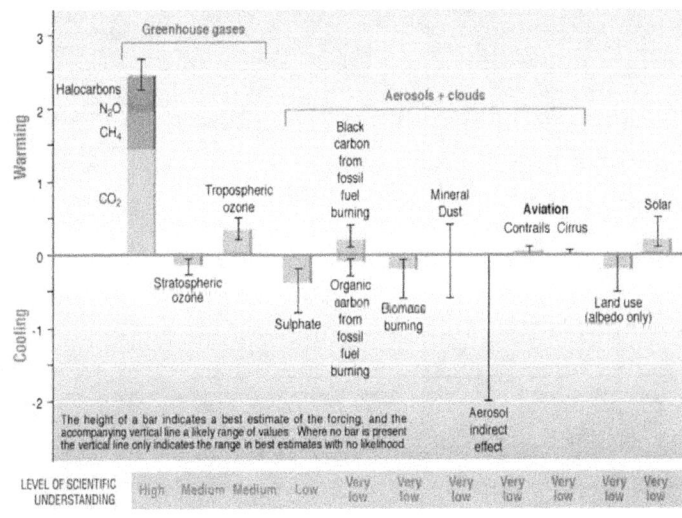

Figure 3-2. Global Mean Radiative Forcing of the Climate System for the Year 2000, Relative to 1750 (Source: IPCC [6])

---

5 A large body of work has been undertaken to understand the influence of external factors on climate using the concept of changes in radiative "forcing" due to changes in the atmospheric composition, alteration of surface reflectance by land use, and variations in solar input. Some of the radiative forcing agents, such as CO_2 are well mixed over the globe, thereby perturbing the global heat balance. Others, such as aerosols, represent perturbations with stronger regional signatures because of their spatial distribution. For this and other reasons, a simple sum of the positive and negative bars cannot be expected to yield the net effect on the climate system.

6 Available at http://www.ipcc.ch/present/graphics/2001syr/large/06.01.jpg.

projected rate of emissions growth is dependent on many factors that cannot be predicted with certainty. Studies conducted by organizations, including the IPCC,[7] the Stanford Energy Modeling Forum (EMF),[8] and others,[9] indicate that among the more significant factors expected to drive future GHG emissions growth are demographic changes (e.g., regional population growth), social and economic development (e.g., gross world product and standard of living), fossil fuel use (i.e., coal, oil, and natural gas), and land use changes. The most important factors limiting increases in future GHG emissions include improvements in energy efficiency; increases in nuclear, renewable, and non-CO_2-emitting fossil

energy supply; decreases in GHG emissions from industry, agriculture, and forestry; and rapid technological change that results in reducing GHG emissions.

CO_2 Emissions from Energy Consumption

The International Energy Agency estimates that 1.6 billion people lack access to electricity. The United Nations estimates that 2 billion people are without clean and safe cooking fuels, relying instead on traditional biomass (UNDP 2000). Over the course

Primary Energy Use Projections Using Various Energy-Economic Models and Assumptions

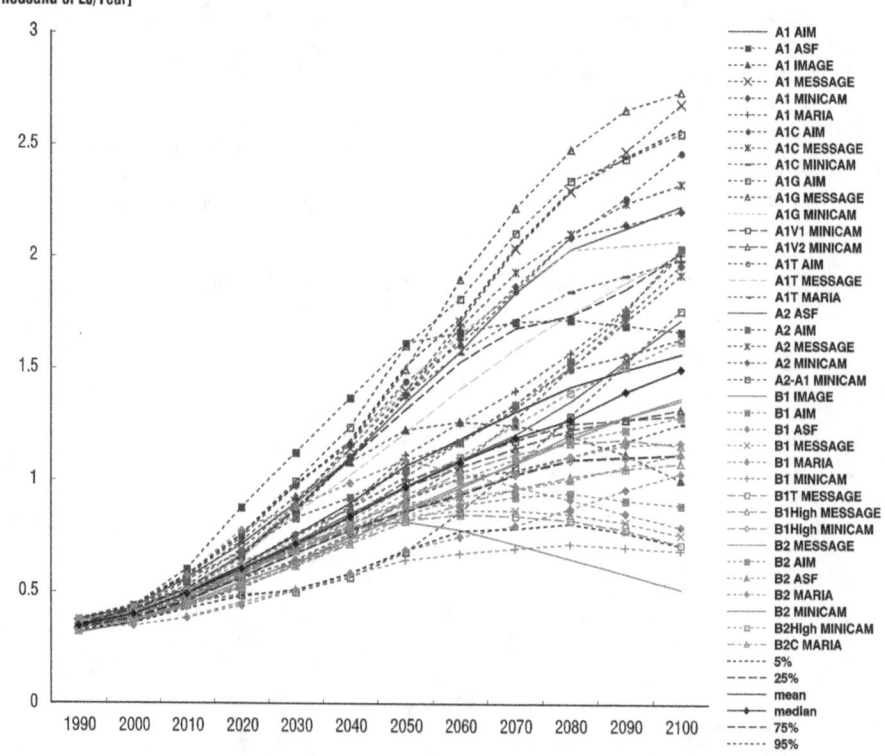

Figure 3-3. Primary Energy Use Projections Using Various Energy-Economic Models and Assumptions
Note: The mean, median, and percentile bands in the figure are based on the range of projections and do not represent probabilities of the projections. (Source: IPCC 2000)

---

7   One study that examined emissions growth in the absence of special initiatives directed at climate change is the *Special Report on Emissions Scenarios (SRES)* by the Intergovernmental Panel on Climate Change (IPCC 2000), in which six leading energy-economic models were used to explore a suite of scenarios that projected growth in global energy and GHG emissions.

8   See http://www.stanford.edu/group/EMF/publications/index.htm.

9   See for example, *Direct and Indirect Human Contributions to Terrestrial Carbon Fluxes: A Workshop Summary* (2004) and *Human Interactions with the Carbon Cycle: Summary of a Workshop* (2002), both available from the National Academies Press (Coppock and Johnson 2004 and Stern 2002).

of the 21st century, a greater percentage of the world's population is expected to gain access to electricity and commercial fuels, as well as experience major improvements to quality of life. These changes are expected to result in increased per capita energy use. In addition, world population is expected to grow, which will further increase overall demand for energy.

Estimates of projected energy demand vary considerably. The Energy Information Administration (EIA 2005) projects world primary energy demand will increase from about 411.5 exajoules (EJ) in 2002 to 680 EJ in 2025. Most of that increase will come in demand from developing countries. EIA forecasts primary energy use in the developed world will rise just 1.1 percent annually while that in the developing world will rise 3.2 percent annually. Energy use in the emerging

economies of developing Asia, driven largely by demand in China and India, is projected to more than double over the course of the quarter century.

In the IPCC's *Special Report on Emissions Scenarios* (SRES) (IPCC 2000), projected world primary energy use in 2100 fell within a range of 600 to 2800 EJ for 90 percent of the scenarios explored (Figure 3-3). Among the scenarios surveyed in SRES, the average annual growth rates for global energy demand over the period from 2000 to 2100 ranged from 2.4 percent per year to -0.1percent per year, with a median value of 1.3 percent per year.[10]

As about four-fifths of GHG emissions are energy related, energy generation and consumption are key determinants of CO_2 emissions. The scenarios with the highest CO_2 emissions are those that assume the highest energy demand along with the highest

CO_2 Emissions Projections from Energy Use Using Various Energy-Economic Models and Assumptions

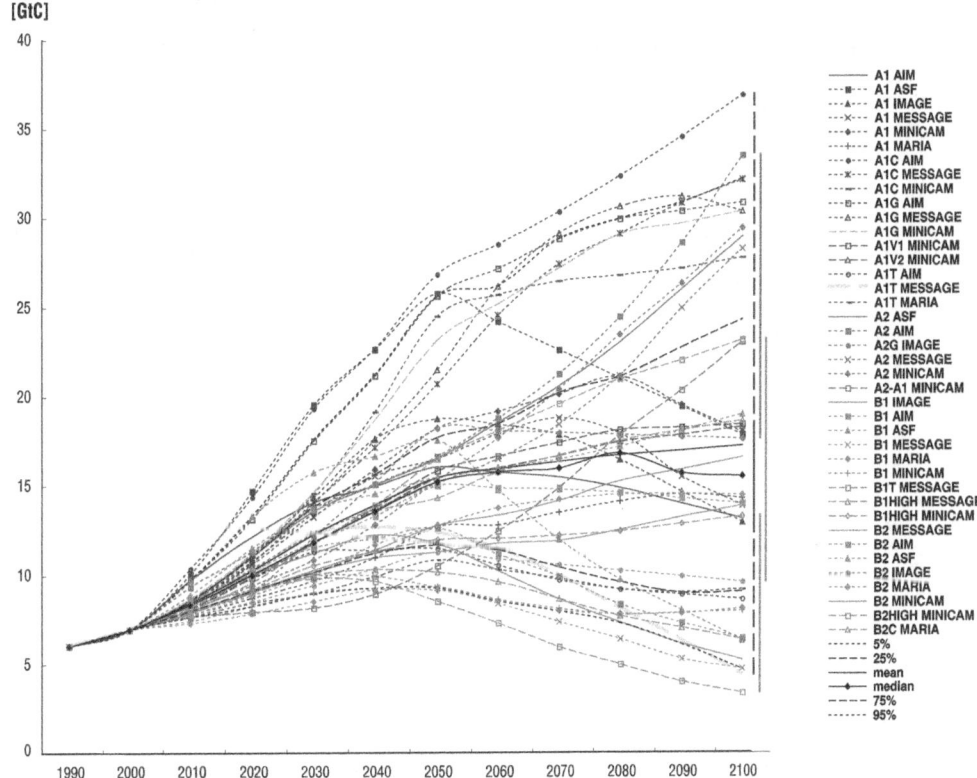

Figure 3-4. *CO_2 Emissions Projections from Energy Use Using Various Energy-Economic Models and Assumptions*
Note: *The mean, median, and percentile bands in the figure are based on the range of projections, and do not represent probabilities of the projections. (Source: IPCC 2000)*

---

[10] Scenarios that show low or negative energy consumption growth rates over time represent cases where technological improvement is projected to be very rapid and where population and GDP growth rates lie at the lower bounds of the projections.

proportion of fossil fuel use unaccompanied by CO_2 capture and storage. Since 1900, global primary energy consumption has, on average, increased at more than two percent per year, similar to the 20-year trend from 1975 to 1995.

Without constraints, energy-related CO_2 emissions are expected to increase significantly over the next 100 years at rates similar to those for the growth in energy consumption. Specific projections vary depending on assumptions. According to EIA (2005), annual global CO_2 emissions may increase by about 60 percent between 2002 and 2025. Higher growth rates are expected in the developing regions of the world, where CO_2 emissions may increase by a factor of two or more by 2025.

In 2025, global use of petroleum products, primarily in the transportation sector, is expected to continue to account for the largest share of global emissions of CO_2. This is followed in importance by the use of coal, primarily for electricity generation, and natural gas. Natural gas is a versatile fuel, used for power generation, residential and commercial fuel, and many other uses. CO_2 emissions from fossil fuel combustion by end-use sector for 2002 can be broken down as follows: transportation, 24 percent; electricity and heat, 35 percent; and industrial and other end uses, 41 percent.

IPCC's SRES examined projections of CO_2 emissions from energy use based on multiple reference scenarios from six long-term modeling efforts. It found that different assumptions about the driving forces led to widely divergent emissions trajectories (Figure 3-4). Ninety percent of the CO_2 emissions projections fall within the upper and lower bounds shown in Figure 3-4. The mean, median, and percentage bands shown were calculated based on the range of projections across the full set of scenarios and do not represent probabilities associated with the projections.

The upper bounds in Figures 3-3 and 3-4 are formed by scenario results that assume very high world economic growth, high per-capita energy use, and continued dominance of fossil fuels. At this upper bound, world CO_2 emissions from energy use are projected to grow from about 6 GtC/year in 2000 to more than 30 GtC/year in 2100—a five-fold increase.

The lower bounds in Figures 3-3 and 3-4 are formed by scenarios that assume less population growth, changes in the composition of economic activity away from energy-intensive output, lower per-capita energy use, more energy efficiency, and considerably more use of carbon-neutral fuels, compared to the upper bound. At this lower bound, CO_2 emissions are projected to grow for the first half the century, but then to decline to levels about equal to those in 2000, representing no net growth by 2100. Assumptions for the various scenarios are described in Box 3-1.[11] The models used in this study include AIM,[12] ASF,[13] IMAGE,[14] MARIA,[15] MESSAGE,[16] and MiniCAM.[17]

Recent studies have explored the uncertainty in future emissions using a probabilistic approach (see for example, Webster et al. 2002).[18] While there are some differences in the upper and lower bounds of the emissions projections between the SRES scenarios and these more recent probabilistic-based analyses, the range of the SRES scenarios overlaps to a large degree with the range of emissions estimated using these probabilistic approaches.

---

[11] The range of CO_2 emissions in the SRES has been compared to scenarios done later (post-SRES). In general, the ranges are not very different. The estimated CO_2 emissions in post-SRES scenarios have a higher lower bound, a similar median, and a higher upper bound of the distribution. The post-SRES scenarios use lower population estimates, both in range and median. The post-SRES economic development projections (based on market exchange rates) have approximately the same lower bound and median but a lower upper bound of the distribution. A comprehensive database of emissions scenarios is available at http://www-cger.nies.go.jp/cger-e/db/enterprise/scenario/scenario_index_e.html.

[12] Asian Pacific Integrated Model (AIM) from the National Institute of Environmental Studies in Japan (Morita et al. 1994).

[13] Atmospheric Stabilization Framework Model (ASF) from ICF Consulting in the USA (Lashof and Tirpak 1990; Pepper et al. 1992, 1998; Sankovski et al. 2000).

[14] Integrated Model to Assess the Greenhouse Effect (IMAGE) from the National Institute for Public Health and Environmental Hygiene (RIVM) (Alcamo et al. 1998; de Vries, Olivier et al. 1994, de Vries, Janssen et al. 1999; de Vries, Bollen et al. 2000), used in connection with the Dutch Bureau for Economic Policy Analysis (CPB) WorldScan model (de Jong and Zalm 1991), the Netherlands.

[15] Multiregional Approach for Resource and Industry Allocation (MARIA) from the Science University of Tokyo in Japan (Mori and Takahashi 1999; Mori 2000).

[16] Model for Energy Supply Strategy Alternatives and their General Environmental Impact (MESSAGE) from the International Institute of Applied Systems Analysis (IIASA) in Austria (Messner and Strubegger 1995; Riahi and Roehrl 2000).

[17] Mini Climate Assessment Model (MiniCAM) from the Pacific Northwest National Laboratory (PNNL) in the USA (Edmonds et al. 1994, 1996a, 1996b).

[18] Use of uncertainty analysis and probabilistic forecasting can help identify and quantify critical but uncertain parameters (such as demographic or technology trends over time). Multiple simulations are performed by sampling from those distributions to construct probability distributions of the outcomes (such as GHG emissions). Distributions for factors, such as labor productivity growth, energy efficiency improvement, agricultural and industrial emissions coefficients for various GHGs, etc., are quantified by expert elicitation or from a review of the literature. These distributions are then used in assessment models to generate a distribution of results, such as GHG emissions and/or climate impacts (e.g., temperature change or sea-level rise).

BOX 3-1 THE SRES SCENARIOS

The SRES scenarios are organized around four major storylines, which received the names A1, A2, B1, and B2. Each of these storylines represented different general conceptions of how the world might evolve over time, including the evolution of key drivers such as economic growth (including differences or convergence in regional economic activity), population growth, and technological change. Each driver was interpreted by the participating modeling teams in terms of quantitative assumptions about the evolution of specific model parameters. Some scenario drivers, such as economic growth, final energy, and population growth, were harmonized across many of the models, while others, such as the specific technology assumptions, were developed by the individual modeling teams to be generally consistent with the storylines. For the A1 Scenario, four basic assumptions about technology were also developed, so there are four categories of technology scenarios under the A1. The scenarios are described as follows:

A1. The A1 storyline and scenario family describe a future world of very rapid economic growth, global population that peaks in mid-century and declines thereafter, and the rapid introduction of new and more efficient technologies. Major underlying themes are convergence among regions, capacity building, and increased cultural and social interactions, with a substantial reduction in regional differences in per capita income. The A1 scenario family develops into three groups that describe alternative directions of technological change in the energy system. The four A1 groups are distinguished by their technological emphasis: fossil intensive (A1C – coal- and A1G – gas), non-fossil energy sources (A1T), or a balance across all sources (A1B), where balanced is defined as not relying too heavily on one particular energy source, on the assumption that similar improvement rates apply to all energy supply and end-use technologies.

A2. The A2 storyline and scenario family describe a very heterogeneous world. The underlying theme is self-reliance and preservation of local identities. Fertility patterns across regions converge very slowly, which results in a continuously increasing population. Economic development is primarily regionally oriented, and per capita economic growth and technological change more fragmented and slower than in other storylines.

B1. The B1 storyline and scenario family describe a convergent world with the same global population, which peaks in mid-century and declines thereafter, as in the A1 storyline, but with rapid change in economic structures toward a service and information economy, with reductions in material intensity and the introduction of clean and resource-efficient technologies. The emphasis is on global solutions to economic, social, and environmental sustainability, including improved equity, but without additional climate initiatives.

B2. The B2 storyline and scenario family describe a world in which the emphasis is on local solutions to economic, social, and environmental sustainability. It is a world with continuously increasing global population, at a rate lower than in A2, intermediate levels of economic development, and less rapid and more diverse technological change than in the B1 and A1 storylines. While the scenario is also oriented towards environmental protection and social equity, it focuses on local and regional levels.

The set of harmonized drivers depended both on the scenario and the specific model. Key drivers that characterized the scenarios are summarized qualitatively in the table below. All scenarios assume energy intensity reductions over the coming century at or greater than the historical average over the past few decades. Comparison of the emissions trajectories in Figures 3-5 and 3-6 can be interpreted in terms of the relative evolution of these drivers and the discussion of these drivers above.

DRIVER	A1				A2	B1	B2
	A1C	A1G	A1B	A1T			
Population Growth	low	low	low	low	high	low	medium
GDP Growth	very high	very high	very high	very high	medium	high	medium
Energy Use	very high	very high	very high	high	high	low	medium
Energy Intensity Improvement	high	high	high	high	low	very high	medium
Land-use Changes	low-medium	low-medium	low	low	medium-high	high	medium
Availability of Conventional Oil & Gas	high	high	medium	medium	low	low	medium
Pace of Technological Change	rapid	rapid	rapid	rapid	slow	medium	medium

CO$_2$ Emissions and Sequestration from Changes in Land Use

CO$_2$ emissions in the future will be influenced not only by trends in CO$_2$ emissions from energy use and industrial sources, but also by trends in land use that result in either net CO$_2$ sequestration or release. CO$_2$ emissions and carbon sequestration associated with various land uses will be driven primarily by increasing demand for food. Other important factors include demand for wood products, land management changes, demand for biomass energy and bio-based products, and technological change.

The role of land-use change has received relatively limited consideration (compared to energy use) in prior modeling exercises aimed at developing long-run GHG emissions scenarios. To date, the most comprehensive treatment is contained in the scenarios developed for the IPCC SRES (IPCC 2000). In developing these scenarios, the IPCC assembled a data base of over 400 earlier emissions scenarios. Of these, 26 scenarios (all the work of three modeling groups) explicitly considered the role of land-use change on global CO$_2$ emissions. The projections vary considerably in the near term, with some

scenarios showing increasing and some decreasing net global CO$_2$ emissions from land-use change (Figure 3-5). A key insight to emerge from the IPCC exercise was that the link between land-use change and global CO$_2$ emissions is more complex and uncertain than had been reflected in previous analyses.

Across and within the four storylines described in Box 3.1, the scenarios produced a wide range of land-use paths that included large increases and decreases in the global areas of cropland, grassland, and forest over the course of the century. Differences in land-use patterns resulted primarily from alternative assumptions about population and income growth. Hence, the scenarios indicate that land-use change could become either an important source or sink of global CO$_2$ emissions over the next 100 years, depending on the mix of goods and services the world's population demands from its land resources.

Further, future paths of technological change in today's land-intensive sectors—including agriculture, forestry, energy, construction, and environment quality—will help to define the contribution land-use change makes to net CO$_2$ emissions. Many of the IPCC scenarios show that CO$_2$ emissions from

Net CO$_2$ Emissions from Land-Use Change

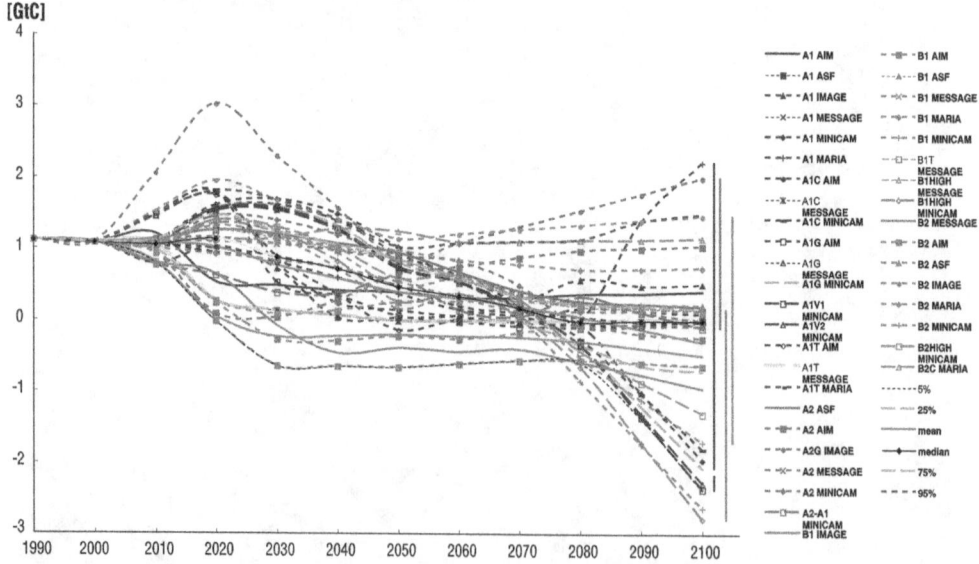

Figure 3-5.   Net CO$_2$ Emissions from Land-Use Change
Note: The mean, median and percentile bands in the figure are based on the range of projections, and do not represent probabilities. (Source: IPCC 2000)[19]

---

[19] The structure of the underlying modeling exercise required harmonization in 2000. Such harmonization in the context of a modeling exercise does not necessarily reflect agreement.

deforestation are likely to peak after several decades and then subsequently decline.[20] Despite the differences among the scenarios' assumptions, most of the scenarios show a net decrease to below current levels by the end of the century.

Sohngen and Mendelsohn (2003) linked a global forestry model with the global optimization model DICE to more explicitly explore the relationships between forestry management, land-use emissions, and global energy systems. They reported a net sequestration potential ranging from 32 to 102 GtC in the coming century, depending on carbon prices. Much of the sequestration occurs through avoided deforestation.

Other Greenhouse Gases

The non-CO_2 GHGs include a diverse group of gases such as methane, nitrous oxide, chlorofluorocarbons, and other gases with high global warming potential (see Chapter 7). Future growth in emissions of non-CO_2 GHGs will depend on the future level of the activities that emit these gases, as well as the amount of emissions control that occurs. The cost-effectiveness of emission controls for mitigating the various GHGs will depend on their relative estimated costs and anticipated climate-related benefits.

Integrated assessment models have only recently begun to project long-term trends in non-CO_2 GHGs. In a recent international modeling exercise conducted by the Stanford Energy Modeling Forum (EMF-21), non-CO_2 emissions and mitigation potential were projected by 18 models of various forms (Weyant and de la Chesnaye 2005).[21] Each model ran a "reference case" scenario, in which non-CO_2 GHGs were allowed to grow in the absence of any constraints or incentives for GHG emissions mitigation.

The results for methane and N_2O are shown in Figures 3-6 and 3-7, respectively. The projections vary considerably among models. On average, emissions of non-CO_2 GHGs were projected to increase from 2.7 gigatons of carbon equivalent emissions (GtC-eq.) in 2000 to 5.1 GtC-eq. in 2100. On average, methane emissions were projected to increase by 0.6 percent/year between 2000 and 2100; nitrous oxide by 0.4 percent/year; and the fluorinated

gases by 1.9 percent/year. (By comparison, in these same scenarios, CO_2 emissions were projected to grow by 1.1 percent/year over the same time period.)

A recent modeling study conducted for CCTP projected the contributions of CO_2 and other GHGs to future radiative forcing (Clarke et al. 2006). In the reference case scenario (without actions specifically targeted toward lowering GHGs), radiative forcing from pre-industrial levels was projected to increase to 6.5 watts per square meter (W/m^2), compared to a level of about 2 W/m^2 in 2000 (Figure 3-8). In this projection, CO_2 contributed approximately 80 percent of the radiative forcing in 2100, with other GHGs (CH_4, N_2O, and the F-gases) contributing about 20 percent.

Some other models show larger contributions from non-CO_2 GHGs. The analysis for CCTP (which used MiniCAM) projects a considerable amount of control of methane in the absence of mandated controls.[22] In results from other models, radiative forcing was projected to increase to over 9 W/m^2, due in part to growth in non-CO_2 GHG emissions (van Vuuren et al. 2006). These studies indicate the importance of non-CO_2 GHGs, especially in the future, as their emissions rise as a result of increased industrial and agricultural activity and population growth.

3.3 Analytical Context for CCTP Planning

For the purposes of CCTP planning and analysis, it is useful to understand the potential contributions that advanced technologies could make to future GHG emissions reductions and to the potential stabilization of atmospheric GHG concentrations and its integrated multi-gas metric, radiative forcing, over a century-long planning horizon. No particular GHG stabilization level is assumed in this plan, nor is there an assumed "best path" for reaching stabilization. Accordingly, the synthesis assessment of various analyses of advanced technology scenarios explores two degrees of freedom (or uncertainty)—one is about the hypothesized levels of GHG concentrations

---

[20] This pattern is tied to declines in the rate of population growth toward the latter half of the century and increases in agricultural productivity.

[21] The models included a variety of model types, including integrated assessment models and general equilibrium models.

[22] Note that certain models (such as MiniCAM) project that some GHG-reducing technologies penetrate the market without incentives or policies. For example, in MiniCAM, technologies for reducing methane emissions from coal and natural gas production would penetrate the market when it is cost-effective to do so, based on the value of the methane (natural gas) collected, which is marketable as a fuel.

Methane Emissions Projections from the EMF-21 Study, With No Explicit Initiatives to Reduce GHG Emissions

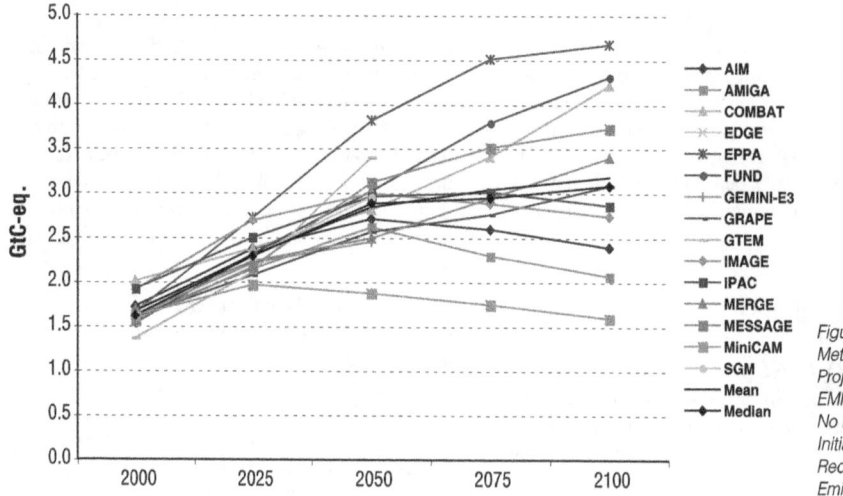

Figure 3-6.
Methane Emissions
Projections from the
EMF-21 Study, With
No Explicit
Initiatives to
Reduce GHG
Emissions

Nitrous Oxide Emissions Projections from the EMF-21 Study, With No Explicit Initiatives to Reduce GHG Emissions

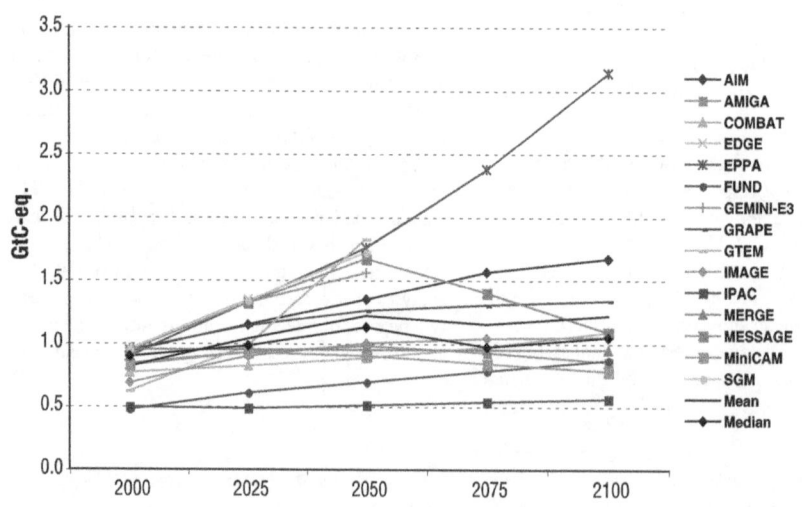

Figure 3-7.
Nitrous Oxide
Emissions
Projections from
the EMF-21
Study, With No
Explicit
Initiatives to
Reduce GHG
Emissions

Radiative Forcing in a Reference Case Scenario

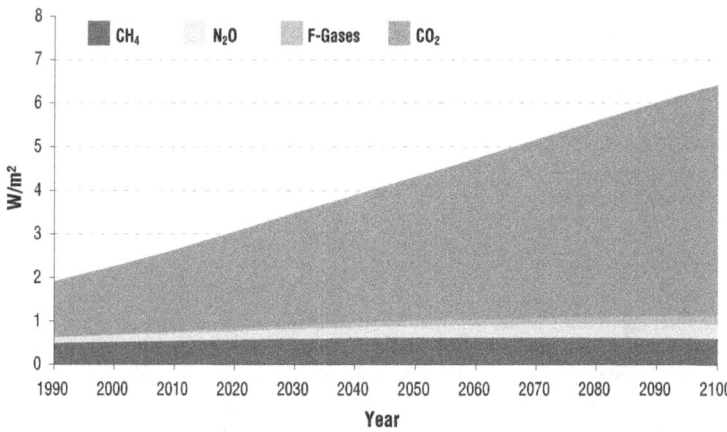

Figure 3-8. Radiative
Forcing in a Reference
Case Scenario
(Source: Clarke et al.
2006)

and the other is about the technologies. Useful insights may be gained by modeling the hypothesized levels across a range of assumptions about, and mixes of, advanced technologies.

Cost-effective means to stabilize radiative forcing would include reductions in emissions of both CO_2 and other GHGs. A recent analysis by Pacific Northwest National Laboratory (PNNL) examined various GHG emissions reduction options associated with a range of radiative forcing levels in the year 2100, each leading to long-run stabilization of radiative forcing (Clarke et al. 20006). The results are presented in Figure 3-9, which compares the estimated radiative forcing in 2000 to projected radiative forcing in 2100 under an unconstrained

emissions scenario (Reference Case) and four emissions-constrained scenarios that lead to lower radiative forcing in 2100.

Other studies also examined integrated multi-gas strategies for reducing radiative forcing, considering possible tradeoffs among CO_2 and other GHG emissions. For example, using results from a variety of different models, an analysis by Weyant and de la Chesnaye (2005) showed that a multi-gas approach results, on average, in CH_4 emissions reductions of almost 50 percent, N_2O reductions of about 40 percent, and a reduction in F-gases of almost 75 percent (on a GtC-eq. basis) by 2100, compared to reference case levels.

Radiative Forcing Levels under Different Degrees of Constraint

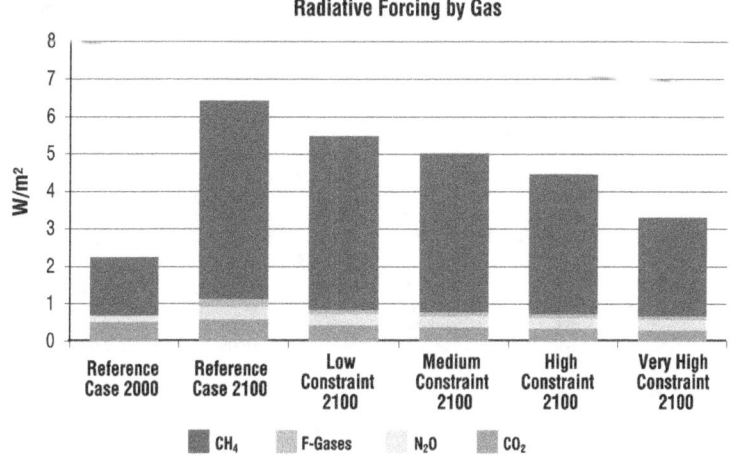

Figure 3-9. Radiative
Forcing Levels under
Different Degrees of
Constraint
(Source: Clarke et al.
2006)

For any assumed constrained radiative forcing scenario, reductions in the non-CO_2 gases lead to the need for smaller reductions in CO_2 emissions. Nevertheless, CO_2 remains by far the most important GHG. Accordingly, a range of potential CO_2 stabilization levels have been explored, using different models and assumptions. Figure 3-10 shows one set of relationships between CO_2 emissions and CO_2 concentrations over time, across a range of CO_2 stabilization levels commonly found among the literature on scenarios.[23]

The set of hypothetical CO_2 emissions scenarios (Figure 3.10-A), shown here across a range of five corresponding CO_2 concentration stabilization levels (Figure 3.10-B), illustrates a general pattern found consistently across the analyses. Emissions scenarios leading to CO_2 stabilization typically show growth of emissions in the near term, but with that growth slowing; the emissions eventually peak and then gradually decline; and ultimately they approach levels that are low or near zero. In almost all stabilization scenarios, emissions must continue to decline beyond 2100 and into the 22nd century and beyond. Over the same time period, as discussed earlier, the world's energy needs can be expected to continue to grow.

An illustration of the scale of the overall emissions reductions needed to achieve stabilization of CO_2 concentrations is shown in Figure 3-11. To meet the CO_2 stabilization level in this hypothetical example, by 2100 annual CO_2 emissions would need to be reduced by almost 15 GtC-eq/year below the level in an otherwise "unconstrained" emissions scenario case.[24] For the example shown, the cumulative emissions reduction would be approximately 600 GtC-eq over the course of the 21st century. For other scenarios (PNNL), the 100-year cumulative CO_2 emissions reductions ranged from about 300 to about 1,000 GtC-eq.[25]

Illustrative CO_2 Emissions Profiles and Corresponding Concentrations

A. Emissions Scenarios from Fossil Fuel Use and Industrial Activities

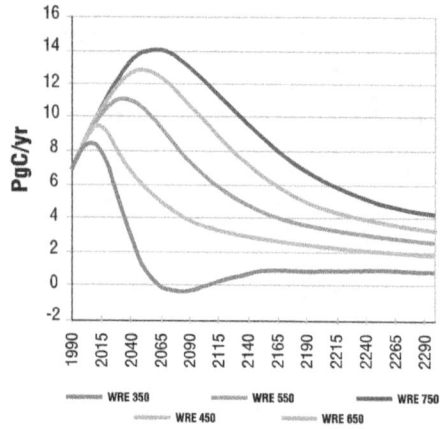

B. Corresponding Atmospheric Concentration Levels

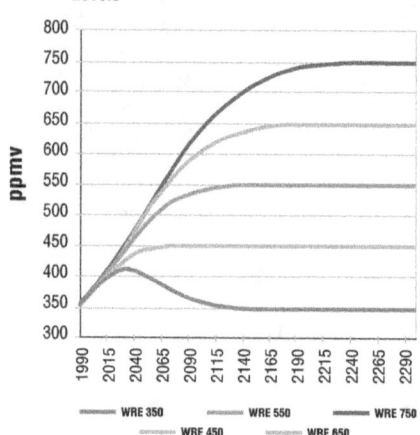

Figure 3-10. Illustrative CO_2 Emissions Profiles and Corresponding Concentrations

[23] Derived from Wigley et al. 1996. The emissions scenarios represent net emissions from fossil fuels (i.e., including emissions reductions from carbon dioxide capture and storage) and industrial sources. They do not include emissions from land use and land-use change. The concentration levels are based on a range of specific assumptions regarding net emissions from land use and land-use change, and about the carbon cycle more generally, including assumptions regarding the rate of ocean uptake. Note that significant uncertainties remain about many aspects of the carbon cycle; reducing these carbon cycle uncertainties is one of the goals of the U.S. Climate Change Science Program (CCSP). Other estimated scenarios showing relationships between emissions and concentrations can be found in the literature on scenarios.

[24] The "unconstrained" case in this illustrative example is based on the reference scenario developed for CCTP by PNNL; see Clarke et al. (2006). Note that this reference scenario includes a considerable amount of energy efficiency as well as increased use of renewable and nuclear energy resulting from improvements in these technologies over time, increased prices for fossil fuels, and hence increased ability of renewable and nuclear technologies to compete in the market. The lower curve represents a reduced-emissions scenario leading to stabilization; it is a slightly modified version of the 550 ppmv trajectory shown in Figure 3-10A.

[25] See Clarke et al. 2006 and IPCC 2001. Estimates of the emissions reductions required to stabilize concentrations are uncertain and vary based on assumptions. Key assumptions include: (1) estimates of future emissions to 2100 in the absence of actions aimed at GHG mitigation (i.e., the "reference" scenario); (2) selection of the stabilization level or levels of atmospheric GHG concentrations; and (3) the nature of baseline and advanced technology scenarios and their associated time-specific emissions trajectories. Results for CCTP are shown as "mitigated emissions" on Figure 3-14.

Potential Scale of CO_2 Emissions Reductions to Stabilize GHG Concentrations: Hypothetical Unconstrained and Constrained Emissions Scenarios

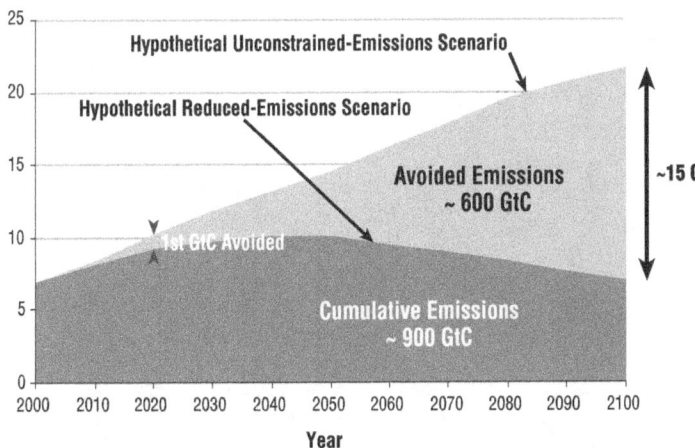

Figure 3-11. Potential Scale of CO_2 Emissions Reductions to Stabilize GHG Concentrations: Hypothetical Unconstrained and Constrained Emissions Scenarios (Source: Clarke et al. 2006)

Box 3-2 provides illustrations of technological measures that could achieve an annual reduction of one GtC-eq./yr. As these illustrations suggest, GHG-reducing technologies would need to be implemented on a significant scale. The costs of achieving such reductions using today's technology options could be high. Developing and deploying advanced technology with significantly improved performance and cost characteristics could substantially lower these costs, thereby facilitating their entry into the marketplace or their expanded adoption. The implication for CCTP and its associated science and technology R&D programs is that advanced technologies, including novel or breakthrough technologies, are important, if not essential, to the CCTP goals of significantly reducing or avoiding GHG emissions, while maintaining economic growth and ensuring safety and overall environmental quality.

3.4 Exploring the Role for Advanced Technology

Reducing or avoiding GHG emissions on the scale hypothesized in Section 3.3 could be facilitated by a variety of advanced scenarios, each characterized by a different mix of technologies, estimated cost, and timing of the emission reductions. Given the diversity of activities and processes that emit GHGs, achieving emissions reductions on such a scale will likely require significant contributions from a combination of existing, improved or transitional, and advanced technologies. At present, the "best" combination of technologies (and other means) is unknown, but insights about the role for technology can be gained through analysis.

Estimating the potential contribution to CCTP strategic goals of any technology is difficult and depends in large part on assumptions about the success of scientific and technical advancements and other factors. These assumptions may be explored in scenario analyses featuring advanced technology. For example, several studies have projected that lower carbon fuels (e.g., natural gas) and lower GHG-intensity technologies (e.g., coal gasification) could bridge the transition to zero- or very low-carbon technologies.[26] Two other themes common among many GHG mitigation technology scenarios are steady improvement in energy efficiency (e.g., lowering of GHG-intensity) and the emergence of bio-based products as an important energy source throughout the 21st century.

In addition to technological considerations, cost considerations are major elements in the mitigation

---

[26] See, for example, studies by van Vuuren et al. (2004) and Manne and Richels (2004), and the the mitigation scenarios studied in the IPCC Working Group III (see http://www.grida.no/climate/ipcc_tar/wg3/084.htm).

BOX 3-2

HOW BIG IS ONE GIGATON PER YEAR OF GHG REDUCTION?

Actions that provide 1 gigaton per year of carbon-equivalent mitigation for the duration of their existence:

- **Coal-Fired Power Plants.** Build 1,000 "zero-emission" 500-MW coal-fired power plants to supplant coal-fired power plants without CO_2 capture and storage. (Current global installed generating capacity is about 2 million MW.)

- **Geologic Storage.** Install 3,700 carbon storage sites like Norway's Sleipner project (0.27 MtC/year).

- **Nuclear.** Build 500 new nuclear power plants, each 1 GW in size, to supplant an equal capacity of coal-fired power plants without CO_2 capture and storage. This would more than double the current number of nuclear plants worldwide.

- **Electricity from Landfill Gas Projects.** Install 7,874 "typical" landfill gas electricity projects (typical size being 3 MW projects at non-regulated landfills) that collect landfill methane emissions and use them as fuel for electric generation.

- **Efficiency.** Deploy 1 billion new cars at 40 miles per gallon (mpg), instead of new cars at 20 mpg.

- **Wind Energy.** Install 650,000 wind turbines (1.5 MW each, operating at 0.45 capacity factor) to supplant coal-fired power plants without CO_2 capture and storage.

- **Solar Photovoltaics.** Install 6 million acres of solar photovoltaics to supplant coal-fired power plants without CO_2 capture and storage (assuming 10% cell DC efficiency, 1700 kWhr/m² solar radiance, and 90% DC-AC conversion efficiency).

- **Biomass Fuels from Plantations.** Convert a barren area about 15 times the size of Iowa's farmland (about 33 million acres) to biomass crop production.

- **CO_2 Storage in New Forest.** Convert a barren area about 40 times the size of Iowa's farmland to new forest.

Note: SRES (IPCC 2000) scenarios assume that all of these technologies will be used extensively prior to 2100.

---

scenarios. When projected declines in the costs of low-carbon-emitting technologies make them attractive economically, they play major roles in many scenarios. Different technologies may mature and become cost-competitive at different times over the course of a 100-year planning horizon. For example, increased energy efficiency (using today's technologies), mitigation of non-CO_2 GHGs such as methane, and terrestrial sequestration may be cost-effective options in the near term. Transformative supply-side and end-use technologies with greatly reduced GHG emissions profiles could become more cost-effective later, as these technologies advance.

Several landmark multi-model studies,[27] as well as various scenario analysis efforts based on individual models, have explored a range of emission reduction scenarios. In most of these analyses, advanced technology scenarios are modeled under a range of hypothetical GHG emissions constraints (e.g., low,

medium, high, and very high). These advanced technology scenarios, in turn, can be compared against baseline scenarios, where the given GHG emissions constraints are met, but with less optimistic assumptions about the advancement of technology. The results can suggest what roles various technologies may play, if assumptions about their advancement could be realized. The results can also, by inference, provide insights about various R&D programmatic goals, suggesting what technological progress would be needed, and by when, in order to achieve the hypothesized results.

Reference Case, Baseline, and Advanced Technology Scenarios

A number of analytical approaches have been pursued to explore the potential contributions of advanced

---

[27] For example, the IPCC "Post-SRES" report on mitigation (IPCC 2001) and the EMF studies (Weyant 2004).

[28] The low- or no-cost suite of technologies generally includes improvements to current systems and various cost-effective energy conservation strategies. These are usually modeled as a general rate of energy-efficiency (or intensity) improvement, and are often included in the business-as-usual (or "reference case") emissions projections.

technologies. One approach is to focus on a particular technology or genre of technology, one at a time, and estimate what could be achieved if it were to be fully adopted by a certain time in the future. For example, Brown et al. (1996) estimated the amount of mitigation that could be achieved through use of a variety of individual technologies. More recently, Pacala and Socolow (2004) discussed technology "wedges," each of which represents the mitigation of one gigaton of carbon emissions in the year 2050 (Box 3-2), some of the examples of which were inspired by Pacala and Socolow). Hoffert et al. (2002) examined technologies needed to deliver a certain amount of carbon-free energy by the end of the 21st century. Such approaches are useful for understanding the technical potential of various technologies.

Other analytical approaches address important underlying factors that may influence a technology's ability, within a larger competitive context, to achieve its technical potential. In this context, advanced technologies would need to meet an array of conditions before they could be successfully adopted. They would need to be cost-competitive, in the context of the future global economy and the world energy market, compared to other available technologies. Other considerations include ease of use, reliability, public safety, and acceptance; and still others include policy, environmental, or regulatory factors. Taking these considerations into account requires an integrated assessment approach using models, which typically require competition among technologies to meet certain exogenously imposed emissions constraints, in conjunction with other emissions-related factors and considerations.

Such models simulate, for each step in time, the competitive deployment of technologies and approaches that would be needed to achieve a given amount of emissions reductions at the lowest cost in that time period. Depending on the level of emissions constraint assumed, low or no-cost approaches may supply a large portion of the emissions reduction.[28] More costly advanced technologies may be adopted more widely in scenarios that require moderate to high levels of emissions reduction. Expensive, undeveloped, or undemonstrated technologies, or technologies that face difficult challenges to wide-scale deployment,

may enter the market later in the mitigation period. Hence, the mix of technologies in any given scenario depends on assumptions about technical readiness, costs, and barriers to adoption for each technology, as well as the level of emissions constraint assumed.

In a recent model-based integrated assessment, sponsored by CCTP and conducted by PNNL, a set of 17 scenarios was developed to explore and compare advanced technology options for achieving significant GHG emission reductions. The 17 scenarios include a Reference Case and four sets of GHG-emissions-constrained scenarios (each set has four different levels of emission constraint). One set of emissions-constrained scenarios (the Baseline Scenarios) assumes reference case technologies are available to meet the emissions constraints, and three sets of emissions-constrained scenarios assume advanced technologies become available.[29] The scenarios are summarized as follows:

◆ A Reference Case scenario represents a hypothetical technological future, where GHG emissions are not constrained, but where significant technical improvements are achieved in a broad spectrum of currently known or available technologies for supplying and using energy.[30] This scenario results in improvements in global GHG-intensity over time, but not in lower GHG emissions. It provides reference level energy and emissions projections to which the energy and emissions in the emissions-constrained scenarios can be compared.

◆ A set of four Baseline Scenarios use the same Reference Case technology assumptions described above but applies four hypothesized GHG emissions constraints. Because these scenarios require emission reductions from the Reference Case, low- or zero-emission technologies and other means to reduce GHG gases are deployed at higher rates in these baseline emission-constrained scenarios than in the Reference Case. These baseline scenarios provide energy and mitigation cost projections to which the energy mix and costs in the advanced technology scenarios can be compared.

◆ Each of the three advanced technology scenarios includes a distinct set of technology advancements, beyond those in the Reference Case.[31] Each of

---

[29] The four hypothesized GHG emissions constraints (i.e., Very High, High, Medium, and Low) were designed to stabilize, over the long term, the aggregated radiative forcing of the following GHGs: CO_2, CH_4, N_2O, and the so-called "F-gases" (hydrofluorocarbons [HFCs], perfluorocarbons [PFCs], and sulfur hexafluoride [SF_6]). A range of additional substances, such as aerosols, also have important effects on radiative forcing. These substances are not included in this analysis.

[30] The reference case assumes energy efficiency improvements occur, as well as cost decreases in renewable and nuclear technologies that bring their costs to below today's levels.

[31] In the PNNL analysis, Scenarios 1, 2, and 3 are given representative labels of "Closing the Loop on Carbon," "A New Energy Backbone," and "Beyond the Standard Suite," respectively.

these, in turn, is also applied under the four GHG emissions constraints (for a total of twelve advanced technology cases). The advanced technology scenarios include:

Scenario 1, which assumes successful development of carbon capture and storage technologies for use in electricity, as well as in applications such as hydrogen and cement production.

Scenario 2, which assumes additional technological improvement and cost reduction for carbon-free energy sources, such as wind power, solar energy systems, and nuclear power.[32]

Scenario 3, which assumes major advances in fusion energy and/or novel energy applications for solar energy and biotechnology such that they can provide zero-carbon energy at competitive costs in the second half of this century.[33]

A number of common features cut across all three of the advanced technology scenarios:

◆ Additional gains in energy efficiency beyond the reference case occur.

◆ Additional technologies for managing non-CO_2 GHGs become available.

◆ Terrestrial carbon sequestration increases.

◆ The full potential of conventional oil and gas is realized.

◆ Hydrogen production technology advances.

None of the advanced technology scenarios is intended to represent a preferred "path" to the future. Rather, each is designed with unique features, distinct from the others, so as to explore orthogonally a wide range of technology options. The current CCTP portfolio of technology R&D is diversified and includes elements of all three of the advanced technology scenarios. The scenarios are not necessarily intended to represent every component of the CCTP portfolio, however. Instead, they illustrate several possible pathways to lower emissions that

could be the outcome of successful technology R&D.

Figures 3-12 and 3-13 provide illustrative results across the three advanced technology scenarios for a high emissions constraint case. Figure 3-12 shows the mix of technologies and their associated contributions to total global energy demand for the three advanced energy scenarios. Figure 3-13 shows the CO_2 emissions reduction contributions from the various energy sources and technologies, under the same set of assumptions.

Although each scenario assumes advances in particular classes of technology, all scenarios result in a mix of energy efficiency and energy supply technologies. The overall results show the extent of the variation possible in the mix of emissions-reducing technologies under a variety of assumptions and planning uncertainties.

Economic Benefits of Advanced Technologies

The purpose of CCTP is to accelerate the development of promising technologies that can reduce, avoid, capture, or sequester emissions of GHGs at greatly reduced cost compared to current technologies. Providing advanced technology options can enable progress toward CCTP's strategic goals through greater choice and competition. Stated in other terms, the same amount of GHG emissions could be reduced with advanced technology options at costs significantly lower than would otherwise be the case had they not been developed or made available.

In the PNNL analysis (Clarke et al. 2006), the estimated costs of achieving a range of emission reductions were compared for cases with and without the use of advanced technology.[34] The scenarios described above were supplemented by an additional scenario in which the advanced technologies in all three scenarios were combined in one model run. The resulting cost estimates (Figure 3-14) show that the cumulative cost for meeting the hypothetical carbon constraints is significantly lower in all three advanced technology scenarios than in the Baseline Scenarios that use reference case technology

---

[32] Note that renewables and nuclear energy increase over time in the reference case scenario, but increase more in Scenario 2 due to more significant decreases in cost and performance.

[33] Advanced biotechnologies (sometimes called "Bio-X") are those that combine the biosciences with fields such as nanotechnology, chemistry, computers, medicine, and others to create novel solutions for technology challenges. This could lead to innovative concepts such as the use of "enzyme machines" or even new materials (e.g., bio-nano hybrids) that could replace traditional technology altogether.

[34] The term "advanced technologies" is used to represent major improvement in current technologies, as well as novel technologies currently not in use. In this study, technology advancement was assumed to lead to more efficient energy technologies with lower capital and operating costs. Details on the assumptions can be found in Clarke et al. (2006). The resulting cost reductions do not consider the cost associated with performing any R&D that might be necessary to achieve the improved technology performance.

World Primary Energy Demand for Three Advanced Technology Scenarios Under a High GHG-Emissions-Constraint Case

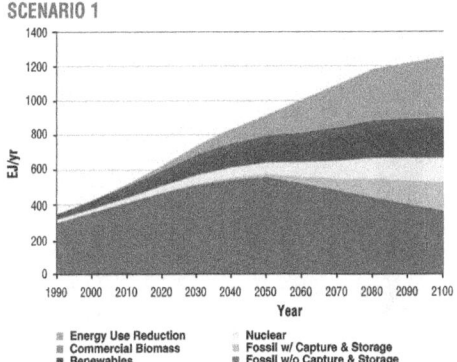

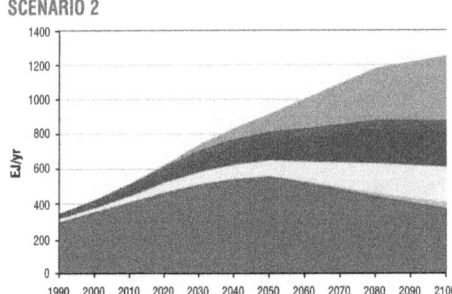

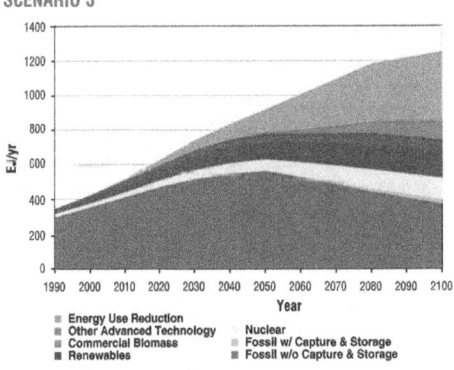

World Carbon Dioxide Emissions for Three Advanced Technology Scenarios Under a High GHG-Emissions-Constraint Case

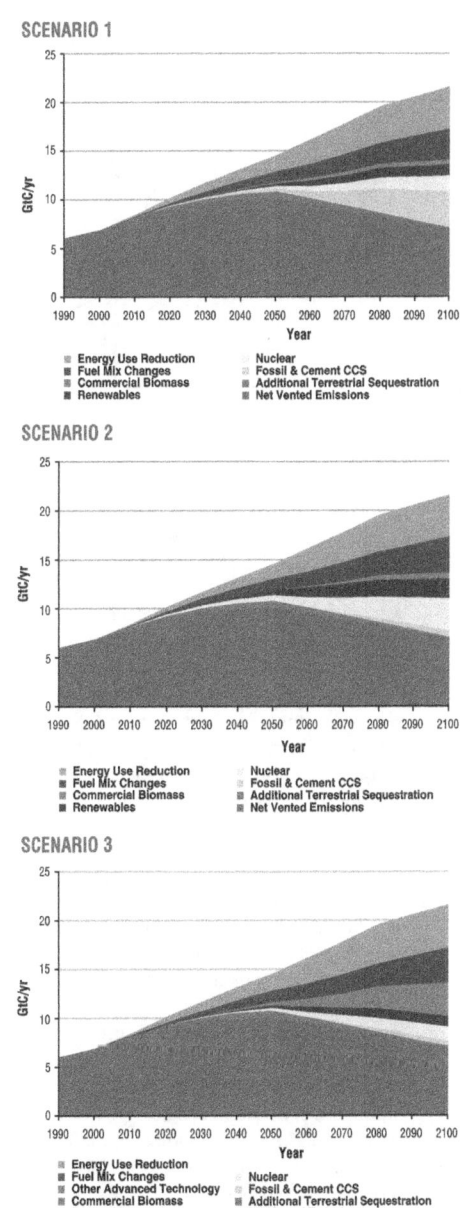

Figure 3-12. World Primary Energy Demand for Three Advanced Technology Scenarios Under a High GHG-Emissions-Constraint Case Note: "Energy Use Reduction" is the amount of energy conserved or saved through advanced energy-efficient end-use technologies compared to a Reference Case, which also includes a considerable increase in energy efficiency compared to today's level. (Source: Clarke et al. 2006)

Figure 3-13. World Carbon Dioxide Emissions for Three Advanced Technology Scenarios Under a High GHG-Emissions-Constraint Case Note: "Energy Use Reduction" is the amount of energy conserved or saved through advanced energy-efficient end-use technologies compared to a Reference Case, which also includes a considerable increase in energy efficiency compared to today's level. (Source: Clarke et al. 2006)

Cost Reductions Associated with Advanced Technology Scenarios, Compared to a Baseline Case without Advanced Technologies

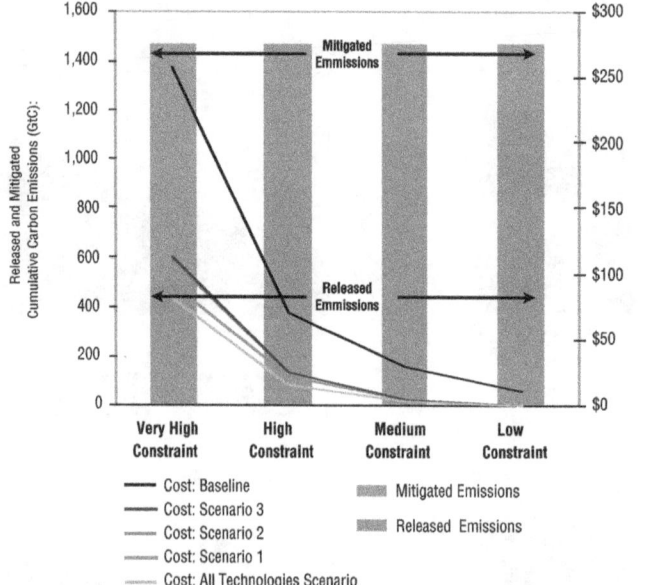

Figure 3-14. Cost Reductions Associated with Advanced Technology Scenarios, Compared to a Baseline Case without Advanced Technologies (Source: Clarke et al. 2006)

Note: Cumulative released emissions (shown in the bar graphs) are highest and mitigated emissions are lowest when the emissions constraint is least stringent (Low Constraint). Costs (line graphs) are highest when the emissions constraint is most stringent (Very High Constraint). Costs are lower when advanced technology was assumed to be available (purple, red, green, and blue lines), than when technology was assumed to advance only incrementally (black line).

assumptions. Accumulated over the course of the 21st century, the potential economic benefits of an advanced technology strategy are likely to be significant, independent of which technologies eventually emerge as most successful. Furthermore, the cumulative cost for meeting the carbon constraints was even lower in the case in which all technology advances were combined, implying that benefits will be greatest if many types of advanced technologies achieve success.

Numerous other studies reach similar conclusions. For example, Manne and Richels (2004) examined limiting global temperature rise using scenarios with "optimistic" technological assumptions. They assumed advanced technologies, such as fuel cells and integrated gasification combined cycle with CO_2 capture and storage, are available in the future. They compared these to more "pessimistic" scenarios without such advanced technologies. The costs[35] were estimated to be 2.5 times lower in the advanced technology cases than in the scenarios without advanced technologies. In another study, Edmonds et al. (2004) report that when a suite of advanced technologies (such as carbon capture and storage, biotechnology, and hydrogen energy systems) are available to be deployed at a large scale, the inferred added dollar value to GHG emissions that would be

required to achieve the assumed reduction was 60 percent lower than when the advanced technologies were not available.

The economic benefits of accelerated technical advances are also identified by a class of studies that explore the dynamics of technical change (e.g., Edenhofer et al. (2005), Manne and Richels (2004), van Vuuren et al. (2004), Löschel (2002), Gerlagh and van der Zwaan (2003) and Goulder and Mathai (2000)). These analyses model technical change through mechanisms such as "technology learning" or "learning by doing," where costs of technologies decline with experience as a function of investment in R&D or other mechanisms. The mechanisms are then used to examine the effects of the technological change on the costs of complying with various climate policy and emissions constraints.

Although these studies vary in design, model platform, and methodology, a common conclusion can be drawn from all of them. That is, technological change can significantly reduce the costs of emissions reduction policies. In these studies, the costs were typically reduced by over 50 percent when technological change was introduced in a portfolio of mitigation options. One of the major conclusions drawn at the recent IPCC Expert Meeting on

---

[35] In the study, costs included those associated with fuel switching (to fuels or technologies with lower emissions), changes in domestic and international fuel prices, and price-induced conservation activities.

Emission Scenarios was that technological change can have a significant impact on reducing costs of GHG stabilization.[36]

In addition, the studies indicate that the benefits are greatest when technology advancement is pervasive. A study conducted using the MESSAGE model (Roehrl and Riahi 2000) concluded that mitigation costs are highest in scenarios with static technologies and lowest in scenarios where technology improvements span both supply and end-use sectors.

A multi-gas, rather than a CO_2-only, approach is another important dimension to lowering the costs of stabilizing radiative forcing from GHGs. Weyant and de la Chesnaye (2005) showed that using a multi-gas approach results in a cost reduction of 30 to 50 percent, compared to CO_2-only approaches.

Timing of Advanced Technology Market Penetration

CCTP's planning activities must also consider the timing of the commercial readiness of the advanced technologies included in its R&D portfolio. However, the time at which certain technologies would need to be ready for large-scale deployment in the marketplace is uncertain and would vary,

depending on the level of the GHG reductions to be achieved. Understanding how timing varies with varying GHG concentration levels and with varying assumptions about the technology mix provides insights for R&D planning and related technology development strategies. Clearly, R&D programs must complete their contributions well before the time when large-scale deployment of the technologies is expected.

As an illustration, in the PNNL scenario analysis under a "high" emissions constraint, advanced technologies for reducing emissions for energy end use and infrastructure begin achieving emissions reductions at a significant level (one GtC/year) between 2030 and 2040. Under a similar constraint, technologies effecting capture, storage, and sequestration of CO_2 begin making significant contributions (one GtC/year) around 2040 or later. Variations among these dates result from varying assumptions about technologies. A summary of the insights about timing, shown for each CCTP strategic goal, is presented in Table 3-1, across a wide range of hypothesized GHG emissions constraints. In general, the higher the emissions constraint, the sooner the advanced technologies are needed and deployed.

Estimated Timing of the First GtC-eq./Year of Reduced or Avoided Emissions (Compared to the Reference Case) for Advanced Technology Scenarios

CCTP STRATEGIC GOAL	VERY HIGH CONSTRAINT	HIGH CONSTRAINT	MEDIUM CONSTRAINT	LOW CONSTRAINT
Goal #1. Reduce Emissions from Energy End Use & Infrastructure	2010 - 2020	2030 - 2040	2030 - 2050	2040 - 2060
Goal #2. Reduce Emissions from Energy Supply	2020 - 2040	2040 - 2060	2050 - 2070	2060 - 2100
Goal #3. Capture and Store or Sequester Carbon Dioxide	2020 - 2050	2040 or Later	2060 or Later	Beyond 2100
Goal #4. Reduce Emissions of Non-CO₂ GHGs	2020 - 2030	2050 - 2060	2050 - 2060	2070 - 2080

Note: The years shown in the table represent the period in which the first GtC (or GtC-eq.) of incremental emissions reduction (compared to the Reference Case) is projected to occur due to penetration of each class of advanced technology in any one of the advanced technology scenarios. The Reference Class includes significant penetration of energy-efficient end use technologies, nuclear, renewable, and biomass energy, terrestrial sequestration, and non-CO_2 emission reductions.

Table 3-1. Estimated Timing of the First GtC-eq./Year of Reduced or Avoided Emissions (Compared to the Reference Case) for Advanced Technology Scenarios (Source: Clarke et al. 2006)

---

[36] Meeting Report of the IPCC Expert Meeting on Emission Scenarios, 12-14 January 2005, Washington D.C. http://www.ipcc.ch/meet/washington.pdf.

3.5 CCTP Goals for Advanced Technology

Review of scenario analyses indicates that, given the scale of the challenge, no single technology or class of technology would be likely to provide, by itself, the quantity of GHG emissions reductions needed to achieve stabilization of GHG concentrations, or its integrated multi-gas metric, radiative forcing, at most of the levels typically hypothesized and examined in the technology literature on scenarios. Instead, these studies show that, under a wide range of differing assumptions and planning uncertainties, technological advances aimed at the following four broad areas are likely to be needed in combination in order to contribute to the needed GHG emissions reductions:[37]

1. Energy End-Use Efficiency and Infrastructure

2. Low- and Zero-CO_2 Emissions Energy Supply

3. CO_2 Capture/Storage and Sequestration

4. Non-CO_2 GHGs

Energy End-Use Efficiency and Infrastructure

Ultimately, global GHG emissions are driven by the demand for services (heating, cooling, transportation, agriculture, industrial process activities, etc.) that require energy or other services and consumables with embodied GHG emissions. Technological advances that can reduce the energy and services required to meet these needs are one of the key means for reducing GHG emissions from energy end-use and infrastructure. Scenario analyses suggest that increased use of highly energy-efficient technologies and other means of reducing energy end

use could play a major role in contributing to cost-effective emissions reduction.

In published scenarios, increasing demand for energy services, driven by population and economic growth, drives growth in GHG emissions over the 21st century. If gross world product were to grow by 2.0 percent per year over the span of the 21st century, and the demand for energy services were to grow at a commensurate 2.0 percent per year, then energy demand would grow seven-fold by 2100. Many published scenarios assume gross world product growth above these rates. For example, at the top of the range of the IPCC's *Special Report on Emissions Scenarios* (SRES) scenarios, gross world product grows at more than 3.0 percent per year from 1990 through 2100.

In virtually all published scenarios, however, the demand for final energy[38] and associated emissions of CO_2 grows at rates lower than the growth rates of gross world product. This is because improvements in end-use efficiency, along with structural changes in economic activity, tend to drive up economic value-added and drive down energy inputs associated with increasing global prosperity.[39] In 1990, global final energy intensity (energy used per constant dollar of gross world product) was roughly 17 million joules per dollar. In the IPCC's SRES scenarios, final energy intensities in 2100 ranged from 1.4 million to 5.9 million joules per dollar of gross world product.[40] Without these reductions in energy intensity, which are significant, energy demand growth and associated GHG emissions would be significantly higher. This point is illustrated in Figure 3-15, which shows the relationship between global CO_2 emissions and energy intensities in 2100 in various published scenarios. Although Figure 3-15 shows variation across multiple scenarios due to differences in energy mix and the levels of deployment of low or zero-emitting technologies, the general pattern indicates that lower energy use per unit of economic output leads to lower CO_2 emissions per unit of economic output.

---

[37] CCTP also includes two supporting technology areas. These are measuring and monitoring technologies, and application of basic science to applied technology R&D. These supporting areas are not discussed in this chapter, although they are integral elements of the overall CCTP technology strategic plan.

[38] Final energy refers to energy delivered to the point of end use (e.g., to buildings or gas stations) as opposed to energy used as an input to, for example, electricity generation. However, final energy does not represent the actual energy ultimately delivered in the way of services to end-users (e.g., building cooling or vehicle miles traveled) because final energy must be converted to these final services by equipment such as air conditioners and automobiles. Hence, changes in final energy should be interpreted as combining changes both in the level of services demanded and in the efficiency at which final energy is used to supply those services.

[39] See the GHG Emissions Scenario Database at http://www-cger.nies.go.jp/cgere/db/enterprise/scenario/scenario_index_e.html.

[40] Range based on the illustrative scenarios from IPCC (2000). Changes in final energy intensity incorporate both changes in end-use efficiency and changes in the relationship between economic output and the demand for services such as transportation and air conditioning (see footnote 39). Hence, improvements in energy intensity should not be interpreted strictly as improvements in end-use efficiency.

Global CO$_2$ Emissions Intensity versus Global Energy Intensity

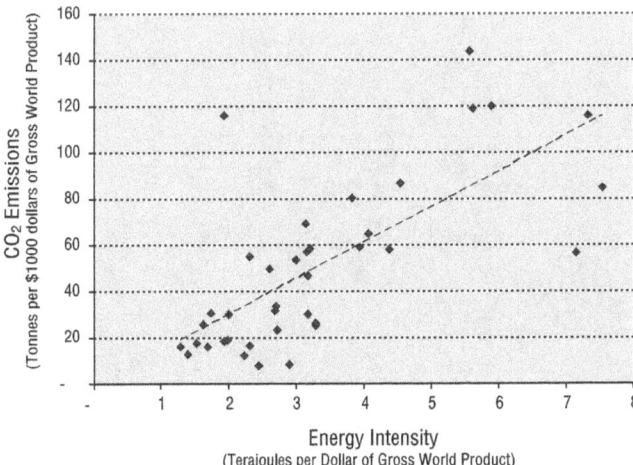

Figure 3-15. Global CO$_2$ Emissions Intensity versus Global Energy Intensity

(Source: Data from the Greenhouse Gas Emissions Scenario Database[41])

This relationship suggests the significance of the emissions reduction benefits that would accrue from increasing the efficiency of end-use technologies. If R&D efforts were to increase the rate of final energy intensity improvement by one-quarter of one percent per year over the span of the 21st century (leading to an increase in energy efficiency by the end of the century of roughly 25 percent), scenario analyses indicate that CO$_2$ emissions would be reduced by 3.5 GtC/year by 2100. For perspective, this is roughly half of the total global CO$_2$ emissions in 2006.[42]

This point is further illustrated by the advanced technology scenarios in the PNNL report on climate change technology strategies (Clarke et al. 2006). In this study, advanced energy-efficiency technologies were assumed to lower final energy requirements by 17 to 32 percent globally in 2100. These reductions were responsible for a cumulative decrease of between 130 to 270 GtC of CO$_2$ emissions over the course of the 21st century. Energy-efficiency improvements also contributed to the lowering of the costs for achieving stabilization of GHG concentrations across a range of varied assumptions.

Similarly, one of the IPCC's SRES scenarios featuring advanced end-use technologies (A1T) indicates that reductions from end-use efficiency (through reduced final energy intensity) could be responsible for roughly 4.0 GtC/yr by 2100.[43] In addition, Hanson and Laitner (2004) developed an advanced technology scenario and projected that approximately one-third of the U.S. carbon emissions reductions in 2050— roughly one GtC/yr—were attributable to the deployment of more efficient end-use technologies.[44]

Providing technological options to improve energy end-use efficiency can provide a fundamental way to achieve GHG emissions reductions and lower the need for CO$_2$-free energy supply. This insight is robust across a full spectrum of varied technology futures—whether these futures emphasize fossil fuels combined with CO$_2$ capture and storage, renewable or nuclear power, or novel technologies such as fusion and advanced bio-technology.

Low- and Zero-CO$_2$-Emissions Energy Supply Technologies

Supplying the world's energy needs while achieving substantial reductions in GHG emissions may also

---

[41] See http://www-cger.nies.go.jp/cger-e/db/enterprise/scenario/scenario_index_e.html.

[42] This calculation is based on the illustrative B2 scenario from IPCC (2000). It assumes that lower final energy requirements would not alter the relative proportions of energy provided from different sources.

[43] Result based on the illustrative scenarios for the A1 set. It was calculated based on a comparison of the illustrative A1T scenario with the illustrative A1B scenario, assuming no change in the primary energy mix between the two. While not identical to A1T, A1B is similar in terms of the emissions per unit of primary energy and therefore serves as an effective reference. The particular scenario cited above used the AIM (model).

[44] Note that many of the assumptions in this study followed from the study, *Scenarios for a Clean Energy Future* (Brown et al. 2000).

Global CO$_2$ Emissions Intensity versus Percentage of Renewable and Nuclear Energy in the Energy Supply Mix

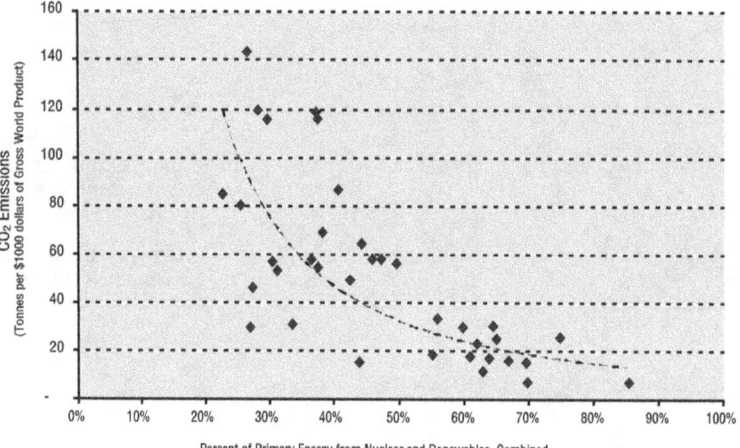

Figure 3-16. Global CO$_2$ Emissions Intensity versus Percentage of Renewable and Nuclear Energy in the Energy Supply Mix

(Source: Data from the Greenhouse Gas Emissions Scenario Database[45])

require large contributions from energy supply technologies with low or near-zero emissions profiles. These include renewable energy sources for electricity, such as wind, solar, and hydroelectric power; biomass-based energy systems; nuclear power; and the use of these and other technologies to produce the energy carrier, hydrogen. These could also include other advanced technologies such as fusion energy, solar energy from space or remote desert locations, and novel biotechnologies.

In integrated assessment models, low- and near-zero-emissions energy supply technologies are modeled at varying levels of technological detail. While some models explicitly model various low- and near-zero-emissions energy supply technologies, others use one or more generic technology classes to represent these technologies. In either case, the low- and near-zero-emissions technologies prove to be important components to technology strategies aimed at stabilizing GHG concentrations. Figure 3-16 shows the strong correlation between the CO$_2$ emissions intensity (tonnes of CO$_2$ emissions per constant dollar of GWP) of the global economy and the percentage

of renewable and nuclear energy found in the energy supply mix, as projected in a wide number of scenarios.

A number of scenario analyses give quantitative significance to the importance of low- and near-zero-emissions energy supply technologies in reducing emissions. For example, Akimoto et al. (2004)[46] show that for a hypothetical climate policy where emissions are constrained, the share of the world's primary energy in 2100 met by biomass and wind energy increased by more than 70 percent from their reference case contributions of 10 percent and 4 percent, respectively. In addition, solar power supplied almost 5 percent of the world's primary energy demand by 2100.[47] Growth in nuclear energy (fission), biomass, and renewable energy accounted for about 30 percent of the emissions reduction in 2100, in about equal shares. Similarly, Edmonds et al. (2004) reported that contributions from solar and nuclear energy grew under CO$_2$ emissions constraints, especially when no technological advancement was assumed for fossil-based generation technologies and CO$_2$ capture and storage (CCS) technologies.[48]

---

[45] See http://www.cger.nies.go.jp/cger-e/db/enterprise/scenario/scenario_index_e.html.

[46] The study used an updated version of the DNE21 model, an integrated assessment model which hard-links macroeconomic, energy systems, and climate change models, and seeks optimal development of the world's energy system for a given climate policy based on maximizing macroeconomic consumption.

[47] The upper limit of the world total nuclear production assumed in this scenario was 920 GW in 2050 and 1450 GW in 2100, so nuclear energy was not a major contributor in this analysis.

[48] This study used the MiniCAM model and the IPCC SRES B2 Scenario to examine the role of advanced technologies under a climate policy aimed at stabilizing atmospheric CO$_2$ concentrations at 550 ppmv.

As discussed in previous sections, Clarke et al. (2006) examined several advanced technology scenarios to achieve a range of emissions reductions. Low- and near-zero-emissions energy technologies (including solar, wind, biomass, nuclear fission, and advanced concepts such as nuclear fusion and novel biotechnology) contribute between 23 percent and 34 percent of world primary energy demand by 2100. These technologies were projected to contribute between 30 and 340 GtC of CO_2 emissions reductions (cumulative) over the course of the 21st century, under a variety of scenarios aimed at stabilizing GHG concentrations (Clarke et al. 2006).

Finally, in several scenarios that explored the use of hydrogen as a fuel or energy carrier, renewable energy sources were found to be important means for generating hydrogen and other fuels. Edmonds et al. (2004) showed that, under a medium carbon constraint, the preferred feedstock for hydrogen production shifts from fossil fuels to biomass, because the application of CCS to biomass-based H_2 production can result favorably in net negative emissions. Alternatively, Mori and Saito (2004) report that H_2 production from fast breeder reactors can, under certain emissions constraints, cost-effectively supply nearly all of the final energy demand for hydrogen.[49]

Carbon Capture/Storage and Sequestration

Several physical, chemical, geochemical, and biological mechanisms can remove CO_2 from the atmosphere or from point sources, and store or use the resulting CO_2 or chemical derivatives (see e.g., Halmann 1993, Kojima 1997, Inui et al. 1998, Lackner 2002). The CCTP technology thrusts related to capturing/storing and sequestering CO_2 (see Chapter 6) include: (1) engineered capture and storage of CO_2 from power plants and other industrial sources of CO_2 emissions; (2) terrestrial sequestration of CO_2 in trees, soils, and other terrestrial systems; and (3) ocean sequestration via direct injection or other means.

Capture and Storage of Carbon Dioxide

Carbon capture and storage (CCS) refers to the capture, purification, and concentration of molecular carbon dioxide resulting from combustion or other industrial processes, and the subsequent transport to and storage of CO_2 in suitable geologic or ocean reservoirs. CCS has the potential to lower the carbon emissions intensity of fossil energy systems. CCS could also be applied to bio-based electricity-generation systems or to other non-fossil-fuel waste streams, such as those from calcining operations (cement or lime production, for example).

Figure 3-17 shows the amount of carbon dioxide captured and stored, as a function of the amount of primary energy supplied by fossil fuels for various scenarios from the Center for Global Environmental Research data base[50] in the years 2050 and 2100. Both parts of the figure show a relationship between the amount of carbon sequestered and the amount of fossil fuel used. The plots show that by the middle of the century, the deployment of CCS technologies in conjunction with fossil fuel use is occurring in many scenarios, even though lower-carbon fuels (such as natural gas) are still available. By the end of the century, when such fuels are less abundant and/or expensive, CCS is almost always deployed in scenarios with high fossil fuel consumption.

A number of recent studies using integrated assessment models have examined the potential of CCS to lower future CO_2 emissions. Edmonds et al. (2004) report that fossil energy technologies with CCS can supply approximately 55 percent of the global electricity generation by the end of the century in an advanced technology scenario with high emissions reductions.[51] This was more than twice the contribution than in a case where CCS (and other advanced energy technologies) was not assumed to advance. McFarland et al. (2004) find that fossil-based power systems with CCS account for approximately 70 percent of global electricity production by 2100 under a high GHG emissions constraint, where CCS systems and other advanced fossil energy systems are allowed to deploy to their full market potential.[52] Clarke et al. (2006) show fossil systems with CCS contributing up to 50 percent

---

[49] This study used the MARIA integrated assessment model to examine the role of nuclear technology and hydrogen use under different climate policies, and different technology advancement assumptions.

[50] Center for Global Environmental Research, National Institute for Environmental Studies, Tsukuba, Japan. See http://www-cger.nies.go.jp/cger-e/db/enterprise/scenario/scenario_index_e.html.

[51] This analysis employed the PNNL MiniCAM model, using the IPCC SRES B2 scenario as the reference case, and compared that case to an advanced technology case that incorporates more efficient and economical CCS, higher efficiency fossil generation, and hydrogen energy systems.

[52] This study used the MIT EPPA model, a recursive dynamic multi-regional general equilibrium model of the world economy. Bottom-up information about coal- and natural gas-based generation systems with CCS were used in a top-down energy economics model to examine the effect of CCS on different climate policies.

Carbon Dioxide Captured and Stored, as a Function of Primary Energy Supplied from Fossil Fuels for Various IPCC Scenarios

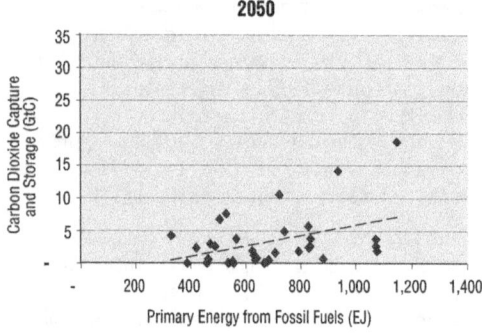
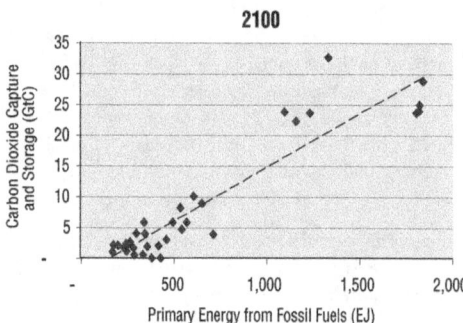

Figure 3-17. Carbon Dioxide Captured and Stored, as a Function of Primary Energy Supplied from Fossil Fuels for Various IPCC Scenarios. Note: The dashed lines in the figures reflect the correlation between fossil energy and CCS deployment. (Source: Various scenario results were extracted from the database maintained by the Center for Global Environmental Research, National Institute for Environmental Studies, Tsukuba, Japan. See http://www-cger.nies.go.jp/cger-e/db/enterprise/scenario/scenario_index_e.html.)

more of the world's total primary energy consumption in 2100 in scenarios featuring technology advancement in CCS, compared to other advanced technology scenarios.[53]

Recent studies have also examined the economic benefits of using CCS in isolation or along with other technological advancements. By allowing fossil energy resources to be used while simultaneously delivering reductions in CO_2 emissions, CCS technologies help to constrain the rate of increase and ultimate peak of carbon prices (an indication of the overall cost of achieving the emission reductions).[54] For example, Edmonds et al. (2004) show that through the large-scale adoption of CCS and other advanced fossil energy technologies, peak carbon permit prices were 62 percent lower than if those technologies were not deployed to their full market potential. In the study by McFarland et al. (2004), CCS reduces carbon prices by 33 percent at the end of the century.

While many studies employ different modeling approaches, technology representations, and assumed climate policies, the synthesis assessment of scenarios shows that CCS has the potential to play a significant role in emissions mitigation during the 21st century. The advancement of CCS-related technologies magnifies this contribution, delivering substantial economic savings. Early technical resolution of the viability of various CCS options could have significant implications for subsequent R&D investment strategies.

Terrestrial Sequestration

Land-use change that results in net CO_2 release to the atmosphere accounts for about 22 percent of today's global CO_2 emissions (IPCC 1996). At the same time, terrestrial systems in many parts of the world are being managed in ways that remove CO_2 (also referred to here simply as "carbon") from the atmosphere and sequester it in soils and biomass. Over the next several decades, the potential exists to offset a significant portion of global CO_2 emissions by managing the world's terrestrial systems to accumulate and store additional carbon. How much of this potential can be realized is uncertain, however, and will depend on the development and diffusion of advanced technologies in a variety of economic sectors.

---

[53] This study used the PNNL MiniCAM model to examine energy and economic implications of different technology futures and different levels of emissions reductions. One future assumes that CCS technologies meet aggressive technical, economic, and environmental goals for application on fossil and biomass-based energy systems, along with higher-efficiency fossil generation and greater end-use efficiency gains.

[54] Since the cost of compliance is the total area under the marginal abatement curve, the last two metrics are strongly correlated, i.e., the greater the reduction in the carbon price, the greater the reduction in the cost of compliance.

Globally, the goods and services derived from land resources—including food, water, shelter, energy, and recreation—are basic to human existence and quality of life. Future changes in cropland, grassland, and forest land areas—regionally and globally—will be driven by the ability of land resources to provide these basic goods and services. Hence, the potential to use terrestrial systems to sequester carbon and mitigate global GHG emissions will be directly affected by the development of advanced technologies that reduce human pressures on land by increasing land productivity across a range of economic sectors—including (but not limited to) agriculture, forestry, and energy.

In agriculture, advanced technologies could enhance terrestrial carbon sequestration by enabling the development of new food and fiber products, production processes, and distribution systems that reduce the amount of land needed to feed and clothe the world's population. In forestry, advanced technologies could accelerate the processes of reforestation and afforestation, as well as increase the quantity of wood products that could be obtained from a unit of forest land. Advanced energy technologies could increase terrestrial sequestration by reducing deforestation pressures in developing countries and shifting cropland to bioenergy crop systems that not only increase soil carbon levels but also shift energy production toward technologies that recycle atmospheric CO_2.

Sohngen and Mendelsohn (2003) suggest that global forests have a net sequestration potential ranging from 32 to 102 GtC in the coming century, depending on carbon prices (see Section 3.2.1.2). In a more recent study performed as part of EMF-21, the net sequestration was projected to increase from today's level by 48 to 148 GtC by 2100 under different climate policies (Sohngen and Sedjo, 2006). The cost of land-use and forest sequestration has been estimated to range between $10 and $200 per ton of carbon stored (Richards and Stokes 2004).

Non-CO_2 GHG Emissions

Non-CO_2 GHGs play an important role in the CCTP analytical framework because of their high potential to reduce overall radiative forcing, both in the near term and over the next 100 years, and to reduce the overall cost of GHG stabilization. As shown in Figure 3-1, combined emissions from non-CO_2 gases accounted for about one-quarter of all GHG emissions (in terms of global warming

potential) in the year 2000. These gases are particularly important because a variety of scenario analyses show that a significant level of reduction is achievable in the first half of the 21st century.

Potential reductions and cost savings are illustrated in the Energy Modeling Forum multi-gas scenario study—EMF-21 (Weyant and de la Chesnaye 2005)—and other long-term multi-gas studies (e.g., Manne and Richels 2000, 2001; Reilly et al. 2002). The various models exercised in the EMF-21 study used a range of assumptions about technology development, leading to a range of reductions of non-CO_2 GHGs. The studies suggest that, between 2000 and 2100, emissions of non-CO_2 "well mixed" gases (methane, nitrous oxide, and the F-gases) in a moderately constrained emissions case[55] could be reduced by as much as 48 percent, and the cost of GHG stabilization could be lowered by 30 to 60 percent compared to a CO_2-only scenario.

In addition to the long-term EMF-21 multi-gas scenarios, two other studies illustrate maximum technology potential of non-CO_2 mitigation options over the medium term. Delhotal and Gallaher (2005) projected the reduction potential of technological improvements out to 2030 in the three major methane emitting sectors—landfills, natural gas, and coal—for selected countries. By 2030, cost-effective technologies could reduce methane emission to less than 50 percent of current levels in the United States, and could potentially reduce emissions by a factor of two in countries such as China, Mexico, and Russia in the same time frame. Another study by the International Institute for Applied Systems Analysis (IIASA) (Cofala et al. 2005) shows the "maximum potential reductions" to 2030. This study concluded that if all currently available technologies were applied to landfills, agriculture, the natural gas sector, the coal sector, and oil and gas extraction, without consideration of cost, global CH_4 emissions would stop increasing and remain constant through 2030.

The scenarios described above do not explicitly include new or highly advanced mitigation technologies for non-CO_2 GHGs. An analysis conducted by PNNL in cooperation with the U.S. Environmental Protection Agency assumed the development of advanced technologies in areas such as methane emissions from waste and energy sectors, methane and nitrous oxide emissions from agriculture, and high-GWP emissions from the industrial sector (Clarke et al. 2006). Compared to a reference scenario with no emissions constraints and no new non-CO_2 mitigation technologies, the

---

[55] The constrained case was defined as 4.5 W/m² stabilization target by 2100.

advanced technology scenarios showed that reductions in emissions of non-CO_2 GHGs could potentially contribute 91 to 165 GtC-eq. in cumulative emissions reductions over the 21st century. The assumptions underlying the advanced technology scenario are based on the currently known methods to achieve emissions reduction, as well as detailed "bottom-up" analyses of the technical potential to reduce non-CO_2 GHGs further. Results from this analysis for a high GHG-constrained case are shown in Figure 3-18.

Summary of Relative Contributions of Four CCTP Goals

As described in the sections above, a variety of scenario analyses conducted by different research groups show the importance of technology advancement consistent with each of the four core CCTP emissions-reduction goals:

1. Reduce emissions from energy end-use and infrastructure.

2. Reduce emissions from energy supply.

3. Capture and sequester CO_2.

4. Reduce emissions of non-CO_2 GHGs.

In general, scenario analyses indicate that no single technology option, as presently envisioned, is able to provide sufficient emissions reductions to meet

stabilization objectives. Rather, even when assumptions vary, the analyses strongly indicate that a portfolio of technologies is required, with each technology contributing significantly to the GHG emission reductions required.

This point is illustrated by the results of a recent PNNL study (Clarke et al. 2006) in which each of the four technology areas was shown to make contributions toward stabilizing concentrations. Based on the assumptions used in this set of scenarios, no one area was markedly more or less important than others. Figure 3-19 shows the contributions of four technology categories (directly linked to the four CCTP goals stated above) to cumulative GHG emissions reductions between 2000 and 2100. The figure represents one set of possible outcomes from scenarios that are based on a particular set of assumptions about advanced technologies over the next century. It offers a glimpse of the range of emissions reductions new technologies might make possible through reduced energy end use; low- or zero-emission energy supply; carbon capture, storage, and sequestration; and reduction of other GHGs—on a 100-year scale and across a range of uncertainties. Given the magnitude of the CO_2 challenge and the uncertainties in cost, efficacy, impacts, and ultimate design of the mitigation technologies being considered, pursuit of new technological advances and alternative approaches may prove beneficial to the formulation of GHG stabilization strategies.

World Non-CO_2 GHG Emissions Under High Emissions Constraints[56]

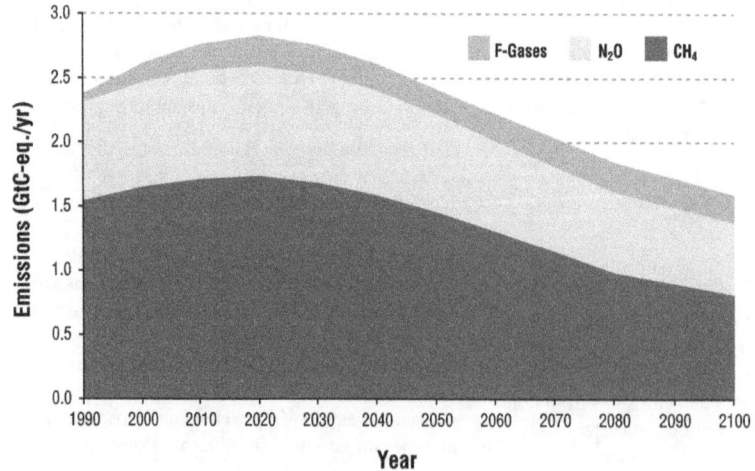

Figure 3-18. World Non-CO_2 GHG Emissions Under High Emissions Constraints[56] (Source: Clarke et al. 2006)

[56] This figure was based on the A New Energy Backbone scenario (Scenario 2).

Cumulative Contributions Between 2000 and 2100 to the Reduction, Avoidance, Capture, and Sequestration of Greenhouse Gas Emissions for the Three Advanced Technology Scenarios, Under Varying Carbon Constraints[57]

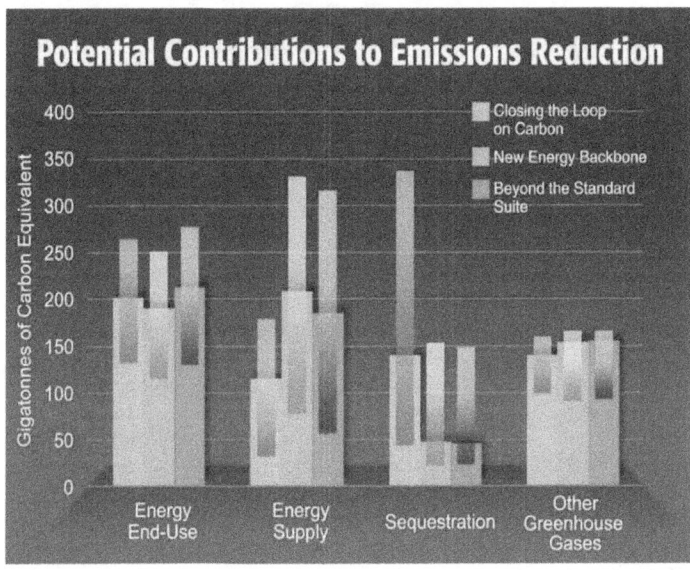

Figure 3-19. Cumulative Contributions Between 2000 and 2100 to the Reduction, Avoidance, Capture, and Sequestration of Greenhouse Gas Emissions for the Three Advanced Technology Scenarios, Under Varying Carbon Constraints[57]

Note: The thick bars show the contribution in the high emission reduction case and the thinner bars show the variation in the contribution between the very high emission reduction case and the low emission reduction case.

3.6 Summary of Insights

Many studies have examined long-term GHG emissions trends under a range of assumptions about the rate of change of population, economic growth, and technology change; and the potential role for advanced technology in mitigating emissions growth. Although the rate of GHG emissions growth over the 21st century is uncertain and dependent on many variables, the scenarios presented in this chapter suggest that significant increases in GHG emissions are projected through the end of the 21st century if there are no constraints on emissions.

Further scientific study is required to understand the quantities and timing of GHG emissions reductions that would be needed to stabilize GHG concentrations at a level that would prevent dangerous anthropogenic interference (DAI) with the climate system. In the approach adopted by CCTP

to explore the potential roles for and benefits of advanced technologies, four levels of GHG concentrations were assumed, with results presented for each.

Regarding the scale of the challenge, the scenarios analysis conducted for CCTP suggests that mid-range estimates of the cumulative global emissions reductions needed to result in progress over the course of the 21st century toward eventual stabilization, across a range of GHG concentrations, would be on the order of 300 to 1,000 GtC-eq.[58] Analyses using different assumptions may result in different values, but a number of mid-range analyses indicate 100-year cumulative reductions of similar magnitude.[59] These reductions (or avoidances) would be in addition to the GHG emissions avoided by the substantial energy-efficiency improvements and CO_2 emission-free energy sources already assumed (embedded) in their respective reference case scenarios. Technology advancements could make such reductions much more feasible in the context of economic growth.

---

[57] The figure shows the cumulative contributions between 2000 and 2100 to the reduction, avoidance, capture/storage, and sequestration of GHG emissions under the three Advanced Technology Scenarios, based on varying emissions constrained cases. The thick bars show the contribution under the high emission constraint and the thinner, semi-transparent bars show the variation in the contribution between the very high emissions constraint and the low emissions constraint. "Energy End-Use" includes emission reductions due to energy-efficiency measures. "Energy Supply" includes emissions reductions from the substitution of non-fossil energy supply technologies with low or zero CO_2 emissions for fossil-based power generation without capture and storage of CO_2. "Sequestration" includes carbon capture and storage from fossil-based technologies, as well as terrestrial sequestration.

The synthesis assessment of a large number of scenario analyses conducted by different research groups indicates that emissions reductions of the scale needed to achieve stabilization of GHG concentrations can be achieved through various combinations of many different technologies. An important insight that can be drawn from these studies is that, under a wide range of differing assumptions, advanced technologies associated with CCTP Strategic Goals 1 through 4 could all contribute significantly to overall GHG emission reductions.

While many technologies may reduce or avoid GHG emissions, scenario analyses can suggest roles for advanced technologies that could result in potentially large relative economic benefits. When the costs of achieving different levels of emissions reductions were compared for cases with and without advanced technologies, the cumulative cost savings of the former were projected to be 60 percent or more over the course of the 21st century. Further, by including the non-CO_2 gases in a multi-gas GHG reduction strategy aimed at stabilizing at various DAI levels of radiative forcing, overall costs of goal attainment were reduced, potentially by 30 to 50 percent when compared to CO_2-only approaches.

Finally, scenario analyses indicate that the timing of the commercial readiness of advanced technology options is an important R&D planning consideration, and particularly so for R&D planning under scenarios with the higher GHG emissions constraints. Table 3-1 is one set of representative results in this regard, showing when the first GtC/year of reduced or avoided emissions would be needed, depending on the range of GHG emissions constraints. Looking over a 100-year planning horizon, and allowing for capital stock turnover and other inertias, inherent in the global energy system and infrastructure, technologies with low or near-zero net emissions characteristics would need to be available and moving into the marketplace years before the periods shown on Table 3-1.

The following chapters focus in depth on various technological means for making progress toward, and eventually achieving, each CCTP strategic goal. Guided in part by the insights gained through the review and synthesis of the scenario analyses, each chapter's discussion addresses the rationale and technology strategy that would guide investments in the current technology portfolio, explains potential R&D progress, and identifies candidate areas for future research directions that could enrich and broaden the overall portfolio.

---

[58] Estimates of emissions reductions required to stabilize GHG concentrations are uncertain and vary based on assumptions. See Section 3.3 and footnote 25. See also "mitigated emissions" in Figure 3-14.

[59] Manne and Richels 2004, Weyant 2004, and Roehrl and Riahi 2000.

3.7   References

Akimoto, K., T. Tomoda, Y. Fujii, and K. Yamaji. 2004. Assessment of global warming mitigation options with integrated assessment model DNE21. *Energy Economics* 26(4):635-653. Special Issue EMF 19 Alternative technology strategies for climate change policy, John P. Weyant, ed.

Alcamo, J., R. Leemans, and E. Kreileman, eds. 1998. *Global change scenarios of the 21ᵗ century. Results from the IMAGE 2.1 model.* London: Elsevier Science.

Ausubel, J.H., A. Grübler, and N. Nakicenovic. 1988. Carbon dioxide emissions in a methane economy. *Climatic Change* 12:245-263.

Brown, M.A., M.D. Levine, and W.D. Short. 2000. *Scenarios for a clean energy future*. Interlaboratory Working Group. Published by Oak Ridge National Laboratory, Oak Ridge, TN; and Lawrence Berkeley National Laboratory, Berkeley, CA; ORNL/CON-476 and LBNL-44029. http://www.ornl.gov/sci/eere/cef/

Brown, S., J. Sathaye, M. Cannell, and P.E. Kauppi. 1996. Mitigation of carbon emissions to the atmosphere by forest management. *Commonwealth Forestry Review* 75:80-91.

Clarke, L., M. Wise, M. Placet, C. Izaurralde, J. Lurz, S. Kim, S. Smith, and A. Thomson. 2006. *Climate Change Mitigation: An Analysis of Advanced Technology Scenarios*. Richland, WA: Pacific Northwest National Laboratory.

Cofala, J. Markus Amann, and Reinhard Mechler. 2005. Scenarios of world anthropogenic emissions of air pollutants and methane up to 2030. Laxenburg, Austria: International Institute for Applied Systems Analysis. http://www.iiasa.ac.at/rains/global_emiss/Global%20emissions%20of%20air%20pollutants%20.pdf

Coppock, R. and S. Johnson. 2004. *Direct and indirect human contributions to terrestrial carbon fluxes: a workshop summary*. Washington, DC: National Academies Press.

Delhotal, K.C., and M. Gallaher. 2005. *Estimating technical change and potential diffusion of methane abatement technologies for the coal-mining, natural gas, and landfill sectors*. Conference proceedings, IPCC Expert Meeting on Industrial Technology Development, Transfer and Diffusion. http://arch.rivm.nl/env/int/ipcc/docs/TTDT/TTDT%20Meeting%20Report%20Open%20website%20version.pdf

De Jong, A., and G. Zalm. 1991. Scanning the future: a long-term scenario study of the world economy 1990-2015. In *Long-term prospects of the world economy*, 27-74. Paris: OECD.

De Vries, H.J.M., J.G.J. Olivier, R.A. van den Wijngaart, G.J.J. Kreileman, and A.M.C. Toet. 1994. Model for calculating regional energy use, industrial production and greenhouse gas emissions for evaluating global climate scenarios. *Water, Air Soil Pollution* 76:79-131.

De Vries, B., M. Janssen, and A. Beusen. 1999. Perspectives on global energy futures—simulations with the TIME model. *Energy Policy* 27:477-494.

De Vries, B., J. Bollen, L. Bouwman, M. den Elzen, M. Janssen, and E. Kreileman. 2000. Greenhouse gas emissions in an equity-, environment- and service-oriented world: An IMAGE-based scenario for the next century. *Technological Forecasting & Social Change* 63(2-3).

Edenhofer, O., N. Bauer, and E. Kriegler. 2005. The impact of technological change on climate protection and welfare: Insights from the model MIND. *Ecological Economics* 54:277– 292.

Edmonds, J., J. Clarke, J. Dooley, S.H. Kim, and S.J. Smith. 2004. Stabilization of CO_2 in a B2 world: insights on the roles of carbon capture and disposal, hydrogen, and transportation technologies. *Energy Economics* 26(4):517-537. Special Issue EMF 19 Alternative technology strategies for climate change policy, John P. Weyant, ed.

Edmonds, J., M. Wise, and C. MacCracken. 1994. *Advanced energy technologies and climate change. an analysis using the Global Change Assessment Model (GCAM)*. PNL-9798, UC-402. Richland, WA: Pacific Northwest Laboratory. http://sedac.ciesin.columbia.edu/mva/MCPAPER/mcpaper.html

Edmonds, J., M. Wise, H. Pitcher, R. Richels, T. Wigley, and C. MacCracken. 1996a. An integrated assessment of climate change and the accelerated introduction of advanced energy technologies: an application of MiniCAM 1.0. *Mitigation and Adaptation Strategies for Global Change* 1(4):311-339.

Edmonds, J., M. Wise, R. Sands, R. Brown, and H. Kheshgi. 1996b. *Agriculture, land-use, and commercial biomass energy: a preliminary integrated analysis of the potential role of biomass energy for reducing future greenhouse related emissions*. PNNL-11155. Washington, DC: Pacific Northwest National Laboratories.

Energy Information Administration (EIA). 2005. *International energy outlook* 2005. DOE/EIA-0484 (2005). Washington, DC: U.S. Department of Energy. http://www.eia.doe.gov/oiaf/ieo/index.html

EPA (See U.S. Environmental Protection Agency)

Gerlagh, R. and B. van der Zwaan. 2003. Gross world product and consumption in a global warming model with endogenous technological change. *Resource and Energy Economics* 25:35–57.

Goulder, L.H., and K. Mathai. 2000. Optimal CO Abatement in the Presence of Induced Technological Change. *Journal of Environmental Economics and Management* 39:1-38.

Grübler, A., and N. Nakicenovic. 1996. Decarbonizing the global energy system. *Technological Forecasting and Social Change* 53(1):97-110.

Halmann, H.M. 1993. *Chemical fixation of carbon dioxide*. Boca Raton, Florida: CRC Press.

Hanson, D., and J. Laitner. 2004. An integrated analysis of policies that increase investments in advanced energy-efficient/low-carbon technologies, *Energy Economics* 26(4):739-755. Special Issue EMF 19 Alternative technology strategies for climate change policy, John P. Weyant, ed.

Hoffert, Martin L. et al. 2002. Advanced technology paths to global climate stability: energy for a greenhouse planet. *Science* 298:981-987.

IIASA-WEC (International Institute for Applied Systems Analysis-World Energy Council). 1995. *Global energy perspectives to 2050 and beyond*. London: WEC.

Intergovernmental Panel on Climate Change (IPCC). 1996. *Climate change 1995: the science of climate change. Contribution of Working Group I to the Second Assessment Report*. Cambridge, UK: Cambridge University Press.

Intergovernmental Panel on Climate Change (IPCC). 2000. *Special report on emissions scenarios*. Cambridge, UK: Cambridge University Press. http://www.grida.no/climate/ipcc/emission/index.htm

Intergovernmental Panel on Climate Change (IPCC). 2001. *Climate Change 2001. Mitigation: a report of Working Group III of the Intergovernmental Panel on Climate Change*. Bert Metz, Ogunlade Davidson, Rob Swart and Jiahua Pan, eds. Cambridge, UK: Cambridge University Press. http://www.grida.no/climate/ipcc_tar/wg3/index.htm

Inui, T., M. Anpo, K. Izui, S. Yanagida, and S. Yamaguchi. 1998. *Advances in chemical conversions for mitigating carbon dioxide*. Amsterdam: Elsevier.

IPCC (See Intergovernmental Panel on Climate Change).

Kojima, T. 1997. *The carbon dioxide problem: Integrated energy and environmental policies for the 21ˢᵗ century*. Amsterdam: Gordon and Breach.

Lackner, K.S. 2002. Carbonate chemistry for sequestering fossil carbon. *Annual Review of Energy and Environment* 27:193-232.

Lashof, D., and D.A. Tirpak. 1990. *Policy options for stabilizing global climate*. 21P-2003. Washington, DC: U.S. Environmental Protection Agency.

Löschel, A. 2002. Technological change in economic models of environmental policy: a survey. *Ecological Economics* 43:105-126.

Manne, A., and R. Richels. 2000. *A multi-gas approach to climate policy—with and without GWPs*. http://www.stanford.edu/group/MERGE/multigas.pdf

Manne, A., and R. Richels. 2001. An alternative approach to establishing trade-offs among greenhouse gases. *Nature* 410:675-677.

Manne, A., and R. Richels. 2004. The impact of learning-by-doing on the timing and costs of CO_2 abatement. *Energy Economics* 26(4):603-619. Special Issue EMF 19 Alternative technology strategies for climate change policy, John P. Weyant, ed.

McFarland, J.R., J.M. Reilly, and H.J. Herzog. 2004. Representing energy technologies in top-down economic models using bottom-up information. *Energy Economics* 26(4):685-707. Special Issue EMF 19 Alternative technology strategies for climate change policy, John P. Weyant, ed.

Messner, S., and M. Strubegger. 1995. *User's guide for MESSAGE III*. WP-95-69. Laxenburg, Austria: International Institute for Applied Systems Analysis.

Mori, S. 2000. The development of greenhouse gas emissions scenarios using an extension of the MARIA model for the assessment of resource and energy technologies. *Technological Forecasting & Social Change* 63(2-3).

Mori, S., and M. Takahashi. 1999. An integrated assessment model for the evaluation of new energy technologies and food productivity. *International Journal of Global Energy Issues* 11(1-4):1-18.

Mori, S., and T. Saito. 2004. Potentials of hydrogen and nuclear towards global warming mitigation—expansion of an integrated assessment model MARIA and simulations. *Energy Economics* 26(4):565-578. Special Issue EMF 19 Alternative technology strategies for climate change policy, John P. Weyant, ed.

Morita, T., Y. Matsuoka, I. Penna, and M. Kainuma. 1994. *Global carbon dioxide emission scenarios and their basic assumptions: 1994 survey*. CGER-1011-94. Tsukuba, Japan: Center for Global Environmental Research, National Institute for Environmental Studies.

Pacala, S., and R. Socolow. 2004. Stabilization wedges: solving the climate problem for the next 50 years with current technologies. *Science* 305:968-972.

Pepper, W.J., J. Leggett, R. Swart, J. Wasson, J. Edmonds, and I. Mintzer. 1992. Emissions scenarios for the IPCC. An update: assumptions, methodology, and results. Support document for Chapter A3. In *Climate change 1992: supplementary report to the IPCC scientific assessment*, J.T. Houghton, B.A. Callandar, and S.K. Varney, eds. Cambridge, UK: Cambridge University Press.

Reilly, J., M. Mayer, and J. Harnisch. 2002. The Kyoto Protocol and non-CO_2 greenhouse gases and carbon sinks. *Environmental Modeling and Assessment* 7(4):217-229.

Riahi, K., and R.A. Roehrl. 2000. Greenhouse gas emissions in a dynamics-as-usual scenarios of economic and energy development. *Technological Forecasting & Social Change* 63:175-206.

Richards, K. and C. Stokes. 2004. A review of forest carbon sequestration cost studies: A dozen years of research. *Climatic Change* 63 (1-2): 1-48.

Richels, R.G., A.S. Manne, and T.M.L. Wigley. 2004. *Moving beyond concentrations: the challenge of limiting temperature change*. Working Paper 04-11. AEI-Brookings Joint Center for Regulatory Studies. http://www.aei-brookings.org/admin/authorpdfs/page.php?id=937

Roehrl, R.A., and K. Riahi. 2000. Technology dynamics and greenhouse gas emissions mitigation: a cost assessment. *Technological Forecasting and Social Change* 63, 231–261

Sankovski, A., W. Barbour, and W. Pepper. 2000. Quantification of the IS99 emission scenario storylines using the atmospheric stabilization framework (ASF). *Technological Forecasting & Social Change* 63(2-3).

Sohngen, B., and R. Sedjo (2006). Carbon sequestration costs in global forests. *Energy Policy*. Special Edition on Multi-Gas Scenarios and Climate Change (to be published).

Sohngen, B. and R. Mendelsohn. 2003. An optimal control model of forest carbon sequestration. *American Journal of Agricultural Economics*, 85(2): 448-457.

Stern, P.C. 2002. *Human interactions with the carbon cycle: summary of workshop.* Washington, DC: National Academies Press.

United Nations Development Program (UNDP). 2000. *World Energy Assessment.* New York.

U.S Climate Change Science Program. 2003. *Strategic plan for the U.S. Climate Change Science Program.* http://www.climatescience.gov/Library/stratplan2003.default.htm

U.S. Environmental Protection Agency. 2005. *U.S. emissions inventory: inventory of U.S. greenhouse gas emissions and sinks: 1990-2003.* EPA 430-R-05-003. Washington, DC: U.S. Environmental Protection Agency. http://yosemite.epa.gov/oar/globalwarming.nsf/content/ResourceCenterPublicationsGHGEmissionsUSEmissionsInventory2005.html

van Vuuren, D.P., B. de Vries, B. Eickhout, and T. Kram. 2004. Responses to technology and taxes in a simulated world. *Energy Economics* 26(4):579-601. Special Issue EMF 19 Alternative technology strategies for climate change policy, John P. Weyant, ed.

Webster, M.D., M. Babiker, M. Mayer, J.M. Reilly, J. Harnisch, R. Hyman, M.C. Sarofim, and C. Wang. 2002. Uncertainty in emissions projections for climate models. *Atmospheric Environment* 36 (2002) 3659–3670.

Wexler, L. 1996. *Improving population assumptions in greenhouse emissions models.* WP-96-099. Laxenburg, Austria: International Institute for Applied Systems Analysis. http://www.iiasa.ac.at/Publications/Documents/WP-96-099.pdf

Weyant, J.P., ed. 2004. *Energy Economics* 26(4): Special Issue EMF 19 Alternative technology strategies for climate change policy.

Weyant, J.P., and F. de la Chesnaye. 2005. Multigas scenarios to stabilize radiative forcing. *Energy Journal.* Special Edition on Multi-gas Scenarios and Climate Change.

Wigley T., R. Richels, and J. Edmonds. 1996. Economic and environmental choices in the stabilization of atmospheric CO_2 concentrations. *Nature* 379(6562): 240-243.

Reducing Emissions from Energy End Use and Infrastructure

The potential for advanced technology to enable and facilitate accelerated reductions in greenhouse gas (GHG) emissions, mainly carbon dioxide (CO_2), from energy end use and infrastructure is believed to be significant—on a par with, or greater than, that of each of the other main elements of CCTP's technology strategy. Emissions reductions can be achieved from all end-use sectors of the global economy, including industry, residential and commercial buildings, and transportation, through conservation practices,[1] technological and other economic productivity improvements that lead to increased energy efficiency, and shifts in the composition of economic activity toward lower energy- and GHG-intensive outputs.

Historically, global energy productivity—loosely measured in terms of economic output per unit of energy input—has shown steady increases, averaging gains of about 0.9 percent per year over the period 1971 to 2002 (IEA 2004). Use of more energy-efficient processes and replacement of older, less-efficient capital stock are important contributors to these gains. Another factor in increasing individual country measures of energy productivity, especially in industrialized countries, has been a shift over the past several decades in the composition of economic output toward less energy-intensive goods and services.

In published scenarios of CO_2 emissions over the 21st century, increasing demand for energy services, driven by population and economic growth, results in growth of CO_2 emissions. By contrast, almost all scenarios that explore various future paths toward significant emissions reductions show that energy end-use reduction[2] plays a key role in achieving those reductions. In one set of scenarios, as shown in Figure 3-19 and highlighted above, energy end-use reductions led to a cumulative decrease (over 100 years) of between 7 and 22 thousand exajoules (EJ) in global energy use, and between 110 and 270 gigatons of reduced or avoided global emissions of carbon (GtC), compared to a reference case used in the study (see Chapter 3). Although bracketed by

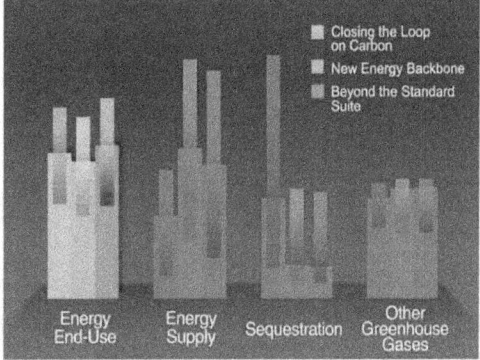

**Energy End-Use
Potential Contributions to Emissions Reduction**

Potential contributions of Energy End-Use reduction to cumulative GHG emissions reductions to 2100, across a range of uncertainties, for three advanced technology scenarios. See Chapter 3 for details.

uncertainties, this figure suggests both the potential role for advanced technology and a long-term goal for contributions from this sector of the global economy.

In the United States, the largest end-use sources of CO_2 emissions (Table 4-1) are the following:

◆ Electricity and fuel use in buildings;

---

[1] In this context, "conservation" refers to practices that conserve resources or reduce waste, such as in the case of energy, turning off lights, equipment, etc., when not in use.

[2] End use reduction includes improvements in energy efficiency in the end-use sectors, as well as improvements in efficiency of energy conversion, e.g., increased efficiency in electricity generation.

◆ Transportation fuels;

◆ Electricity and fuel use in industry; and

◆ A few industrial processes not related to combustion.

This chapter explores energy end-use and carbon emission-reduction strategies and opportunities within each of these end-use categories. Sections 4.1 through 4.3 address transportation, buildings, and industry, respectively. Section 4.4 deals with technology strategies for the electric grid and infrastructure that can enable and facilitate CO_2 emissions reductions in all sectors. Each section provides information on its respective sector, explores the potential role for advanced technology in reducing emissions, outlines a strategy for technology development, describes the current portfolio, and identifies potential directions for future research. This chapter focuses primarily on reducing and avoiding CO_2 emissions. Many industrial processes and energy end uses produce significant quantities of other non-CO_2 GHGs. These other GHGs are addressed in Chapter 7, "Reducing Emissions of Other Greenhouse Gases." The descriptions of the technologies and deployment programs in this section include active Internet links to an updated version of the CCTP report *Technology Options in the Near and Long Term* (CCTP 2005).[3]

Transportation

The transport of people, goods, and services accounts for a significant share of global energy demand, mostly in the form of petroleum, and is among the fastest growing sources worldwide of emissions of GHGs, mainly CO_2. In the developing parts of Asia and the Americas, emissions from transportation-related use of energy are expected to increase dramatically during the next 25 years. In the United States, from 1991 to 2000, vehicle miles traveled, a measure of highway transportation demand, increased at an average rate of 2.5 percent per year (DOT 2002a), outpacing population growth. In 2003, the U.S. transportation sector accounted for 39 percent of total CO_2 emissions, with the highway modes accounting for more than 82 percent of these (Table 4-2). Through 2025, future growth in U.S. transportation energy use and emissions is projected to be strongly influenced by the growth in light-duty trucks (pickup trucks, vans, and sport utility vehicles, under 8,500 lb gross vehicle weight rating) (Figure 4-1). According to the Federal Highway Administration's Freight Analysis Framework, freight tonnage will grow by 70 percent during the first two decades of the 21ˢᵗ century (DOT 2002b).

CO_2 Emissions in the United States by End-Use Sector, 2003 (GtC)

END-USE SECTOR	EMISSIONS FROM ELECTRICITY	EMISSIONS FROM COMBUSTION OF FUELS	EMISSIONS, TOTAL	% OF TOTAL
Transportation	0.009	0.485	0.493	31.1%
Residential and Commercial Buildings	0.410	0.169	0.579	36.5%
Industrial Energy Use	0.211	0.258	0.468	29.5%
Industrial Processes			0.040	2.5%
Waste Disposal Activities		0.005	0.005	0.3%
Total	**0.630**	**0.957**	**1.586**	**100.0%**

Table 4-1. CO_2 Emissions in the United States by End-Use Sector, 2003 (GtC)

Source: EPA 2005, Tables 2-16, 3-44, and 4-1.
Note: Values may not sum to total due to independent rounding of values.

---

[3] See http://www.climatetechnology.gov/library/2005/tech-options/index.htm.

Potential Role of Technology

Advanced technologies can make significant contributions to reducing CO_2 emissions from transportation activity. In the near term, advanced highway vehicle technologies, such as electric-fuel-engine hybrids ("hybrid-electric" vehicles) and clean diesel engines, could improve vehicle efficiency and, hence, lower CO_2 emissions. Other reductions might result from modal shifts (e.g., from cars to light rail) higher load factors, improved overall system-level efficiency, or reduced transportation demand. Improved intermodal connections could allow for better mode-shifting and improved efficiency in freight transportation. Application of developing technology will reduce idling and the concomitant emissions from heavy-duty vehicles, including vessels, trains, and long-haul trucks. Intelligent transportation systems can reduce congestion, resulting in decreases in fuel use. In the long term, technologies such as cars and trucks powered by hydrogen, bio-based fuels, and electricity show promise for transportation with either no highway CO_2 emissions or no net-CO_2 emissions.

CO_2 Emissions in the United States from Transportation, by Mode, in 2003 (GtC)

	EMISSIONS	% OF TOTAL
Passenger Cars	0.173	35.6%
Light-Duty Trucks	0.131	26.9%
Other Trucks	0.093	19.2%
Aircraft [a]	0.047	9.6%
Other [b]	0.013	2.6%
Boats & Vessels	0.016	3.2%
Locomotives	0.012	2.4%
Buses	0.002	0.5%
Total [c]	**0.477**	**100.0%**

(a) Aircraft emissions consist of emissions from all jet fuel (less bunker fuels) and aviation gas consumption.

(b) "Other" CO_2 emissions include motorcycles, pipelines, and lubricants.

(c) Percentages may not sum to 100 percent due to independent rounding of values.

Source: EPA 2005.

Table 4-2. CO_2 Emissions in the United States from Transportation, by Mode, in 2003 (GtC)

Transportation Sector Energy Use by Mode and Type

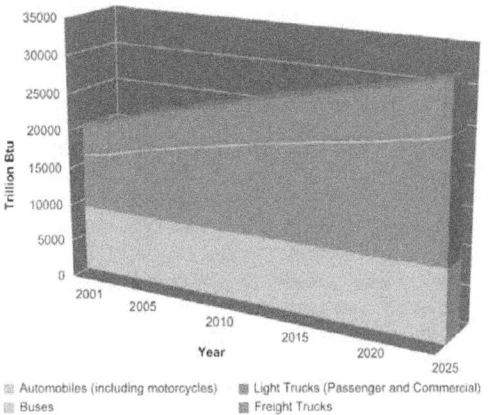

Legend: Automobiles (including motorcycles), Light Trucks (Passenger and Commercial), Buses, Freight Trucks

Figure 4-1. Projected Energy Consumption in U.S. Highway Vehicles.
Source: EIA 2004

The focus of CCTP is on technology developments that may reduce or capture GHG emissions, but it should be noted that, in the transportation sector particularly, there are non-technological options for policymakers to consider as well. They will not be discussed in the rest of this chapter, but they are worth mentioning here. For instance, local investments in bicycle lanes and paths and in pedestrian-friendly development planning can help reduce total vehicle-miles traveled. Urban and suburban planning that is well integrated with public transportation can similarly reduce fuel use per person-mile of travel. In addition, new communications technologies may alter the concepts about individual mobility. Work locations may be centered near or in residential locations, and work processes and products may be more commonly communicated or delivered via digital media. With global trends toward increasing urbanization in both population concentrations and opportunities for employment, people may rely in the future on improved modes of local, light-rail or intra-city passenger transport, coupled with other advances in electrified intercity transport that would curb the growth of fuel use and emissions from transportation.

Technology Strategy

Realizing these opportunities requires a research portfolio that embraces a combination of advanced vehicle, fuel, and transportation system technologies. Within constraints of available resources, a balanced portfolio needs to address major sources of CO_2 emissions in this sector, including passenger cars, light trucks, and other trucks; key modes of transport, including highway, aviation, and urban transit; system-wide planning and enhancements; and both near- and long-term opportunities.

In the near term, CO_2 emissions and transportation energy use can be reduced through improved vehicle efficiency, clean diesel engines, hybrid propulsion, and the use of hydrogenated low-sulfur gasoline. Other fuels, such as ethanol, natural gas, electricity with storage, and biodiesel, can also provide attractive means for reducing emissions of CO_2. These efficiency gains and fuel alternatives also provide other benefits, such as improving urban and regional air quality and enhancing energy security.

In aviation, emissions could be lowered through new technologies to improve air traffic management. An example is Reduced Vertical Separation Minimums (RVSM). RVSM has been used for transatlantic flights since 1997, and it became standard in U.S. airspace in January 2005. Full implementation of RVSM may reduce fuel use by approximately 500 million gallons each year.

In the long term, hydrogen may prove to be a low- or no-net-carbon energy carrier, if it can be cost-effectively produced with few or no GHG emissions, such as with renewable or nuclear energy, or with fossil fuels in conjunction with carbon capture and storage. Hydrogen and biofuels as substitutes for petroleum-based fuels in the transportation and other sectors also offer significant national security benefits. Hydrogen and alternative fuels are discussed in more depth in Chapter 5, "Reducing Emissions from Energy Supply." Hydrogen can be used in internal combustion engines, but its use in highly efficient fuel-cell-powered vehicles is considered a very important future option (Figure 4-2). In aviation, new engines and aircraft will feature enhanced engine cycles, more efficient aircraft aerodynamics, and reduced weight— thereby improving fuel efficiency. Research sponsored by the Federal Government through NASA, in collaboration with the Next Generation Air Transportation System (NGATS) plan, could enable these enhancements. NGATS is a multi-agency-integrated effort to ensure that the future air transportation system meets air transportation security, mobility, and capacity needs, while reducing environmental impacts.

Current Portfolio

Across the current Federal portfolio of transportation-related R&D, activities are focused on a number of major programs.

◆ Research on **light vehicles,** organized primarily in support of the FreedomCAR and Fuel Partnership, focuses on materials; power electronics; hybrid vehicles operating on gasoline, diesel, or alternative fuels; high-efficiency, low-emission advanced combustion engines, enabled by improved fuels; and high-volume, cost-effective production of lightweight materials. Beginning in Fiscal Year 2007, the Department of Energy is increasing the funding for advanced batteries, power electronics, and systems analyses specifically needed to accelerate the introduction of "plug-in" hybrid vehicles, which provide their owners with the option of operating as a normal hybrid vehicle or in all-electric mode, consuming no petroleum at all. With appropriate power-aggregation infrastructure, these vehicles can potentially also act as grid-attached storage when they are parked, contributing to electric grid stability.

The vehicle technologies research programs have a number of specific goals: (a) electric propulsion systems with a 15-year life capable of delivering at least 55 kW for 18 seconds and 30 kW continuous at a system cost of $12/kW peak; (b) internal combustion engine powertrain systems costing $30/kW, having peak brake engine efficiency of 45 percent, and that meet or exceed emissions standards; (c) electric drivetrain energy storage with a 15-year life at 300 Wh with discharge power of 25kW for 18 seconds and $20/kW; (d) material and manufacturing technologies for high-volume production vehicles, which enable/support the simultaneous attainment of 50 percent reduction in the weight of vehicle structure and subsystems, affordability, and increased used of recyclable/renewable materials; and (e) internal combustion engine powertrain systems, operating on hydrogen with a cost target of $45/kW by 2010 and $30/kW in 2015, having a peak brake engine efficiency of 45 percent, and that meet or exceed emissions standards.[4]

◆ Research areas for **heavy vehicles,** organized primarily under the 21ˢᵗ Century Truck Partnership, include lightweight materials, aerodynamic drag, tire rolling resistance,

---

[4] For more information, see Section 1.1.1 (CCTP 2005): http://www.climatetechnology.gov/library/2005/tech-options/tor2005-111.pdf. See also: http://www.eere.energy.gov/hydrogenandfuelcells/fuelcells/transportation.html, and http://www.epa.gov/otaq/technology.

electrification of ancillary equipment, advanced high-efficiency combustion propulsion systems (including energy-efficient emissions reduction), fuel options (both petroleum- and non-petroleum-based), hybrid technologies for urban driving applications, and onboard power units for auxiliary power needs. The research objectives are to: (1) reduce energy consumption in long-haul operations, (2) increase efficiency and reduce emissions during stop-and-go operations, and (3) develop more efficient and less-polluting energy sources to meet truck stationary power requirements (i.e., anti-idling). By 2010, the goals include a laboratory demonstration of an emissions-compliant engine system that is commercially viable for Class 7-8 highway trucks, and an engine that improves the system efficiency to 53 percent by 2010 and to 55 percent by 2013, from the 2002 baseline of 40 percent. By 2012, the goals include advanced technology concepts that reduce the aerodynamic drag of a Class 8 tractor-trailer combination by 20 percent.[5]

Figure 4-2. *In the longer term, hydrogen fuel cell vehicles may provide for a transportation future that has much lower CO_2 emissions.*

Courtesy: DOE/NREL, Credit: SunLine Transit Agency

◆ **Fuels research** encompasses the development of new fuel blend formulations that will enable more efficient and cleaner combustion and the development of renewable and non-petroleum-based fuels that could displace 5 percent of petroleum used by commercial vehicles.[6]

◆ Research on **intelligent transportation systems** infrastructure includes sensors, information technology, and communications to improve efficiency and ease congestion. Intelligent transportation systems (ITS) goals include improved analysis capabilities that properly assess the impact of ITS strategies and strategies that will improve travel efficiency resulting in lower delays, thereby reducing emissions.[7]

◆ Research on **aviation fuel efficiency** includes engine and airframe design improvements. Aviation fuel efficiency goals include improved aviation fuel efficiency per revenue plane-mile by 1 percent per year through 2008, and new technologies with the potential to reduce CO_2 emissions from future aircraft by 25 percent within 10 years and by 50 percent within 25 years.[8]

◆ **Best Workplaces for Commuters** program is a voluntary employer-adopted program that increases commuter flexibility by expanding transportation mode options, using flexible scheduling, and increasing work location choices.

◆ The **Clean Cities** program supports efforts to deploy alternative fuel vehicles (AFVs) and develop the necessary supporting infrastructure. Clean Cities works through a network of more than 80 volunteer, community-based coalitions, which develop public/private partnerships to promote the use of alternative fuels and vehicles, expand the use of fuel blends, encourage the use of fuel economy practices, increase the acquisition of hybrid vehicles by fleets and consumers, and advance the use of idle reduction technologies in heavy-duty vehicles.

◆ **Congestion Mitigation and Air Quality Improvement Program**, administered by the U.S. Department of Transportation, in consultation with the EPA, provides states with funds to reduce congestion and to improve air quality through transportation control measures and other strategies.

◆ **Mobile Air Conditioning Climate Protection Partnership** strives to reduce GHG emissions from vehicle air conditioning systems through voluntary approaches. The program promotes cost-effective designs and improved service procedures that minimize emissions from mobile air conditioning systems.

5 See Section 1.1.2 (CCTP 2005): http://www.climatetechnology.gov/library/2005/tech-options/tor2005-112.pdf. See also: http://www.epa.gov/otaq/technology.

6 See Section 1.1.3 (CCTP 2005): http://www.climatetechnology.gov/library/2005/tech-options/tor2005-113.pdf.

7 See Section 1.1.4 (CCTP 2005): http://www.climatetechnology.gov/library/2005/tech-options/tor2005-114.pdf.

8 See Section 1.1.5 (CCTP 2005): http://www.climatetechnology.gov/library/2005/tech-options/tor2005-115.pdf.

◆ **SmartWay Transport Partnership** program is designed to reduce emissions from the freight sector through the implementation of innovative technology and advanced management practices. To date, 133 companies have joined the Partnership and have committed to reduce the CO_2 emissions associated with their respective freight operations. Additionally, there are now over 50 diesel truck and locomotive engine idling reduction projects around the country.

Future Research Directions

The current portfolio supports the main components of the technology development strategy and addresses the highest priority current investment opportunities in this technology area. For the future, CCTP seeks to consider a full array of promising technology options. From diverse sources, suggestions for future research have come to CCTP's attention. Some of these, and others, are currently being explored and under consideration for the future R&D portfolio. These include:

◆ **Freight Transport.** Strategies and technologies to address congestion in urban areas and freight gateways by increasing freight transfer and movement efficiency among ships, trucks and rail in anticipation of large growth in freight volumes.

◆ **Advanced Urban Concepts.** Studies of advanced urban-engineering concepts for cities to evaluate alternatives to urban sprawl. Such engineering analysis would consider the co-location of activities with complementary needs for energy, water, and other resources; and would enable evaluation of alternative configurations that could significantly reduce vehicle-miles traveled and GHG emissions.

◆ **Integrated Urban Planning**. Concept and engineering studies for large-scale institutional and infrastructure changes required to manage CO_2, electricity, and hydrogen systems reliably and securely. Analysis of the infrastructure requirements for plug-in hybrid electric vehicles is needed. By being plugged into the power grid at night, when electricity is cheapest and most available, and by operating solely on battery power for the first 10-60 miles, this technology could significantly reduce oil consumption.

◆ **Large-Scale Hydrogen Storage.** Technologies for large-scale hydrogen storage and transportation and low-cost, lightweight electricity storage including advanced batteries and ultracapacitors.

In addition, supporting or crosscutting areas for future research include the following:

◆ **Advanced Thermoelectric Concepts.** Advanced thermoelectric concepts to convert temperature differentials into electricity, made more affordable through nanoscale manufacturing.

◆ **Battery and Fuel Cell Systems.** Basic electrochemistry to produce safe, reliable battery and fuel cell systems with acceptable energy and power density, cycle life, and performance under temperature extremes.

◆ **New Combustion Regimes.** Advanced combustion research on new combustion regimes in conventional vehicle propulsion technologies, using conventional fuels as well as alternatives such as cellulosic ethanol and biodiesel where near-zero regulated emissions and lowered carbon emissions can be achieved.

4.2 Buildings

The built environment—consisting of residential, commercial, and institutional buildings—accounts for about one-third of primary global energy demand (IPCC 2000) and represents a major source of energy-related GHG emissions, mainly CO_2. Growth in global energy demand in buildings has averaged 3.5 percent per year since 1970 (IPCC 2001).

Over the long term, buildings are expected to continue to be a significant component of increasing global energy demand and a large source of CO_2 emissions. Energy demand in this sector will be driven by growth in population, by the economic expansion that is expected to increase the demand for building services (especially electric appliances, electronic equipment, and the amount of conditioned space per person), and by the continuing trends toward world urbanization. As urbanization occurs, energy consumption increases, because urban buildings usually have electricity access and have a higher level of energy consumption per unit area than buildings in more primitive rural areas. According to a recent projection by the United Nations, the percentage of the world's population living in urban areas will increase from 49 percent in 2005 to 61 percent by 2030 (UN 2005).

In the United States, energy consumption in residential buildings has been increasing proportionately with increases in population, while energy consumption in commercial buildings has tended to follow the level of economic activity, or real Gross Domestic Product (GDP).[9] This trend masks significant increases in efficiency in some building components that are being offset by new or increased energy uses in others. In the United States in 2003, CO_2 emissions from this sector, including those from both fuel combustion and use of electricity derived from CO_2-emitting sources, accounted for nearly 37 percent of total CO_2 emissions (Table 4-1). These emissions have been increasing at 1.9 percent per year since 1990 (EIA 2005). Table 4-3 shows a breakdown of emissions from the buildings sector, by fuel type, in the United States.

Potential Role of Technology

Many opportunities exist for advanced technologies to make significant reductions to energy-related CO_2 emissions in the buildings sector. In the near term, widespread adoption of advanced commercially available technologies, such as ENERGY STAR® compliant equipment, can improve efficiency of energy-using equipment in the primary functional areas of energy use. In residential buildings, these functional areas include space heating, appliances, lighting, water heating, and air conditioning. In commercial buildings, functional areas are lighting, space heating, cooling and ventilation, water heating, office equipment, and refrigeration. Through concerted research, major technical advances have occurred during the past 20 years, with many application areas seeing efficiency gains of 15 percent to 75 percent. Figure 4-3 gives an example of technological improvements that have occurred in refrigerators as an illustration of the kind of gains that have been achieved. Over the longer term, more advances can be expected in these areas, and significant opportunities also lie ahead in the areas of new buildings design, retrofits of existing buildings, and the integration of whole building systems and multi-building complexes through use of sensors, software, and automated maintenance and controls.

By 2025—with advances in building envelopes, equipment, and systems integration—it may be possible to achieve up to a 70 percent reduction in a building's energy use, compared to the average energy use in an equivalent building today (DOE 2005). If augmented by on-site energy technologies (such as photovoltaics or distributed sources of combined heat and power), buildings could become net-zero GHG

Residential and Commercial CO_2 Emissions in the United States, by Source, in 2003 (GtC)

	EMISSIONS	% OF TOTAL
RESIDENTIAL		
Electricity	0.2121	66.9%
Natural Gas	0.0756	23.8%
Petroleum	0.0291	9.2%
Coal	0.0003	0.1%
Total Residential	**0.3171**	**100.0%**
COMMERCIAL		
Electricity	0.2005	76.2%
Natural Gas	0.0466	17.5%
Petroleum	0.0147	5.4%
Coal	0.0025	0.9%
Total Commercial	**0.2643**	**100.0%**

Note: Percentages may not sum to 100 percent, due to independent rounding of values.
Source: EPA 2005, Tables 2-16 and 3.3.

Table 4-3. Residential and Commercial CO_2 Emissions in the United States, by Source, in 2003 (GtC)

emitters and net energy producers.

Technology Strategy

While the built environment is a complex mix of heterogeneous building types (commercial, service, detached dwelling, apartment buildings) and functional uses, all have common features, each of which may benefit from technological research, both as individual components and as integrated systems. Within constraints of available resources, a balanced portfolio needs to address three important aspects of buildings that affect their CO_2 emissions: (1) integrated building design, construction, and operation, including the optimal integration of components; (2) advanced building envelope components, including windows and walls; and (3) advanced, building equipment, appliances, and lights. The portfolio should look at both near- and long-term opportunities.

---

9   According to the Annual Energy Review 2004, residential primary energy use increased 53 percent from 1970 to 2004, while the nation's population increased 44 percent over the same period. Commercial energy use increased 111 percent over this period, driven (in part) by the nearly 200 percent increase in real Gross Domestic Production (GDP).

In the near term, building energy use and CO_2 emissions could be lowered in several ways. Especially in new construction, design strategies that incorporate energy- and material-saving strategies from the very start of the building process can result in significant avoided carbon. Intelligent building systems (such as load balancing and automated sensors and controls) can also be included to help ensure the comfort, health, and safety of residents, as well as aid in the reduction of CO_2. In the building envelope, application of advanced materials such as high R-value insulation, foams, vacuum panels, and spectrally selective windows can reduce space conditioning loads significantly. Choosing highly recyclable materials such as aluminum can reduce the end-of-life impact of building design and contribute to sustainable building practices. Technologies to improve the efficiency of lighting, appliances, heating, cooling, and ventilation are other options.

Refrigerator Energy Efficiency

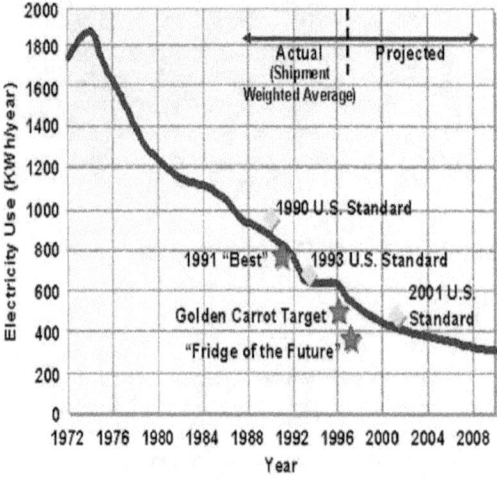

Note: The curve applies to 18-20 cu. ft. top-mount refrigerator/freezers, which capture the largest market share in the United States. The term,"1991 Best" stands for the 1991 top-mount model with lowest energy use. "Golden Carrot Target" was an EPA/electric utility program in the early 1990s to develop a model that was 25 percent more efficient than the then current technology. "Fridge of the Future" is a refrigerator that had a target energy use of 365 kWh/yr or 1 kWh/day for 18-20 cu. ft. top-mount models based on an cooperative research agreement between Oak Ridge National Laboratory (ORNL) and the Association of Home Appliance Manufacturers; this target was exceeded in a test unit (0.93 kWh/day) in FY 1996.

Figure 4-3. Refrigerator Energy Efficiency
(Source: Brown 2003)

In the long term, more advanced research on the building envelope—including dynamic switchable window glazings and dynamic walls, panelized housing construction, façade and roof integration of photovoltaics, and new storage technologies—can drive CO_2 emissions even lower. Distributed power systems, advanced refrigeration and cooling technologies, integrated heat pumps that serve space conditioning and water heating, and solid-state lighting technology are among some of the more promising options for equipment. Among the alternatives, building integration should focus on including sensors and controls, community-scale integration tools, and urban engineering.

Current Portfolio

The current Federal portfolio focuses on four major thrusts. In combination, these activities aim to achieve net-zero-energy residential buildings by 2020 and commercial buildings by 2025.

◆ Research on the **building envelope** (the interface between the interior of a building and the outdoor environment) focuses on systems that determine or provide control over the flow of heat, air, moisture, and light in and out of a building; and on materials that can affect energy use, including insulation, foams, vacuum panels, optical control coatings for windows and roofs, thermal storage, and related controls (such as electrochromic glazings). Research program goals in the building envelope area include the following: (1) by 2008, demonstrate dynamic solar control windows (electrochromics) in commercial buildings, and by 2010 demonstrate market-viable windows with R6 insulation performance for homes; (2) by 2025, develop marketable and advanced energy systems capable of achieving "net-zero" energy use in new residential and commercial buildings; (3) for the long term, and consistent with achieving net zero energy performance, achieve a 30 percent decrease in the average envelope thermal load of existing residential buildings and a 66 percent decrease in the average thermal load of new buildings.[10]

◆ Research on **building equipment** focuses on means to significantly improve efficiency of heating, cooling, ventilating, thermal distribution, lighting (Figure 4-4), home appliances, and on-site energy and power devices. This area also includes a number of crosscutting elements, including advanced refrigerants and cycles, solid-state

---

[10] See Section 1.2.2 (CCTP 2005): http://www.climatetechnology.gov/library/2005/tech-options/tor2005-122.pdf.

lighting, smart sensors and controls, small power supplies, microturbines, heat recovery, and other areas.

Specific goals include: (1) for distributed electricity generation technologies (including microturbines), by 2008, enable a portfolio of equipment that shows an average 25 percent increase in efficiency; (2) for solid-state lighting in general illumination applications, by 2008, develop equipment with luminous efficacy of 79 lumens per watt (LPW); and, for laboratory devices by 2025, a luminous efficacy of 200 LPW. The long-term goals are: (1) by 2025, develop and demonstrate marketable and advanced energy systems that can achieve "net-zero" energy use in new residential and commercial buildings through a 70 percent reduction in building energy use; and (2) by 2030, enable the integration of all aspects of the building envelope, equipment, and appliances with on-site micro-cogeneration and zero-emission technologies.[11]

Figure 4-4. Many opportunities exist for advanced technologies, such as the sub-compact fluorescent coil light bulb, to make significant reductions of CO_2 emissions in the building sector.

Courtesy: DOE/NREL

◆ Research on **residential systems integration**, focuses on progressively higher levels of energy performance over time, and follows a three-phase approach. Phase 1, systems evaluation, involves the design, construction and testing of subsystems for whole house designs (by climate zone) in research houses to evaluate how components perform. In Phase 2, prototype houses (Figure 4-5), the successful Phase 1 subsystems are designed and constructed by production builders to evaluate the ability to implement the systems on a production basis. Phase 3, community evaluations, provides technical support to builder partners to advance from the production prototypes to evaluation of production houses in a subdivision.

◆ Research on **commercial buildings integration** includes development of advanced design "packages" of system-integrated design strategies and operational methodologies for high-performance buildings, which can be used by architects and others to design, build, and operate commercial buildings in an integrated manner. This approach is targeted at new construction, because the opportunities for aggressive performance are much greater than in existing buildings. In addition, R&D includes load balancing and automated sensors and controls, sometimes referred to as intelligent building systems. Such systems continuously monitor building performance, detect anomalies or

degradations, optimize operations across all building systems, guide maintenance, and document and report results. They can also be extended to coordinate on-site energy generation and internal loads, with external power (grid) demands and circumstances, allowing responsiveness to time-variant cost savings, system efficiencies, and grid contingencies. They also ensure occupant comfort, health, and safety, met at the lowest possible cost.

◆ Whole building integration goals include fully and seamlessly integrated building design tools, such as **EnergyPlus**, that support all aspects of design and provide rapid analysis of different pathways to improved energy performance. Also included are the development of automatic operation of buildings systems that require little operator attention; and highly efficient combined cooling, heating, and power systems that use waste heat from small-scale, on-site electricity generation to provide heating and cooling for the buildings, as well as export excess electricity to the grid.[12]

◆ The **ENERGY STAR®** is a joint program of EPA and DOE that enables businesses, organizations, and consumers to realize the cost savings and environmental benefits of energy efficiency investments. Introduced in 1992, the ENERGY

---

[11] See Section 1.2.1 (CCTP 2005): http://www.climatetechnology.gov/library/2005/tech-options/tor2005-121.pdf.

[12] See Section 1.2.3 (CCTP 2005): http://www.climatetechnology.gov/library/2005/tech-options/tor2005-123.pdf.

Figure 4-5. Design strategies that incorporate energy- and material-saving strategies can result in significant reduction of CO_2 emissions.

Courtesy: DOE

STAR® program focuses on four areas of energy efficiency that include products, home improvement, new homes, and business improvement. By providing resources, working with local partners, and labeling ENERGY STAR® qualified-products, the program makes it easy for consumers and businesses to make smart energy-efficient choices that can save about a third on energy bills with similar savings of GHG emissions, without sacrificing features, style or comfort.

♦ **Federal Energy Management Program** aims to reduce energy use in Federal buildings, facilities, and operations by advancing energy efficiency and water conservation, promoting the use of renewable energy, and managing utility choices of Federal agencies. The program accomplishes its mission by leveraging both Federal and private resources to provide Federal agencies the technical and financial assistance they need to achieve their goals.

♦ **Rebuild America** works with a network of hundreds of community-based partnerships across the nation that are saving energy, improving building performance, easing air pollution through reduced energy demand, and enhancing the quality of life through energy efficiency and renewable energy technologies.

♦ **Residential Building Integration: Building America** is intended to design, build, and evaluate energy-efficient homes that use from 30 to 40 percent less total energy than comparable traditional homes with little or no increase in construction costs; and to encourage industry to adopt these practices for new home construction. In addition, ongoing research focuses on integrating onsite power systems, including renewable energy technologies. Through the Building America program, DOE and its more than 470 industry partners are conducting research to develop advanced building energy systems to make homes and communities much more energy efficient.

♦ **State Energy Program** extends grants from DOE to states for a variety of energy-efficiency and renewable-energy activities. Most states have a state energy office designated to help reduce energy waste and increase the use of renewables in the state.

♦ **Weatherization Assistance Program** provides cost-effective energy-efficiency improvements to low-income households through the weatherization of homes. Priority is given to the elderly, persons with disabilities, families with children, and households that spend a disproportionate amount of their income on energy bills (utility bills make up 15 to 20 percent of household expenses for low income families, compared to five percent or less for all other Americans).

Future Research Directions

The current portfolio supports many of the components of the technology development strategy and addresses the highest priority current investment opportunities in this technology area. For the future, CCTP seeks to consider a full array of promising technology options. From diverse sources, suggestions for future research have come to CCTP's attention. Some of these, and others, are currently being explored and under consideration for the future R&D portfolio. These include:

♦ **Building Envelope.** Improved panelized housing construction; methods for integrating photovoltaic systems in building components such as roofs, walls, skylights, and windows, and with building loads and utilities; and exploration of fundamental properties and behaviors of novel materials to maximize moisture resistance and the storage and release of energy. Through nanotechnology advances, research will develop smart roofs and walls that reflect infrared solar radiation in summer and absorb it in winter, with substantial energy savings.

◆ **Building Equipment.** Advances in on-site power production (including fuel cells, microturbines, and reciprocating engines); ultra-efficient Heating, Ventilating, and Air-Conditioning-Residential (HVACR) (including magnetic or solid state cooling systems, advanced desiccants and absorption chilling systems, integrated heat pump systems for space conditioning and water heating, and highly efficient geothermal heat pumps); improved hot water circulation systems; and solid state lighting technology and improved lighting distribution systems. Low-power ubiquitous sensors with wireless communications can optimize building operations through anticipatory and selective heating, cooling, and lighting that manage loads in response to occupant requirements and real-time energy prices.

◆ **Whole Building Integration.** Further development and widespread implementation of building design tools for application in new and retrofit construction; tools and technologies for systems integration in buildings, with a particular focus on sensors and controls for supply and end-use system integration; development of pre-engineered, optimized net-zero energy buildings; community-scale design and system integration tools; and urban engineering to reduce transport energy use and congestion.

4.3 Industry

Industrial activities were estimated to account for about 41 percent of primary global energy consumption in 1995 (IPCC 2000) and a commensurate share of global CO_2 emissions. Certain activities are particularly energy-intensive, including metals industries, such as iron, steel, and aluminum; petroleum refining; basic chemicals and intermediate products; fertilizers; glass; pulp, paper, and other wood products; and mineral products, including cement, lime, limestone, and soda ash. Others are less energy-intensive, including the manufacture or assembly of automobiles, appliances, electronics, textiles, food and beverages, and others. Each regional or national economy varies in the structure, composition, and growth rates of these

industries, shaped in part by its state of economic development and in part by regional advantages in international trade. The industrial sector worldwide is expected to expand in the future and will likely continue to account for a substantial portion of future CO_2 emissions. As competitive environments grow and energy demand increases, industrial firms increasingly will have a financial incentive to invest in improving their energy efficiency.

In the United States in 2003, industry accounted for about one-third of total U.S. CO_2 emissions (Table 4-1). These are attributed to combustion of fuels (51 percent), use of electricity derived from CO_2-emitting sources (41 percent), and industrial processes, including non-combustion processes, that emit CO_2 (8 percent) (Table 4-4).[13]

Potential Role of Technology

The industrial sector presents numerous opportunities for advanced technologies to make significant contributions to the reductions of CO_2 emissions to the Earth's atmosphere. In the near term, advanced technologies can increase the efficiency with which process heat is generated, contained, transferred, and recovered. Process and design enhancements can improve quality, reduce waste, minimize reprocessing, reduce the intensity of material use (with no adverse impact on product or performance), and increase in-process material recycling. Cutting-edge technologies can significantly reduce the intensity with which energy and materials (containing embedded energy) are used. Industrial facilities can implement direct manufacturing processes, which can eliminate some energy-intensive steps, thus both avoiding emissions and enhancing productivity. On the supply side, industry can self-generate clean, high-efficiency power and steam; and create products and byproducts that can serve as clean-burning fuels. The sector can also make greater use of coordinated systems that more efficiently use distributed energy generation, combined heat and power, and cascaded heat.

In the long term, fundamental changes in energy infrastructure could effect significant CO_2 emissions reductions. Revolutionary changes may include novel heat and power sources and systems, including renewable energy resources, hydrogen, and fuel cells. Innovative concepts for new products and high-efficiency processes may be introduced that can take full advantage of recent and promising developments in nanotechnology, micro-manufacturing, sustainable

---

[13] Emissions of GHGs other than CO_2 from industry and agriculture are discussed in Chapter 7, "Reducing Emissions of Other Greenhouse Gases."

biomass production, biofeedstocks, and bioprocessing. As global industry's existing, capital-intensive equipment stock nears the end of its useful service life and as industry expands in rapidly emerging economies in Asia and the Americas, this sector will have an opportunity to adopt novel technologies that could revolutionize basic manufacturing. Advanced technologies will likely involve a mix of pathways, such as on-site energy generation, conversion, and utilization; process efficiency improvements; innovative or enabling concepts, such as advanced sensors and controls, materials, and catalysts; and recovery and reuse of materials and byproducts (Figure 4-6). In the United States, the development and adoption of advanced industrial technologies can not only provide GHG benefits, but also help to maintain U.S. competitiveness.

CO_2 Emissions in the United States from Industrial Sources in 2003 (GtC)

	EMISSIONS 10^9 TONNES C	SHARE OF INDUSTRY TOTAL (%)	SHARE OF INDUSTRIAL PROCESSES (%)
Industrial Fuel Combustion	**0.258**	**50.7%**	
Coal	0.034	6.6%	
Petroleum	0.087	17.1%	
Natural Gas	0.111	21.9%	
Industrial Electricity	**0.211**	**41.4%**	
Industrial Processes (excluding fuel combustion emissions above)	**0.040**	**7.9%**	(See Breakout Below)
Total Industrial CO_2	**0.509**	**100.0%**	

Breakout of Emissions from Industrial Processes:			
Iron and Steel Production	0.0147		36.5%
Cement Manufacture	0.0117		29.2%
Ammonia Manufacture & Urea Application	0.0043		10.6%
Lime Manufacture	0.0035		8.8%
Limestone and Dolomite Use	0.0013		3.2%
Aluminum Production	0.0011		2.9%
Soda Ash Manufacture and Consumption	0.0011		2.8%
Petrochemical Production	0.0008		1.9%
Titanium Dioxide Production	0.0005		1.4%
Phosphoric Acid Production	0.0004		1.0%
Ferroalloy Production	0.0004		1.0%
Carbon Dioxide Consumption	0.0004		0.9%
Total Industrial Process CO_2	**0.0402**		**100.0%**

Source: EPA 2005, Tables 2-14, 2-16, 3-44, and 4-1.
Note: Percentages may not sum to 100 percent due to independent rounding of values.

Table 4-4. CO_2 Emissions in the United States from Industrial Sources in 2003 (GtC) (Excludes Indirect Emissions from Industrial Use of Centrally-Generated Electric Power)

Technology Strategy

A research portfolio should address the more important current and anticipated sources of CO_2 emissions in this sector, with careful attention to whether there is an appropriate role for Federal investment. Some of the largest sources of CO_2 emissions today, and expected in the future, arise from energy conversion to power industrial processes, inefficiencies in the processes themselves, and ineffective reuse of materials or feedstocks; and, in some cases, the intensive use of fossil fuels, especially natural gas.

In the near term, industrial energy use and CO_2 emissions could be lowered through improvements in the industrial use of electricity and fuels to produce process heat and steam, including steam boilers, direct-fired process heaters, and motor-driven systems, such as pumping and compressed air systems. Opportunities for reducing emissions in these areas lie with the adoption of best energy-management practices; adoption of more modern and efficient power and steam generating systems; integrated approaches that combine cooling, heating, and power needs; and capture and use of waste heat. Other areas of opportunity include improvements in specific energy-intensive industrial processes, including hybrid distillation systems; process intensification by combining or removing steps, or designing new processes altogether while producing the same or a better product; the recovery and utilization of waste and feedstocks, which can reduce energy and material requirements; and crosscutting opportunities, such as improved operational capabilities and performance.

In the long term, highly efficient coal gasifiers coupled with CO_2 sequestration technology could provide an alternative to natural gas, and even enable the export of electricity and hydrogen to the utility grid and supply pipelines. Bioproducts could replace fossil feedstocks for manufacturing fuels, chemicals, and materials; while biorefineries could utilize fuels from nonconventional feedstocks to jointly produce materials and value-added chemicals. For non-combustion sources of CO_2, gas capture, separation and sequestration could be applied, or alternative processes or materials could be developed as substitutes. Further, integrated modeling of fundamental physical and chemical properties, along with advanced methods to simulate processes, will stem from advances in computational technology.

Four Possible Pathways to Increased Industrial Efficiency

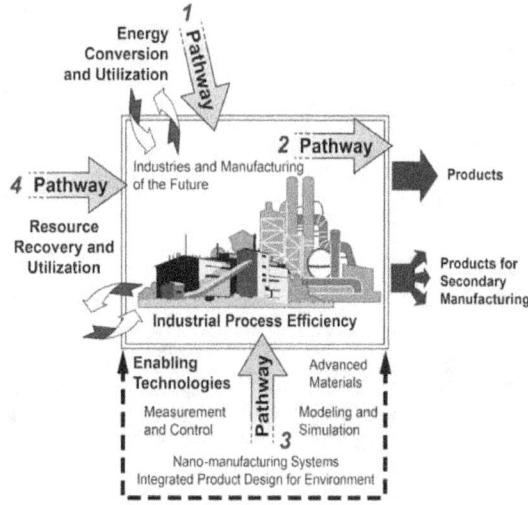

Figure 4-6. Four Possible Pathways to Increased Industrial Efficiency (Source: DOE 1997)

Current Portfolio

The current Federal portfolio focuses on four major thrusts.

◆ Research on **energy conversion and utilization** focuses on a diverse range of advanced and integrated systems. These include advanced combustion technologies, gasification technologies, high-efficiency burners and boilers, thermoelectric technologies (Figure 4-7) to produce electricity using industrial waste heat streams, co-firing with low-GHG fuels, advanced waste heat recovery heat exchangers, and heat-integrated furnace designs. Integrated approaches include combined-cycle power generation, and cogeneration of power and process heat or cooling.

The overall research program goal in this area is to contribute to a 20 percent reduction in the energy intensity (energy per unit of industrial output, as compared to 2002) of energy-intensive industries by 2020. Several specific goals include: (1) by 2006, demonstrate a greater than 94 percent efficient packaged boiler, and, by 2010, have commercially available packaged boilers with thermal efficiencies of 10-12 percent higher than conventional technology; (2) by 2008, demonstrate high-efficiency pulping technology in the pulp and

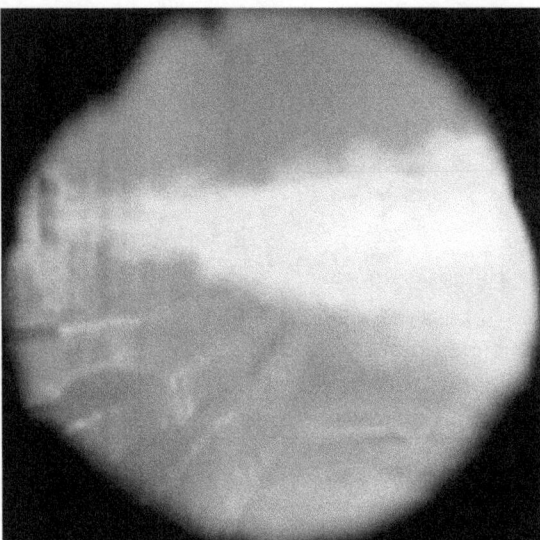

Figure 4-7. Industrial energy use and CO₂ emissions could be lowered through the use of high-efficiency combustion processes, such as oxy-fuel firing for glass manufacturing.

Courtesy of DOE/EERE

paper industry that redirects green liquor to pretreat pulp and reduce lime kiln load and digester energy intensity; and (3) by 2011, demonstrate isothermal melting technology, which could improve efficiency significantly in the aluminum, steel, glass, and metal-casting industries.[14]

◆ Research on **specific, energy-intensive and high-CO₂-emitting industrial processes** focuses on identifying (compared to theoretical minimum energy requirements) and removing process inefficiencies, lowering overall energy requirements for heat and power, and reducing CO_2 emissions. One process under development is a means to produce high-quality iron without the use of metallurgical coke, which—under current methods of steelmaking—is a significant source of CO_2 emissions. Other areas of research focus on processes that may also improve product yield, including oxidation catalysis, advanced processes, and alternative processes that take a completely different route to the same end product, such as use of noncarbon inert anodes in aluminium production.

Industrial process efficiency goals are focused on industry partnerships. The overall research program goal in this area is to realize, before 2020, a 20 percent improvement in energy intensity by the energy-intensive industries through the development and implementation of new and improved processes, materials, and manufacturing practices.[15]

◆ Research on **enabling technologies** includes an array of advanced materials that resist corrosion, degradation, and deformation at high temperatures and pressures; inferential sensors, controls, and automation, with real-time nondestructive sensing and monitoring; and new computational techniques for modeling and simulating chemical pathways and advanced processes.

Research program goals for this area target new enabling technologies that meet a range of cost goals depending on the technologies and on the applications where they are to be used. Specific goals include: (1) by 2010, demonstrate production and application of nano-structured diamond coatings and composites and other ultra-hard materials for use in wear-intensive industrial applications, and develop materials for use in a wide array of severe industrial environments (corrosive, high temperature, and pressure); (2) by 2012, demonstrate the generation of efficient power from high-temperature waste heat using systems with thermoelectric materials; and (3) by 2017, develop and demonstrate integration of sensing technologies with information processing to control plant production.[16]

◆ Research on **resource recovery and utilization** focuses on separating, capturing, and reprocessing materials for feedstocks. Recovery technologies include materials designed for recyclability, advanced separations, new and improved process chemistries, and sensors and controls. Reuse technologies include recycling, closed-loop process and plant designs, catalysts for conversion to suitable feedstocks, and post-consumer processing.

Research program goals in this area target a range of improved recycling/recovery efficiencies. For example, in the chemicals industry the goal is to improve recyclability of materials by as much as 30 percent. Additional goals target new and improved processes to use wastes or byproducts;

---

[14] See Section 1.4.1 (CCTP 2005): http://www.climatetechnology.gov/library/2005/tech-options/tor2005-141.pdf.

[15] See Section 1.4.3 (CCTP 2005): http://www.climatetechnology.gov/library/2005/tech-options/tor2005-143.pdf.

[16] See Section 1.4.4 (CCTP 2005): http://www.climatetechnology.gov/library/2005/tech-options/tor2005-144.pdf.

improve separations to capture and recycle materials, byproducts, solvents, and process water; and identify new markets for recovered materials, including ash and other residuals such as scrubber sludges.[17]

◆ **Climate Leaders** is an EPA partnership encouraging individual companies to develop long-term, comprehensive climate change strategies. Under this program, partners set corporate-wide GHG reduction goals and inventory their emissions to measure progress. The partnership now includes about 70 partners, 38 of whom have already set GHG emissions reduction goals. The US GHG emissions of these partners are equal to more than 7 percent of the U.S. total.

◆ **Climate VISION** assists industry efforts to accelerate the transition to practices, improved processes, and energy technologies that are cost-effective, cleaner, more efficient, and more capable of reducing, capturing, or sequestering GHGs. Already, business associations representing 14 industry sectors and the Business Roundtable have become program partners with the Federal Government and have issued letters of intent to meet specific targets for reducing GHG emissions intensity. These partners represent a broad range of industry sectors: oil and gas production, transportation, and refining; electricity generation; coal and mineral production and mining; manufacturing; railroads; and forestry products. Partnering sectors account for about 90 percent of industrial emissions and 40 to 45 percent of total U.S. emissions.

◆ **Industries of the Future** works in partnership with the nation's most energy-intensive industries, enhancing their long-term competitiveness and accelerating research, development, and deployment of technologies that increase energy and resource efficiency. This program has contributed to the development of hundreds of commercialized industrial technologies, resulting in a cumulative tracked energy savings of over 3,500 trillion Btu.

◆ **Voluntary GHG Emissions Reporting** under 1605(b) provides a means for organizations and individuals that have reduced their emissions to record their accomplishments and share their ideas for action. The enhanced registry will boost measurement accuracy, reliability, and verifiability, working with and taking into account emerging

domestic and international approaches. Currently about 230 U.S. companies and other organizations file GHG reports.

Future Research Directions

The current portfolio supports the main components of the technology development strategy and addresses the highest priority current investment opportunities in this technology area. For the future, CCTP seeks to consider a full array of promising technology options. From diverse sources, suggestions for future research have come to CCTP's attention. Some of these, and others, are currently being explored and considered for the future R&D portfolio. These include:

◆ **Industrial Alternatives to Natural Gas.** Research could be conducted to develop coal gasification systems for large industrial plants (e.g., 100 megawatts). The coal gasifiers would be highly integrated into complex manufacturing plants (e.g., chemical or glass plants). The industrial plant's feedstock, process heat, and power requirements could be accommodated from the coal gasifier, which could also export electricity, hydrogen, or other fuels to the utility grid and gas supply pipelines.

◆ **Industrial Waste Heat Reduction.** Energy conversion and utilization in industry creates large amounts of waste heat. This waste heat could be minimized through beneficial electrification (including infrared drying and induction heating); novel materials that do not require drying or curing; micro-CHP (combined heat and power) systems to convert waste heat to drive other thermal processes; and new ways of boosting low-grade temperatures to more useful high-grade heat.

◆ **Advanced Industrial Materials.** Research on advanced materials can reduce energy requirements and emissions in most industries. Characterization of materials at the nanoscale can lead to engineered materials with improved functionality, including improved properties such as strength, corrosion resistance, and power conversion (e.g., thermoelectrics).

◆ **Cement and Related Products.** Research could focus on various means to reduce or eliminate CO_2 emissions from non-combustion, high-emitting industrial processes, including the

---

[17] For more information, see Section 1.4.2 (CCTP 2005): http://www.climatetechnology.gov/library/2005/tech-options/tor2005-142.pdf.

cement, lime, limestone, and soda ash industries. Worldwide infrastructure building over the 21st century can be expected to create a high demand for these mineral products, the production of which releases CO_2 as a consequence of the calcining process. In the United States in 2003, CO_2 emissions from these sources accounted for 44 percent of the non-energy-related industrial emissions and about 1 percent of total U.S. emissions. Research could be focused on carbon capture and sequestration and on the exploration of substitutes for the end product. Carbon matrixes for construction, for example, might be lighter and stronger than concrete and would provide a means for carbon sequestration.

◆ **Advanced Applications of Biotechnology.** Bioproducts soon could replace fossil feedstocks, such as oil, in current industrial processes for manufacturing fuels, chemicals, and materials. Advances in biosciences and engineering include a better understanding of genomes, proteins, and their functions, including carbon dioxide capture, fixation, and storage; the effects of pre-treatment on biomass feedstock properties; enzymes for hydrolyzing pre-treated biomass into fermentable sugars; micro-organisms used in fermentation; and new tools of discovery such as bio-informatics, high-throughput screening of biodiversity, directed enzyme development and evolution, and gene shuffling. Biorefineries of the future could produce transportation fuels, value-added chemicals, materials, and/or power from non-conventional feedstocks such as agricultural and forest residues, energy crops, and other biomass materials.

◆ **Water and Energy System Optimization.** Energy used by water infrastructure and water used in power and fuel supply systems provide numerous opportunities for improvement with GHG reduction benefits. Opportunities include better matching of water temperatures and purity to end-use requirements; revolutionizing the design of municipal water systems to reduce use, minimize conveyance, and maximize re-use; and integrating water storage and treatment with the intermittency of renewable power supplies.

◆ **Computational Technology.** Process simulation enables more effective design and operation, leading to increased efficiency and improved productivity and product quality. Integrated modeling of fundamental physical and chemical properties can enhance understanding of industrial material properties and chemical processes. For example, modeling of counterflows through structured packings within distillation columns can improve hydrodynamic efficiencies and save a significant portion of the 3 quads of energy used annually in distillation processes, about half in petroleum refineries.

Figure 4-8. Advanced industrial materials, such as those used in high-temperature superconductive wires, can reduce energy losses and improve performance in electric motors.

Courtesy: DOE/NREL, Credit: Reliance Electric Co.

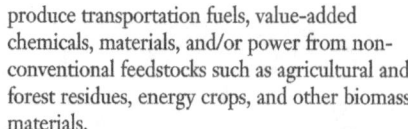

4.4 Electric Grid and Infrastructure

Large reductions in future CO_2 emissions may require that a significant amount of electricity be generated from carbon-free or carbon-neutral sources, including nuclear power and renewable electricity producers such as wind energy, geothermal energy, and solar-based power generating systems. Some renewable energy resources are concentrated in regions of the country that are distant from large urban markets. To accommodate such sources, the future electricity transmission infrastructure (the "grid") would need to extend its capacity and evolve into an intelligent and flexible system that enables the use of a wide and varied set of baseload, peaking, and intermittent generation technologies. The cost of expanding the grid to provide transmission from remote areas to load centers, or to make very long international links between Alaska and Russia, as sometimes proposed, is potentially very expensive, and proper allocation of the investment costs would need to be considered, as well as the overall cost-effectiveness of such projects.

In recent years, the demand for electricity in the United States has increased at a rate such that it could eventually exceed current transmission capacity. Demand is projected to increase by 19 percent from 2003-2012 (EIA 2005); only a 6 percent increase in transmission is planned for 2002-2012 (DOE 2002). There have been few major new investments in transmission during the past 15 years. Outages experienced in parts of the country—including the August 2003 blackout in the Midwest and Northeast—highlight the need to enhance grid reliability.

Enhancements for grid reliability will likely go hand in hand with improved efficiency of electricity transmission. Energy losses in the U.S. transmission and distribution (T&D) system were 5.5 percent in 2003, accounting for 201 billion kilowatt hours of electricity generation and 133 million metric tons of CO_2 emissions (EIA 2005, Table A8; EPA 2005 Table 2-14). About 10 percent of GHG emissions resulting from transmission and distribution are SF_6 emissions from certain high-voltage transmission equipment. The remainder of GHG emissions is from increased operations needed to compensate for energy losses.

Notwithstanding these improvements, the extent to which transmission and distribution losses could be reduced is uncertain, and energy losses on the transmission system could even increase as more long-distance transmission lines are brought into service to transmit power from remote generation sources.

The current portfolio supports the main components of the technology development strategy. Within constrained Federal resources, this portfolio addresses the highest priority current investment opportunities. For the future, CCTP seeks to consider a full array of promising technology options. From diverse sources, suggestions for future research have come to CCTP's attention and others are currently being explored. These include:

◆ **High-Temperature Superconducting Cables and Equipment.** The manufacture of promising HTS materials in long lengths at low cost remains a key program challenge. New, continuously scanning analytical systems are necessary to ensure uniformly high superconductor characteristics over kilometer lengths of wire. Research and development could help develop highly reliable, high-efficiency cryogenic systems to economically cool the superconducting components, including materials for cryogenic insulation and standardized high-efficiency refrigerators and motors (Figure 4-8). Scale-up of national laboratory discoveries for "coated conductors" could be another promising area for the laboratories and their industry partners.

◆ **Materials Science for Energy Conversion.** The digital economy is demanding more low-voltage DC power, which underscores the need for more efficient AC-DC conversion. In addition, on-site sources of photovoltaic (PV) and wind energy would be more efficiently utilized by supplying electricity to DC appliances, which will necessitate a new generation of smart wiring in homes and offices to allow both AC and DC power and end uses.

◆ **Energy Storage.** Energy storage that responds over timescales from milliseconds to hours and outputs that range from watts to megawatts is a critical enabling technology for enhancing customer reliability and power quality, more effective use of renewable resources, integration of distributed resources, and more reliable transmission system operation.

◆ **Real-Time Monitoring and Control.** Introduction of low-cost sensors throughout the power system is needed for real-time monitoring of system conditions. New analytical tools and software must be developed to enhance system observability and power flow control over wide areas.

Potential Role of Technology

There are many T&D technologies that can improve efficiency and reduce GHG emissions. In the near term, these include high-voltage DC (HVDC) transmission, high-strength composite overhead conductors, solid-state transmission controls such as Flexible AC Transmission System (FACTS) devices that include fault current limiters, switches and converters, and information technologies coupled with automated controls (i.e., a "Smart Grid"). High-efficiency conventional transformers—commercially available although not widely used—also could have impacts on distribution system losses.

Advanced conductors integrate new materials with existing materials and other components and

*Figure 4-9.
Superconductor wires
have the ability to
conduct more than 150
times the electrical
current of copper wire of
the same dimension.*

Courtesy: American Superconductor.

subsystems to achieve better technical, environmental, and financial performance—e.g., higher current carrying capacity, more lightweight, greater durability, lower line losses, and lower installation and operations and maintenance costs. Improved sensors and controls, as part of the next-generation electricity T&D system, could significantly increase the efficiency of electricity generation and delivery, thereby reducing the GHG emissions intensity associated with the electric grid. Outfitting the system with digital sensors, information technologies, and controls could further increase system efficiency, and allow greater use of more efficient and low-GHG end-use and other distributed technologies. High-temperature superconductors may be able to be utilized in key parts of the T&D system to reduce or eliminate line losses and increase efficiency. Energy storage allows intermittent renewable resources, such as photovoltaics and wind, to be dispatchable.

Advanced storage concepts and particularly high-temperature superconducting wires and equipment represent longer-term solutions with great promise. Digital sensors, information technologies, and controls may eventually enable real-time responses to system loads. HTS electrical wires might be able to carry 150 times the amount of electricity compared to the same-size conventional copper wires (Figure 4-9). Such possibilities may create totally new ways to operate and configure the grid. Power electronics will be able to provide significant advantages in processing power from distributed energy sources using fast response and autonomous control.

Technology Strategy

Realizing these opportunities requires a research portfolio that focuses on a balance between advanced transmission grid and distributed-generation technologies. Within the constraints of available resources, a balanced portfolio needs to address conductor technology, systems and controls, energy storage, and power electronics to help reduce CO_2 emissions in this sector.

Early research is likely to focus on ensuring reliability, e.g., establishing "self-healing" capabilities for the grid, including intelligent, autonomous device interactions, and advanced communication capabilities. Additional technologies would be needed for wide-area sensing and control, including sensors, secure communication and data management; and for improved grid-state estimation and simulation. Simulation linked to intelligent controllers can lead to improved protection and discrete-event control. Digitally enabled load-management technologies, wireless communications architecture and algorithms for system automation, and advanced power storage technologies will allow intermittent and distributed energy resources to be efficiently integrated.

Longer-term research is likely to focus on the development of fully operational, pre-commercial prototypes of energy-intensive power equipment that, by incorporating HTS wires, will have greater capacity with lower energy losses and half the size of conventional units. Over the long term, the T&D system would also be enhanced by integrating storage and power electronics.

Current Portfolio

Across the current Federal portfolio of electric infrastructure-related R&D, multi-agency activities are focused on a number of major thrusts in high-temperature superconductivity, T&D technologies, distributed generation and combined heat and power, energy storage, sensors, controls and communications, and power electronics. For example:

◆ Research on **high-temperature superconductivity** (HTS) is focused on improving the current carrying capability of long-distance cables; its manufacturability; and cost-effective ways to use the cable in equipment such as motors, transformers, and compensators. More reliable and robust HTS transmission cables that have three to five times the capacity of conventional copper cables and higher efficiency—which is especially useful in congested urban areas—are being developed and built as pre-commercial prototypes. Through years of Federal research in partnership with companies throughout the nation, technology has developed to bond these HTS materials to various metals, providing the flexibility to fashion these ceramics into wires for use in transmission cables; bearings

for flywheels; and coils for power transformers, motors, generators, and the like.

Research program goals in this area include HTS wires with 100 times the capacity of conventional copper/aluminum wires. More broadly, the program aims to develop and demonstrate a diverse portfolio of electric equipment based on HTS, such that the equipment can achieve a 50 percent reduction in energy losses, compared to conventional equipment, and a 50 percent size reduction, compared to conventional equipment with the same rating. Low-cost, high-performance, second-generation coated conductors are expected to become available in 2008 in kilometer-scale lengths. Cost goals include: (1) in 1,000 meter lengths, a wire-cost goal of $50 or less per ampere of current carried by superconducting wire used in power lines cooled to liquid nitrogen temperatures; (2) by 2015, the cost performance ratio for superconducting wires improved by at least a factor of 2.[18]

◆ Research on **transmission and distribution technologies** is focused on real-time information and control technologies; and systems that increase transmission capability, allow economic

A Distributed Energy Future

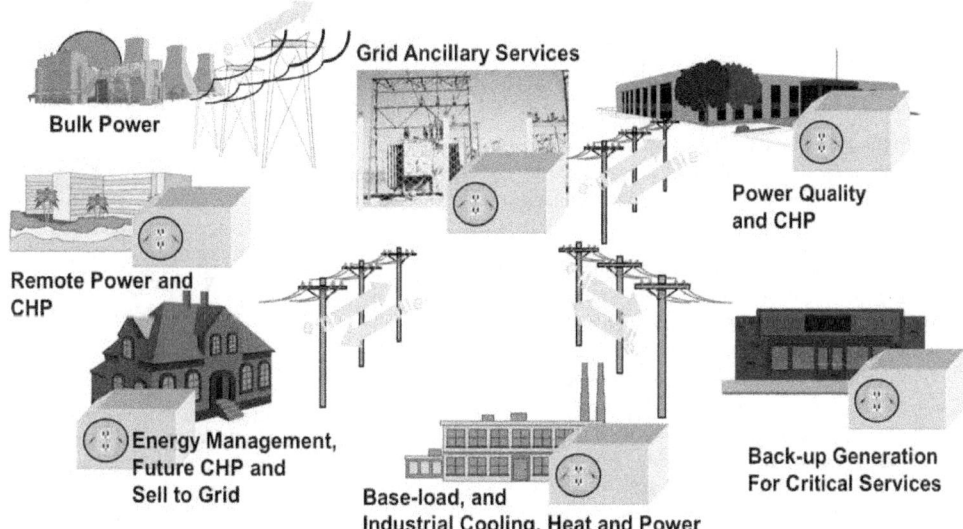

Figure 4-10. A Distributed Energy Future (Source: ORNL, Oak Ridge, Tennessee)

---

[18] See Section 1.3.1 (CCTP 2005): http://www.climatetechnology.gov/library/2005/tech-options/tor2005-131.pdf.

and efficient electricity markets, and improve grid reliability. Examples include high-strength composite overhead conductors, grid-status measurement systems that improve reliability by giving early warning of unstable conditions over major geographic regions, and technologies and regulations that enable the customer to participate more in electric markets through a demand response.[19]

Research program goals in this area include, by 2010, demonstrated reliability of energy-storage systems; reduced cost of advanced conductors systems by 30 percent; and operation of a prototype smart, switchable grid in a region within the U.S. transmission grid.

◆ Research on **distributed generation** (DG) includes renewable resources (e.g., photovoltaics), natural gas engines and turbines, energy-storage devices, and price-responsive loads. These technologies can meet a variety of consumer energy needs, including continuous power, backup power, remote power, and peak shaving. They can be installed directly on the consumer's premises or located nearby in district energy systems, power parks, and mini-grids (Figure 4-10).

Current research focuses on technologies that are powered by natural gas combustion and are located near the building or facility where the electricity is being used. These systems include microturbines, reciprocating engines and larger industrial gas turbines that generate from 25 kW to 10 MW of electricity that is appropriate for hotels, apartment buildings, schools, office buildings, hospitals, etc. Combined cooling, heating, and power (CHP) systems recover and use waste heat from distributed generators to efficiently cool, heat, or dehumidify buildings or make more power.

Research is needed to increase the efficiency and reduce the emissions from microturbines, reciprocating engines, and industrial gas turbines to allow them to be sited anywhere, even in nonattainment areas. These technologies can meet a variety of consumer energy needs, including continuous power, backup power, remote power, and peak shaving. Microturbines and reciprocating engines can also be utilized to burn opportunity fuels such as landfill gases or biogases from wastewater-treatment facilities or other volatile species from industrial processes that

would otherwise be an environmental hazard.[20]

CHP technologies have the potential to take the DG technologies one step further in GHG reduction by utilizing the waste heat from the generation of electricity for making steam, heating water, or producing cooling energy. The average power plant in the United States converts approximately one-third of the input energy into output electricity and then discards the remaining two-thirds of the energy as waste heat. Integrated DG systems with CHP similarly produce electricity at 30 percent to 45 percent efficiency, but then capture much of the waste heat to make steam or heat, to cool water, or to meet other thermal needs and increase the overall efficiency of the system to greater than 70 percent. Research is needed to increase the efficiency of waste-heat-driven absorption chillers and desiccant systems to overall efficiencies well above 80 percent.

The overall research goal of the Distributed Energy Program is to develop and make available, by 2015, a diverse array of high-efficiency, integrated distributed generation and thermal energy technologies, at market-competitive prices, so to enable and facilitate widespread adoption and use by homes, businesses, industry, communities, and electricity companies that may elect to use them. If successful, these technologies will enable the achievement of a 20 percent increase in a building's energy utilization, when compared to a building built to ASHRAE 90.1 standards, using load management, CHP, and energy-storage technologies that are replicable to other localities.

◆ Research on **energy storage** is focused in two general areas. First, research is striving to develop storage technologies that reduce power-quality disturbances and peak electricity demand, and improve system flexibility to reduce adverse effects to industrial and other users. Second, research is seeking to improve electrical energy storage for stationary (utility, customer-side, and renewable) applications. This work is being done in collaboration with a number of universities and industrial partners. This work is set within an international context, where others are investing in high-temperature, sodium-sulfur batteries for utility load-leveling applications and pursuing large-scale vanadium reduction-oxidation battery chemistries.

---

[19] See Section 1.3.2 (CCTP 2005): http://www.climatetechnology.gov/library/2005/tech-options/tor2005-132.pdf.

[20] See Section 1.3.3 (CCTP 2005): http://www.climatetechnology.gov/library/2005/tech-options/tor2005-133.pdf.

The research program goals in this area focus on energy-storage technologies with high reliability and affordable costs. For capital cost, this is interpreted to mean less than or equal to those of some of lower-cost new power generation options ($400-$600/kW). Battery storage systems range from $300-$2,000/kW. For operating cost, this figure would range from compressed gas energy storage (which can cost as little as $1 to $5/kWh) to pumped hydro storage (which can range between $10 and $45/kWh).[21]

♦ Research on **sensors, controls, and communications** focuses on developing distributed intelligent systems to diagnose local faults and coordinate with power electronics and other existing, conventional protection schemes that will provide autonomous control and protection at the local level. This hierarchy will enable isolation and mitigation of faults before they cascade through the system. The work will also help users and electric-power-system operators achieve optimized control of a large, complex network of systems; and will provide remote detection, protection, control, and contingency measures for the electric system.

The initial research program goals for sensors, controls, and communications are to develop, validate, and test computer simulation models of the distribution system to assess the alternative situations. Once the models have been validated on a sufficiently large scale, the functional requirements and architecture specifications can be completed. Then more specific technology solutions can be explored that would conform to the established architecture.[22]

♦ Research on **power electronics** is focused on megawatt-level inverters, fast semiconductor switches, sensors, and devices for Flexible Automated Control Transmission Systems (FACTS). The Office of Naval Research and DOE have a joint program to develop power electronic building blocks. The military is developing more electricity-intensive aircraft, ships, and land vehicles, which are providing power electronic spinoff technology for infrastructure applications.

The research program goal in this area is to build a power electronic system on a base of modules. Each module or block would be a subsystem containing several components, and each one has common power terminals and communication connections.[23]

Future Research Directions

The current portfolio supports the main components of the technology development strategy and addresses the highest priority current investment opportunities in this technology area. For the future, CCTP seeks to consider a full array of promising technology options. From diverse sources, suggestions for future research have come to CCTP's attention. Some of these, and others, are currently being explored and under consideration for the future R&D portfolio. These include:

♦ **High-Temperature Superconducting Cables and Equipment.** The manufacture of promising HTS materials in long lengths at low cost remains a key program challenge. New, continuously-scanning analytical systems are necessary to ensure uniformly high superconductor characteristics over kilometer lengths of wire. Research and development could help develop highly reliable, high-efficiency cryogenic systems to economically cool the superconducting components including materials for cryogenic insulation and standardized high-efficiency refrigerators. Scale-up of national laboratory discoveries for "coated conductors" could be another promising area for the laboratories and their industry partners.

♦ **Energy Storage.** Energy storage that responds over timescales from milliseconds to hours—and outputs that range from watts to megawatts—is a critical enabling technology for enhancing customer reliability and power quality, more effective use of renewable resources, integration of distributed resources, and more reliable transmission system operation.

♦ **Real-Time Monitoring and Control.** Introduction of low-cost sensors throughout the power system is needed for real-time monitoring of system conditions. New analytical tools and software must be developed to enhance system observability and power flow control over wide areas.

[21] See Section 1.3.4 (CCTP 2005): http://www.climatetechnology.gov/library/2005/tech-options/tor2005-134.pdf.

[22] See Section 1.3.5 (CCTP 2005): http://www.climatetechnology.gov/library/2005/tech-options/tor2005-135.pdf.

[23] See Section 1.3.6 (CCTP 2005): http://www.climatetechnology.gov/library/2005/tech-options/tor2005-136.pdf.

4.5 Summary

This chapter reviews various forms of advanced technology, their potential for reducing emissions from end use and infrastructure, and the R&D strategies intended to accelerate their development. Although uncertainties exist about both the level at which GHG concentrations might need to be stabilized in the future and the nature of the technologies that may come to the fore, the long-term potential of advanced end-use and infrastructure technologies is estimated to be significant, both in reducing emissions (as shown in the figure at the beginning of this chapter) and in reducing the costs for achieving those reductions, as suggested by Figure 3-14. Further, the advances in technology development needed to realize this potential, as modeled in the associated analyses, animate the R&D goals for each end-use and infrastructure technology area. It will remain important to continue to coordinate research closely with industrial firms, since in many cases they will have the financial incentive and expertise to make technology advancements without Federal support.

As one illustration among the many hypothetical cases analyzed,[24] when a high-emissions constraint was placed on GHG emissions over the course of the 21st century, the lowest-cost arrays of advanced technology in end use and infrastructure, when compared to a reference case, resulted in 100-year cumulative reductions of emissions of roughly between 190 and 210 GtC. This amounted, roughly, to between 30 and 35 percent of all GHG emissions reduced, avoided, captured and stored, or otherwise withdrawn and sequestered, as needed to attain this level of constrained emission. Similarly, the costs for achieving such emissions reductions, when compared to the reference case, were reduced by roughly a factor of 3. See Chapter 3 for other cases and other scenarios.

As described in this chapter, CCTP's technology development strategy supports potential achievements in this range. The overall strategy is summarized schematically below in Figure 4-11. Advanced technologies are seen entering the marketplace in the near, mid, and long terms, where the long term is sustained indefinitely. Such a progression, if successfully realized worldwide, would be consistent with attaining the end use and infrastructure potential portrayed in Figure 3-19.

The timing and the pace of technology adoption are uncertain and must be guided by science. In the case of the illustration above, the first GtC per year (1GtC/year) of reduced or avoided emissions, as compared to a reference case, would need to be in place and operating, roughly, between 2030 and 2050. For this to happen, a number of new or advanced end-use and infrastructure technologies would need to penetrate the market at significant scale before these dates. Other cases would suggest faster or slower rates of deployment, depending on assumptions. See Chapter 3 for other cases and other scenarios.

Throughout Chapter 4, the discussions of the current activities in each area support the main components of this approach to technology development. The activities outlined in the current portfolio sections address the highest-priority investment opportunities for this point in time. Beyond these activities, the chapter identifies promising directions for future research, identified in part by the end-use and infrastructure technical working group and assessments and inputs from non-Federal experts. CCTP remains open to a full array of promising technology options as current work is completed and changes in the overall portfolio are considered.

---

[24] In Chapter 3, various advanced technology scenarios were analyzed for cases where global emissions of GHGs were hypothetically constrained. Over the course of the 21st century, growth in emissions was assumed to slow, then stop, and eventually reverse in order to ultimately stabilize GHG concentrations in the Earth's atmosphere at levels ranging from 450 to 750 ppm. In each case, technologies competed within the emissions-constrained market, and the results were compared in terms of energy (or other metric), emissions, and costs.

Technologies for Goal #1: Reduce Emissions from End Use and Infrastructure

	NEAR-TERM	MID-TERM	LONG-TERM
Transportation	• Hybrid & Plug-In Hybrid Electric Vehicles • Clean Diesel Vehicles • Alternative and Fuel-Flexible Vehicles • Improved Batteries, Energy Storage • Power Electronics • Engineered Urban Designs • Reduction of Vehicle Miles Traveled • Improved Air Space Operations	• Fuel Cell Vehicles and H_2 Fuels • Efficient, Clean Heavy Trucks • Cellulosic Ethanol Vehicles • Intelligent Transport Systems • Integrated Regional Planning • Low-Emission Aircraft • Intercity Transport Systems	• Zero-Emission Vehicle Systems • Optimized Multi-Modal Intercity & Freight Transport • Widespread Use of Engineered Urban Designs & Regional Planning • Very Low Aviation Emissions (all GHGs)
Buildings	• High-Performance, Integrated Homes • Energy-Efficient Building Materials • High-Efficiency Appliances • Solar Control Windows	• "Smart" Buildings • Solid-State Lighting • Ultra-Efficient HVACR • Intelligent Building Systems • Neural Net Building Controls	• Energy Managed Communities • Low-Power Sensors with Wireless Communications
Industry	• Improved Processes in Energy-Intensive Industries • High-Efficiency Boilers and Combustion Systems • Greater Waste Heat Utilization • Improved Recyclability and Greater Use of Byproducts • Bio-Based Feedstocks	• Transformational Technologies for Energy-Intensive Industries • C&CO_2 Managed Industries • Superconducting Electric Motors • Efficient Thermoelectric Systems • Advanced Separation Technologies • Low-Emission Cement Alternatives • Water and Energy System Optimization	• Integration of Industrial Heat, Power, Processes and Techniques • High-Efficiency, All-Electric Manufacturing • Widespread Use of Bio-Feedstocks • Closed-Cycle Products & Materials
Electric Grid & Infrastructure	• Distributed Generation • Smart Metering & Controls for Peak Shaving • Long-Distance DC Transmission • High-Temperature Superconductivity Demonstrations • Power Electronics • Composite Conductor Cables	• Energy Storage for Load Leveling • Neural Net Grid Systems • Advanced Controls and Power Electronics	• Superconducting Transmission and Equipment • Standardized Power Electronics • Wireless Transmission

Figure 4-11. Technologies for Goal #1: Reduce Emissions from End Use and Infrastructure
(Note: Technologies shown are representations of larger suites. With some overlap, "near-term" envisions significant technology adoption by 10 to 20 years from present, "mid-term" in a following period of 20-40 years, and "long-term" in a following period of 40-60 years. See also List of Acronyms and Abbreviations.)

4.6 References

Brown, M.A. 2003. *What are the administration priorities for climate change technology?* November 6, 2003. http://www.house.gov/science/hearings/energy03/nov06/brown.pdf

Energy Information Administration (EIA). 2004. *Annual energy outlook 2004: with projections to 2025*, DOE/EIA-0383(2004). Washington, DC: U.S. Department of Energy.

Energy Information Administration (EIA). 2005. *Annual energy outlook 2005: with projections to 2025*, DOE/EIA-0383(2005). Washington, DC: U.S. Department of Energy.

Intergovernmental Panel on Climate Change (IPCC). 2000. *Special report on emissions scenarios.* Cambridge, UK: Cambridge University Press. http://www.grida.no/climate/ipcc/emission/

Intergovernmental Panel on Climate Change (IPCC). 2001. *Climate change 2001: mitigation.* Working Group III to the Third Assessment Report. Cambridge, UK: Cambridge University Press. http://www.grida.no/climate/ipcc_tar/wg3/index.htm

International Energy Agency (IEA). 2004. *World energy outlook.* Paris: International Energy Agency.

United Nations (UN) Population Division. 2005. *World population prospects: the 2004 revision population database.* http://esa.un.org/unpp/

U.S. Climate Change Technology Program (CCTP). 2005. *Technology options for the near and long term.* Washington, DC: U.S. Department of Energy. http://www.climatetechnology.gov/library/2005/tech-options/index.htm

U.S. Department of Energy (DOE). 1997. *Technology opportunities to reduce U.S. greenhouse gas emissions.* Washington, DC: U.S. Department of Energy. http://www.ornl.gov/~webworks//cppr/y2003/rpt/110512.pdf

U.S. Department of Energy (DOE). 2002. *National transmission grid study.* Washington, DC: U.S. Department of Energy. http://www.pi.energy.gov/pdf/library/TransmissionGrid.pdf

U.S. Department of Energy (DOE). 2005. *Hydrogen fuel cells and infrastructure technologies multi- year research, development and demonstration plan: 2003-2010* (Draft). Washington, DC: U.S. Department of Energy. http://www.eere.energy.gov/hydrogenandfuelcells/mypp/

U.S. Department of Energy (DOE), Energy Efficiency and Renewable Energy (EERE). 2005. *FY 2006: budget-in-brief.* http://www.eere.energy.gov/office_eere/pdfs/fy05_budget_brief.pdf

U.S. Department of Transportation (DOT). 2002a. *Highway statistics 2001.* Washington, DC: Federal Highway Administration.

U.S. Department of Transportation (DOT). 2002b. *Freight analysis framework.* Washington, DC: Federal Highway Administration.

U.S. Environmental Protection Agency (EPA). 2005. *Inventory of U.S. greenhouse gas emissions and sinks: 1990-2003.* EPA 430-R-05-003. http://yosemite.epa.gov/oar/globalwarming.nsf/UniqueKeyLookup/RAMR69V4ZS/$File/05_complete_report.pdf

Reducing Emissions from Energy Supply

Reducing greenhouse gas (GHG) emissions from energy supply is an important element of CCTP's technology strategy. In part, this is because global energy demand is projected to grow significantly by the year 2100 and, in part, because the infrastructure that will be built to meet this demand will have long-lasting consequences for future GHG emissions.

Some projections show energy demand over the course of the 21st century growing by a factor of more than 6 (from about 400 exajoules [EJ] in 2000 to more than 2400 EJ in 2100). Mid-range scenarios project an increase of about a factor of 3 or more from today's level, even under scenarios in which energy efficiency is assumed to advance steadily over time. Of this growth, global demand for electricity is projected to increase faster than direct use of fuels in end-use applications. Most infrastructure needed to meet this demand, including the replacement of the facilities to be retired, has yet to be built—a circumstance that poses significant opportunities for new and advanced technology to reduce or eliminate much of these future emissions.

In published scenarios of CO_2 emissions to 2100, increasing demand for energy results in concomitant growth in CO_2 emissions. By contrast, almost all scenarios that explore various future paths toward significant emissions reductions show that various forms of low or near net-zero emissions energy supply play a key role in achieving those reductions. In one set of scenarios, as shown in Figure 3-19 and highlighted above, advanced energy supply technologies contribute an additional 20 to 35 thousand EJ toward global energy demand and result in reductions of global carbon emissions between about 30 and 330 gigatons of carbon (GtC), compared to a reference case used in the study (see Chapter 3). Although bracketed by large uncertainties, these figures suggest both the potential role for advanced technology and a long-term goal for contributions from this sector of the global economy.

Today, a range of technologies using fossil fuels, nuclear power, hydroelectric power, and a relatively small (but fast-growing) amount of renewable energy,

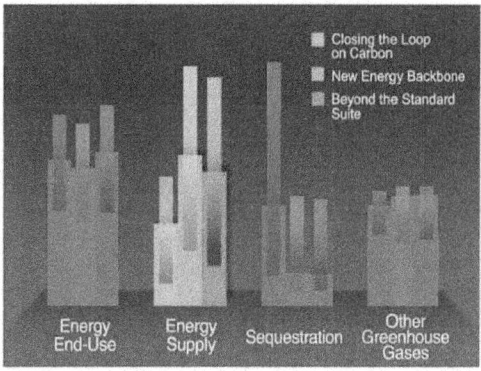

Energy Supply
Potential Contributions to Emissions Reduction

Potential contributions of Energy Supply reduction to cumulative GHG emissions reductions to 2100, across a range of uncertainties, for three advanced technology scenarios. See Chapter 3 for details.

supplies the world's electricity demand. Most of global transportation demand is met with petroleum products (see Figures 5-1 and 5-2).

The development of advanced technologies that can significantly reduce emissions of carbon dioxide (CO_2) from energy supply is a central component of the overall climate change technology strategy. Many opportunities exist for pursuing technological options for energy supply that are characterized by low or near-net-zero emissions and whose development can be facilitated by a coordinated Federal R&D investment plan.

Some advanced energy supply technologies build on the existing energy infrastructure, which is currently dominated by coal and other fossil fuels. One set of technologies that would allow continued use of coal and other fossil fuels—even under scenarios calling for substantial CO_2 emission limitations—is contained in an advanced coal-based production facility. It is based on coal gasification and production of syngas, which can generate electricity, hydrogen, and other valued fuels and chemicals. The facility would be combined with CO_2 capture and storage, and have very low emissions of other pollutants. Some of the emissions-reduction scenarios examined (see Chapter 3) project that if CO_2 capture and storage and improvements in fossil energy conversion efficiencies are achieved, fossil-based energy could continue to supply a large percentage of total energy and electricity in the future (e.g., up to 70 percent of global electricity demand in some scenarios), even under a high carbon constraint. In addition to this mid- to long-term opportunity, lowering CO_2 emissions from fossil fuel combustion in the near term can be achieved by increasing the energy efficiency of combustion technology and by increasing the use of combined heat and power.

Advances in low- and zero-emission technologies have also been identified in a number of scenario analyses as important for reducing GHG emissions. These technologies include advanced forms of renewable energy, such as wind, photovoltaics, solar thermal

applications, and others; biologically based open and closed energy cycles, such as enhanced systems for biomass combustion, biomass conversion to biofuels and other forms of bioenergy; refuse-derived fuels and energy; and various types of nuclear energy, including technologies that employ spent fuel recycling. Variations of these advanced technologies can also be deployed in the production of hydrogen, which may play a big role in reducing emissions from the transportation sector, as well as potentially being used to supply fuel cells for electricity production. Several studies showed that biomass, nuclear, and renewable (solar and wind) energy, combined, would contribute approximately 30 percent of the total reduction in GHG emissions from a "reference case"[1] (see Chapter 3).

Advances in novel or visionary energy supply technologies may also make important contributions toward reduced GHG emissions, including fusion energy; advanced fuel cycles based on combinations of nanotechnology and new forms of bio-assisted energy production, using bioengineered molecules for more efficient photosynthesis; and hydrogen production or photon-water splitting. Other possibilities include advanced technologies for capturing solar energy in Earth orbit, on the moon, or in the vast desert areas

World Electricity Generation

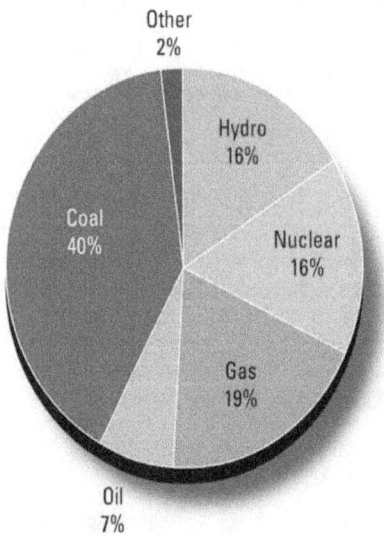

Figure 5-1. World Electricity Generation
(Source: IEA 2004)

World Primary Energy Supply

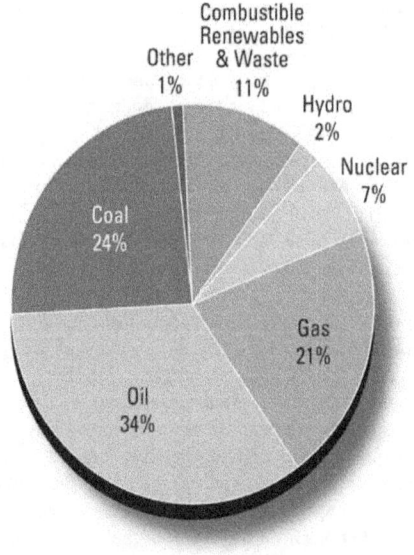

Figure 5-2. World Primary Energy Supply
(Source: IEA 2004)

---

[1] In Chapter 3, the 30 percent value is associated with a hypothesized high emissions constraint.

of Earth—enabled, in part, by new energy carriers and/or low-resistance power transmission over long distances. In one scenario (see Chapter 3), these novel forms of energy were projected to lower cumulative CO_2 emissions by more than 100 GtC over the course of a 100-year period, under a very high emission-constraint scenario.

Because outcomes of various ongoing and planned technology development efforts are not known, a prudent path for science and technology policies in the face of uncertainty is to maintain a diverse R&D portfolio. The current Federal portfolio supports R&D activities important to all three of the general technology areas discussed above. The analysis of the advanced technology scenarios suggests that, through successful development and implementation of these technologies, stabilization trajectories could be met across a wide range of hypothesized concentration levels—and the goal could be accomplished both sooner and at significant cost savings, compared to the case without such dramatic technological advances.

This chapter explores energy supply technologies. For each technology area, the chapter examines the potential role for advanced technology; outlines a technology-development strategy for realizing that potential; highlights the current research portfolio, replete with selected technical goals and milestones; and invites public input on considerations for future research directions. The chapter is organized around the following five energy supply technology areas:

 ◆ Low-Emission, Fossil-Based Fuels and Power.

 ◆ Hydrogen as an Energy Carrier.

 ◆ Renewable Energy and Fuels.

 ◆ Nuclear Fission.

 ◆ Fusion Energy.

Each of these technology sections contains a sub-section describing the current portfolio, where the technology descriptions include an Internet link to the updated version of the CCTP report, *Technology Options for the Near and Long Term* (CCTP 2005).[2]

Low-Emission, Fossil-Based Fuels and Power

Today, fossil fuels are an integral part of the U.S. and global energy mix. Because of its abundance and current relatively low cost, coal now accounts for about half of the electricity generated in the United States, and it is projected to continue to supply one-half of U.S. electricity demands through the year 2030 (EIA 2006). EIA also projects that natural gas will continue to be the "bridge" energy resource, as it offers significant efficiency improvements (and emissions reductions) in both central and distributed electricity generation and combined heat and power (CHP) applications.

Potential Role of Technology

Because coal is America's most plentiful and readily available energy resource, the U.S. Department of Energy has directed a portion of its research and development resources toward finding ways to use coal in a more efficient, cost-effective, and environmentally benign manner, ultimately leading to near-zero atmospheric emissions. Even small improvements in efficiency of the installed base of coal-fueled power stations can result in a significant reduction of carbon emissions. For example, increasing the efficiency of all coal-fired electric-generation capacity in the United States by 1 percentage point would avoid the emission of 14 million tons of carbon per year.[3] That reduction is equivalent to replacing 170 million incandescent light bulbs with fluorescent lights or weatherizing 140 million homes. New U.S. government-industry collaborative efforts are expected to continue to find ways to improve the ability to decrease emissions from coal power generation at lower costs. The objective for future power plant designs is to both increase efficiency and reduce environmental impacts. The focus is on designs that are compatible with carbon sequestration technology, including the development of coal-based, near-zero atmospheric emission power plants.

---

2   The report is available at http://www.climatetechnology.gov/library/2005/tech-options/index.htm.

3   Avoided carbon emissions were calculated based on current coal consumption and power plant efficiencies from the Energy Information Administration's Annual Energy Outlook 2002. Using the published efficiencies, 0.574 quads of energy were saved with a 1 percent improved efficiency, which would result in 14.8 MMT of carbon avoided.

Technology Strategy

The current U.S fossil research portfolio is a fully integrated program with mid- and long-term market-entry offerings. The principal objective is a near-zero atmospheric emission, coal-based electricity generation plant that has the ability to co-produce low-cost hydrogen. In the mid term, that goal is expected to be accomplished through the FutureGen project. This $1 billion venture, cost-shared with industry, will combine electricity and hydrogen production from a single facility with the elimination of virtually all emissions of air pollutants, including sulfur dioxide, nitrogen oxides, mercury, and particulates—as well as 90 percent reduction of atmospheric CO_2 emissions, through a combination of efficiency improvements and carbon capture and storage (called "sequestration" in Figure 5-3). This prototype power plant will serve to demonstrate the most advanced technologies, such as hydrogen fuel cells.[4]

Current Portfolio

The low-emissions, fossil-based power system portfolio has three focus areas:

◆ **Advanced Power Systems:** Advanced coal-fired, power-generation technologies can achieve significant reductions in CO_2 emissions, while providing a reliable, efficient supply of electricity.

Significant reductions in atmospheric CO_2 emissions have been demonstrated via efficiency improvements. Current Integrated Gasification Combined Cycle (IGCC) systems average power plant efficiencies are about 40-42 percent; increasing efficiencies to 48-52 percent in the mid term and 60 percent by 2020 (with the integration of fuel cell technology), will nearly halve emissions of CO_2 per unit of electricity, relative to pulverized-coal-based power plants which have efficiencies of 30-35 percent (Figure 5-4). Development and deployment of CO_2 capture and storage technology could reduce atmospheric emissions to near-zero levels through 2100. Recent R&D activities have focused on IGCC plants, with two U.S. IGCC demonstration plants now in operation.

The research program goal in the Advanced Power Systems area is to increase efficiency of new systems to levels ranging from 48-52 percent by 2010, and 60 percent by 2020, while also achieving an overall electricity production cost that is between 75 percent and 90 percent of current pulverized-coal-based power generation. In addition, emissions of criteria pollutants are targeted to be much less than one-tenth of current new source performance standards.[5]

◆ **Distributed Generation/Stationary Fuel Cells:** The stationary fuel cell (FC) program is focused on reducing the cost of fuel cell technology for distributed generation applications (as opposed to transportation applications) by an order of magnitude.

Coal-Based Energy Complex

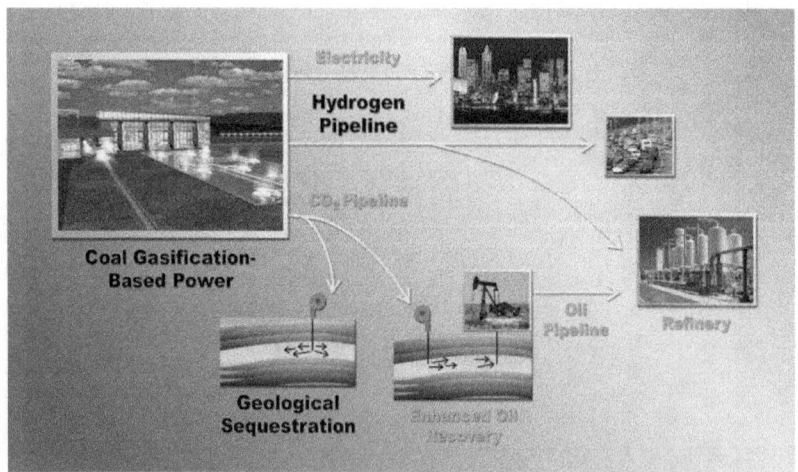

Figure 5-3.
A fully integrated coal-based energy complex will have the ability to co-produce electricity, hydrogen, fuels and chemicals with little or no CO_2 emissions. (Source: DOE 2004)

---

4 See http://www.fossil.energy.gov/programs/powersystems/futuregen/futuregen_report_march_04.pdf.

5 See Section 2.1.2 (CCTP 2005): http://www.climatetechnology.gov/library/2005/tech-options/tor2005-212.pdf.

In the near and mid terms, fuel cell cost reductions could enable the widespread deployment of natural-gas-fueled distributed generation in gas-only, CHP, and fuel cell applications. In the mid- to long-term, this technology, along with others being developed as part of the Distributed Generation effort, will also support coal-based FutureGen/central-station applications. The goal is to develop a modular power system with lower cost and significantly lower carbon dioxide emissions than current plants. Examples of current R&D projects in this area include: (1) low-cost fuel cell systems development; (2) high-temperature fuel cell scale-up and aggregation for fuel cell turbine (FCT) hybrid application; and (3) hybrid systems and component demonstration.

Research program goals in the natural gas fuel cell and hybrid power systems include demonstrating a gas aggregated fuel cell module larger than 250 kW that can run on coal synthetic gas (syngas), while also reducing the costs of the Solid-State Energy Conversion Alliance fuel cell power system to $400/kW by 2010. In addition, by 2012-2015, the program aims to: (1) demonstrate a megawatt-class hybrid system at FutureGen with an overall system efficiency of 50 percent on coal syngas; and (2) integrate optimized turbine systems into zero-emission power plants.[6]

◆ **Co-production/Hydrogen:** This research area focuses on developing technology to co-produce electricity and hydrogen from coal, which could also be relevant to applications using coal and biomass blends, potentially achieving very large reductions in CO_2 emissions when compared to present technologies. This technology will use syngas generated from coal gasification to produce hydrogen.

Near-Zero Atmospheric Emission Power and hydrogen coproduction research goals target a 10-year demonstration project (FutureGen) to create the world's first coal-based, near-zero atmospheric emissions electricity and hydrogen power plant. The near-term goals of the program are to: (1) design, by 2010, a near-term coproduction plant, configured at a size of 275-MW, which would be suitable for commercial deployment; (2) demonstrate pilot-scale reactors using ceramic membranes for oxygen separation and hydrogen recovery; and (3) demonstrate a $400/kW solid-oxide fuel cell.[7]

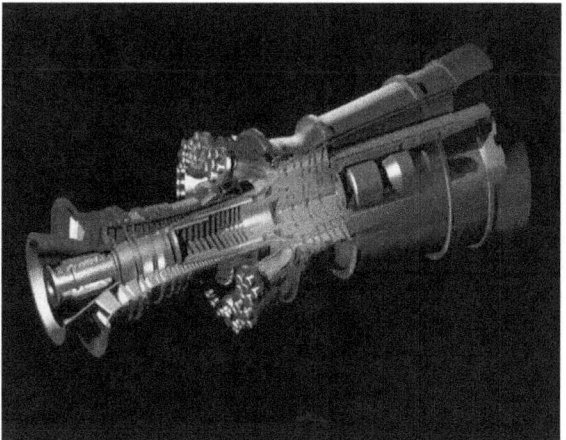

Figure 5-4. *Advanced Gas Turbines can increase efficiencies in combined-cycle systems to 60 percent, while also reducing the cost of electricity production.*

Courtesy: DOE/NETL

Carbon emissions from fossil-fuel-based power systems can be reduced in the near term principally by improving process efficiency and, in the longer term, via more advanced system components, such as high-efficiency fuel cells. In both the near and long terms, incorporating CO_2 capture into the systems' processes, accompanied by long-term CO_2 storage, will be required to achieve low or near-zero atmospheric emissions from these energy sources. Current research activities focus on: (1) ion transport oxygen separation membranes; (2) hydrogen separation membranes; and (3) early-entrance coproduction plant designs. These activities are discussed in more detail in Chapter 6.

Future Research Directions

The current portfolio supports the main components of the technology development strategy and addresses the highest priority current investment opportunities in this technology area. For the future, CCTP seeks to consider a full array of promising technology options. From diverse sources, suggestions for future research have come to CCTP's attention. Some of these, and others, are currently being explored and under consideration for the future R&D portfolio. These include:

◆ **Advanced membranes for gas phase separations.** Separations such as oxygen separation from air, which is required for

---

6 See Section 2.1.3 (CCTP 2005): http://www.climatetechnology.gov/library/2005/tech-options/tor2005-213.pdf.

7 See Section 2.1.1 (CCTP 2005): http://www.climatetechnology.gov/library/2005/tech-options/tor2005-211.pdf.

oxycombustion technology and hydrogen separation from gasified coal streams, are energy intensive and reduce the overall efficiency of advanced combustion and gasification plants. Near-term technologies are under development for improving these gas separations, but in the longer term it is desirable to develop advanced ion transport membranes for oxygen separation and ceramic membranes for hydrogen recovery.

◆ **Solid-oxide fuel cells.** Coal gasification plants that produce hydrogen streams as feed to solid-oxide fuel cells for improving electricity production efficiency are being studied. Future research could examine approaches scaling up low-cost solid oxide fuels cells to multi-hundred megawatt sizes for use as power blocks in near-zero atmospheric emissions systems.

◆ **High-temperature materials and heat transfer technology.** Future generation systems will need to maintain relatively high temperatures between the combustion/gasification stage and the turbine stage to achieve high generation efficiency goals. A primary technology development interest is for high-temperature materials that are stable and resistant to corrosion, erosion, and decrepitation. Advanced materials could improve performance of future heat exchangers, turbine components, particulate filters, and SO_2 removal. Other possible research directions include the use of alternate working fluids and heat-exchange cycles.

◆ **Unconventional combustion systems.** A promising research direction involves use of chemical cycling for CO_2 enrichment. In this concept, CO_2 is continually looped and enriched within the combustion plant and the enriched CO_2/O_2 gas stream is substituted for air in the main combustion chamber.

◆ **Hydrogen co-production.** Technology for hydrogen capture and purification in a coal gasification plant for electricity production may enable hydrogen production for a future hydrogen-powered vehicle economy.

◆ **Advanced hybrid gasification/combustion systems.** These systems appear to offer an alternative path to achieve many of the program goals and may warrant additional study.

◆ **Reduction of N_2O emissions.** Existing post-combustion emissions technology that reduces the emissions of the controlled pollutant NO_X also tends to increase the emissions of N_2O gas, which is a GHG. Post-combustion processes could simultaneously minimize the emissions of both NO_X and N_2O.

5.2 Hydrogen

As discussed above, in a long-term future characterized by low or near-net-zero emissions of GHGs, global energy primary supply can continue its reliance on fossil fuels, provided there are suitable means for capturing and storing the resulting emissions of CO_2. Alternatively, the world could increase reliance on low-carbon and nonfossil energy sources. These approaches share a need for carbonless energy carriers, such as electricity or some alternative, to store and deliver energy on demand to end users. Electricity is increasingly the carbonless energy carrier of choice for stationary energy consumers, but hydrogen could prove to be an attractive carrier for the transportation sector (e.g., highway vehicles and aircraft), as well as stationary applications. If successful, hydrogen could enable reductions in petroleum use and potentially eliminate concomitant air pollutants and CO_2 emissions on a global scale.

Today, hydrogen is used in various chemical processes and is made largely from natural gas, producing CO_2 emissions. However, hydrogen can be produced in a variety of ways that do not emit CO_2, including renewable energy-based electrolysis; various biological and chemical processes; water shift reactions with coal and natural gas, accompanied by CO_2 capture and storage; thermal and electrolytic processes using nuclear energy; and direct photoconversion. Hydrogen can be stored as a pressurized gas or cryogenic liquid, or absorbed within metal hydride powders or physically adsorbed onto carbon-based nanostructures. If progress can be made on a number of technical fronts, so that costs of producing hydrogen are reduced, hydrogen could play a valuable enabling and synergistic role in heat and power generation, transportation, and energy end use.

Potential Role of Technology

As a major constituent of the world's water, biomass, and fossil hydrocarbons, the element hydrogen is ubiquitous. It accounts for 30 percent of the fuel-energy in petroleum, and more than 50 percent of the fuel-energy in natural gas. A fundamental distinction between hydrogen and fossil fuels, however, is that the production of hydrogen, whether from water, methane or other hydrocarbons, is a net-energy consumer. This makes hydrogen not an energy source, per se, but a carrier of energy, similar to electricity.

Like electricity, the life-cycle GHG emissions associated with hydrogen use would vary depending on the method to produce, store, and distribute it. Hydrogen can be generated at various scales, including central plants, fuel stations, businesses, homes, and perhaps onboard vehicles. In principle, the diversity of scales, methods, and sources of production make hydrogen a highly versatile energy carrier, capable of transforming transportation (and potentially other energy services) by enabling compatibility with many primary energy sources. This versatility opens up possibilities for long-term dynamic optimization of CO_2 emissions, technology development lead times, economics, and other factors. In a future "hydrogen economy," hydrogen may ultimately serve as a means of linking energy sources to energy uses in ways that are more flexible, secure, reliable, and responsive to consumer demands than today, while also integrating the transportation and electricity markets.

While its simple molecular structure makes hydrogen an efficient synthetic fuel to produce, use, and/or convert to electricity, the storage and delivery of hydrogen are more challenging than for most fuels. Consequently, most hydrogen today is produced at or near its point of use, consuming other fuels (e.g., natural gas) that are easier to handle and distribute.

Large hydrogen demands at petroleum refineries or ammonia (NH_3) synthesis plants can justify investment in dedicated hydrogen pipelines, but smaller or variable demands for hydrogen are usually met more economically by truck transport of compressed gaseous hydrogen or cryogenic and liquefied hydrogen produced by steam methane reforming. These methods have evolved over decades of industrial experience, with hydrogen as a niche chemical commodity, produced in amounts (8 billion kg hydrogen/yr) equivalent to about 1 percent (approximately 1 EJ per year) of current primary energy use in the United States. For hydrogen use to scale up from its current position to a global carbonless energy carrier (alongside electricity), new energetically and economically efficient technical approaches would be required for hydrogen delivery, storage, and production.

Hydrogen production can be a value-added complement to other advanced climate change technologies, such as those aimed at the use of fossil fuels or biomass with CO_2 capture and storage. As such, hydrogen may be a key and enabling component for full deployment of carbonless electricity technologies (advanced fission, fusion, and/or intermittent renewables).

Figure 5-5. Fuel cells use the chemical energy of hydrogen and oxygen from air to produce electricity without combustion.

Courtesy: DOE/NREL, Credit – Matt Stiveson

In the near term, initial deployment of hydrogen fleet vehicles and distributed power systems may provide early adoption opportunities and demonstrate the capabilities of the existing hydrogen delivery and on-site production infrastructure. This will also contribute in other ways, such as improving urban air quality and strengthening electricity supply reliability. This phase of hydrogen use may also serve as a commercial proving ground for advanced distributed hydrogen production and conversion technologies using existing storage technology, both stationary and vehicular.

In the mid term, light-duty vehicles likely will be the first large mass market (10-15 EJ per year in the United States) for hydrogen. Fuel cells may be particularly attractive in automobiles, given their efficiency versus load characteristics and typical driving patterns (Figure 5-5). Hydrogen production for this application could occur either in large centralized plants or using distributed production technologies on a more localized level—most probably the latter.

In the long term, production technologies must be able to produce hydrogen at a price competitive with gasoline for bulk commercial fuel use in automobiles, freight trucks, aircraft, rail, and ships. This would require efficient production means and large quantities of reasonable-cost energy supplies, perhaps from coal with CO_2 sequestration, advanced nuclear power (high-efficiency electrolysis and thermochemical decomposition of water), fusion energy, renewables (wind-powered electrolysis, direct conversion of water via sunlight, and high-temperature conversion of water using concentrated solar power), or a variety of methods using biomass. Other important factors in the long term include the cost of hydrogen storage and transportation. Finally, advances in basic science associated with direct water-splitting and solid-state hydrogen storage could possibly permit even lower-cost hydrogen production and safer storage, delivery, and utilization in the context of low or near-net-zero emission futures for transportation and electricity generation.

Technology Strategy

Introducing hydrogen into the mix of competitive fuel options and building the foundation for a global hydrogen economy will require a balanced technical approach that not only envisions a plausible commercialization path, but also respects a triad of long-run uncertainties on a global scale: (1) the scale, composition, and energy intensity of future worldwide transportation demand, and potential substitutes; (2) the viability and endurance of CO_2 sequestration; and (3) the long-term economics of carbonless energy sources. The influences of these factors shape the urgency, relative importance, economic status, and ideal end state of a future hydrogen infrastructure.

The International Partnership for the Hydrogen Economy (IPHE) was formed in November 2003 among 15 countries (Australia, India, Brazil, Italy, Canada, Japan, China, Republic of Korea, Norway, France, Russia, Germany, United Kingdom, United States, and Iceland) and the European Commission. The IPHE provides a mechanism to organize, evaluate, and coordinate multinational research, development, and deployment programs that advance the transition to a global hydrogen economy. The partnership leverages national resources, brings together the world's best intellectual skills and talents, and develops interoperable technology standards.

The IPHE has reviewed actions being pursued jointly by participating countries and is identifying additional actions to advance research, development, and deployment of hydrogen production, storage, transport, and distribution technologies; fuel cell technologies; common codes and standards for hydrogen fuel utilization; and coordination of international efforts to develop a global hydrogen economy. More about the IPHE is available at http://www.iphe.net.

The Department of Energy's Hydrogen Fuel Cells and Infrastructure Technologies Program plans to research, develop, and demonstrate the critical technologies (and implement codes and standards for safe use) needed for hydrogen light-duty vehicles (Figure 5-6). The program operates in cooperation with automakers and related parties experienced in refueling infrastructure to develop technology necessary to enable a commercialization decision by 2015 (DOE 2005). Current research program goals call for validation by 2015 of technology for:

◆ Hydrogen storage systems enabling minimum 300-mile vehicle range while meeting identified packaging, cost, and performance requirements;

◆ Hydrogen production to safely and efficiently deliver hydrogen to consumers at prices competitive with gasoline and without adverse environmental impacts; and

◆ Transportation fuel cell power system costs of less than $50/kW (in high-volume production) while meeting performance and durability requirements.

DOE requested a study by the National Research Council (NRC) and the National Academy of Engineering (NAE) to assess the current state of technology for hydrogen production and use, and to review and provide feedback on the DOE RD&D hydrogen program, including recommendations for priorities and strategies to develop a hydrogen economy. The resulting report (NRC/NAE 2004) addressed implications for national goals, R&D priorities, and criteria for transition to a hydrogen economy. It provided recommendations in the areas of systems analysis, fuel cell vehicle technology, infrastructure, transition, safety, CO_2-free hydrogen, carbon capture and storage, and DOE's hydrogen RD&D program. In addition to research being conducted within DOE's Hydrogen, Fuel Cells and Infrastructure Program, the NRC report also addressed DOE's programs for hydrogen production from nuclear and fossil energy sources.

Current Portfolio

Within the constraints of available resources, the current Federal hydrogen technology research portfolio balances the emphasis on near-term technologies that will enable a commercialization decision for hydrogen automobiles by 2015, with the longer-term ultimate development of a mature hydrogen economy founded on advanced hydrogen production, storage, and delivery technologies. Elements of the portfolio include:

◆ **Hydrogen Production from Nuclear Fission.**
High-efficiency, high-temperature fission power plants may one day produce hydrogen economically without producing CO_2 as a byproduct. Hydrogen would be produced by cyclic thermochemical decomposition of water or high-efficiency electrolysis of high-temperature steam.

Hydrogen production from nuclear power RDD&D goals target high-temperature, high-efficiency fission to produce electricity to generate hydrogen from water. Major research areas include support for the development of high-temperature materials, separation membranes, advanced heat exchangers, and supporting systems relating to hydrogen production using the sulfur-iodine (S-I) thermochemical cycle and high-temperature electrolysis. Alternative processes having significantly more technical risk (because less is known about them) continue to be evaluated because their expected lower temperature requirements and, in some cases, reduced complexity could render them more economical in the longer term.[8]

◆ **Hydrogen Production and Distribution Using Electricity and Fossil/Alternative Energy.**
Research and development of small-scale steam reformers, alternative reactor technologies, and hydrogen membrane/separation technologies are aimed at improving the economics of hydrogen production from fossil fuels. Demonstration of on-site electrolysis integrated with renewable electricity and laboratory-scale direct water-splitting by photoelectrochemical and photobiological methods are planned.

Near-term research program goals in this area include, by 2006: (1) completion of research of small-scale steam methane reformers with a projected cost of $3.00/kg hydrogen at the pump; and (2) development of alternative reactors, including auto-thermal reactors; and, by 2007, evaluation of whether renewable energy—when integrated with hydrogen production by water electrolysis—can achieve 64 percent net energy efficiency at a projected cost of $5.50/kg, delivered at 5,000 psi. Midterm goals call for demonstrating, by 2010, at the pilot-plant scale: (1) membrane separation and reactive membrane separation technology for hydrogen production from coal; and (2) distributed hydrogen production from natural gas with a projected cost of $2.50/kg hydrogen at the pump. Longer-term goals call for demonstrating, at laboratory-bench scale: (1) a photo-electrochemical water-splitting system; and

Possible Hydrogen Pathways

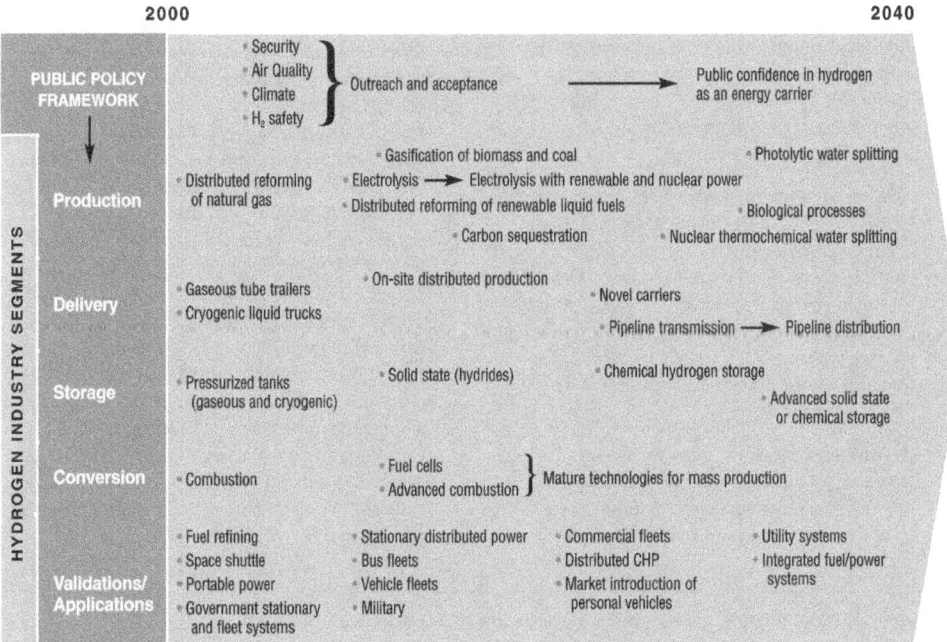

Figure 5-6. Possible Hydrogen Pathways (Source: U.S. Department of Energy Hydrogen Program)

8   See Section 2.2.1 (CCTP 2005): http://www.climatetechnology.gov/library/2005/tech-options/tor2005-221.pdf.

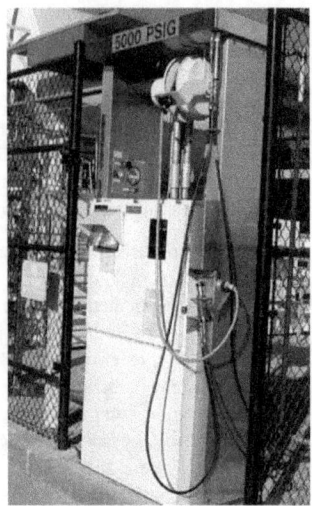

Figure 5-7. A systems approach is needed for integrated hydrogen production, delivery, and storage, as well as for refueling hydrogen vehicles.

Courtesy: DOE/NREL, Credit: Warren Gretz

(2) a biological system for water-splitting (or other substrates) that shows potential to achieve long-term costs that are competitive with conventional fuels.[9]

◆ **Hydrogen Storage.** Four methods of high-density, energy-efficient storage of hydrogen are being researched: (1) composite pressure vessels, which will contain the hydrogen as a compressed gas or cryogenic vapor; (2) physical adsorption on high-surface-area lightweight carbon structures; (3) reversible metal hydrides; and (4) chemical hydrides. Improving hydrogen compression and/or liquefaction equipment—as well as evaluating the compatibility of the existing natural gas pipeline infrastructure for hydrogen distribution—is also planned.

The research program goals of hydrogen storage are: (1) by 2010, develop and verify hydrogen storage systems with 6 weight-percent, 1,500 watt-hrs/liter energy density, and at a cost of $4/kWh of stored energy; and (2) by 2015, develop associated technologies and verify hydrogen storage systems with 9 weight-percent, 2,700 watt-hrs/liter energy density, and at a cost of $2/kWh of stored energy.[10]

• **Hydrogen Use.** DOE aims to demonstrate high-efficiency, solid-oxide fuel cell/turbine hybrid-electric generation systems operating on gasified coal with carbon capture and storage, and to

develop efficient and durable polymer electrolyte membrane (PEM) fuel cells appropriate for automotive and stationary applications. Additional research is underway to use hydrogen in auxiliary power units in heavy vehicles to supplant diesel engine power use for refrigeration and housekeeping tasks.

The research program goals in this area are: (1) by 2010, develop a 60 percent peak-efficient, durable, PEM fuel cell power system for transportation at a cost of $45/kW; and a distributed generation (50-250 kW) PEM fuel cell system operating on natural gas or propane that achieves 40 percent electrical efficiency and 40,000 hours durability at $400-750/kW; and (2) by 2015, reduce the cost of PEM fuel cell power systems to $30/kW for transportation systems.[11]

◆ **Hydrogen Systems Technology Validation.** A systems approach is needed to demonstrate integrated hydrogen production, delivery, and storage, as well as refueling of hydrogen vehicles and use in stationary fuel cells (Figure 5-7). This could involve providing hydrogen in gaseous and liquid form.

The overall goal in this area is to validate, by 2015, integrated hydrogen and fuel cell technologies for transportation, infrastructure, and electric generation in a systems context under real-world operating conditions. Specific goals include the following: (1) by 2006, complete development of a laboratory-scale distributed natural-gas-to-hydrogen production and dispensing system that can produce 5,000 psi hydrogen for $3.00/gge; (2) by 2008, demonstrate stationary fuel cells with a durability of 20,000 hours and 32 percent efficiency; (3) by 2009, demonstrate vehicles with greater than 250-mile range and 2,000-hour fuel cell durability; and (4) by 2009, demonstrate hydrogen production at $3/gge. By 2015, the research program aims to provide critical statistical data that demonstrate that fuel cell vehicles can meet targets of 5,000-hour fuel cell durability, storage systems can efficiently meet 300+ mile range requirements, and hydrogen fuel costs between $2.00 and $3.00/gge. The technology-validation effort also provides information in support of technical codes and standards development of infrastructure safety procedures.[12]

9 See Section 2.2.3 (CCTP 2005): http://www.climatetechnology.gov/library/2005/tech-options/tor2005-223.pdf.

10 See Section 2.2.4 (CCTP 2005): http://www.climatetechnology.gov/library/2005/tech-options/tor2005-224.pdf.

11 See Section 2.2.5 (CCTP 2005): http://www.climatetechnology.gov/library/2005/tech-options/tor2005-225.pdf.

12 See Section 2.2.2 (CCTP 2005): http://www.climatetechnology.gov/library/2005/tech-options/tor2005-222.pdf.

◆ **Hydrogen Infrastructure Safety.** Ensuring the safety of hydrogen infrastructure technologies largely depends on the development of sound, internationally agreed upon codes and standards. DOE will study the flammability and explosive, reactive, and dispersion properties of hydrogen and will subject components, subsystems, and systems to environmental conditions that could result in failure in order to verify design practice and to develop failure-mode models and risk-analysis methodologies. DOE is also compiling a hydrogen safety-incident database, and will develop potential accident scenarios, and draft a handbook on Best Management Practices for Safety to be published in 2008. These technical data and models will be provided to the appropriate organizations (i.e., International Code Council, National Fire Protection Association) to write and publish applicable codes and standards for hydrogen production and delivery processes as well as for hydrogen storage and fuel cell systems. The goal is to have all data and testing completed by 2010 to finalize U.S. technical standards for preparation of a Global Technical Regulation. The Department of Transportation (DOT) has regulatory or codes-and-standards responsibilities, including future Federal Motor Vehicle Safety Standards for fuel cell vehicles. The Federal government will facilitate the development of other necessary standards by standards organizations through R&D and support for appropriate technical representation in working groups. In support of on-vehicle safety, DOE will develop hydrogen safety sensor technology that can meet technical targets for response time and accuracy.[13]

Future Research Directions

The current portfolio supports the main components of the technology development strategy and addresses the highest priority current investment opportunities in this technology area. For the future, CCTP seeks to consider a full array of promising technology options. From diverse sources, suggestions for future research have come to CCTP's attention. Some of these, and others, are currently being explored and under consideration for the future R&D portfolio. These include:

◆ **Commercial Transportation Modes.** If efficient hydrogen-fueled or hybrid-electric vehicles begin to dominate the light-duty passenger vehicle market (beyond 2025), commercial transportation modes (freight trucks, aircraft, marine, and rail) may become the dominant sources of transportation-related CO_2 emissions later in the 21st century. Therefore, the future CCTP portfolio should aim at reducing the cost of hydrogen production and liquefaction of hydrogen for these modes and explore the infrastructure implications of hydrogen production and/or liquefaction on-site at airports, harbors, rail yards, etc. In the case of hydrogen-powered aircraft, the average length of future flights and whether significant demand for hypersonic passenger aircraft develops over the 21st century will be important factors in determining the relative fuel economy advantages of hydrogen, if any, over conventional jet fuel. Scenarios that include a worldwide shift toward hydrogen-powered aircraft and like substitutes for shorter trips (high-speed rail) could be considered.

◆ **Integration of Electricity and Hydrogen Transportation Sectors.** Eventual full deployment for optimal use of solar, wind, biomass, and nuclear electricity may require significant hydrogen storage or increased flexibility in electricity demand. Electrolytic coproduction of hydrogen for transportation fuel would provide such a demand profile. This important possibility needs to be examined to determine the economic and technical parameters for electricity demand, generation, and storage; and for hydrogen production, storage, and use to achieve a synergistic effect between hydrogen vehicles and carbonless electricity generation.

◆ **Vehicle-to-Grid Options.** Fuel cell vehicles represent a potential new power generation source, supplying electricity to homes and to the grid in emergencies or periods of exceptional demand. In this scenario, fuel-cell vehicles would represent new installed peak or backup power generation capacity. Depending on the source of the fuel used, this could also reduce GHG emissions. Better understanding and modeling of the potential benefits of vehicle-to-grid options could inform integrative strategies regarding the energy sector and climate change.

◆ **Develop Fundamental Understanding of the Physical Limits to Efficiency of the Hydrogen Economy.** The fundamental electrochemistry and material science of electrolyzers, fuel cells, and reversible devices need to be fully explored. For example, the theoretical limits on electrolyte conductivity bound the power density and efficiency of both fuel cells and electrolyzers.

---

[13] See Section 2.2.6 (CCTP 2005): http://www.climatetechnology.gov/library/2005/tech-options/tor2005-226.pdf.

Advancing the knowledge of these limits should allow efficiency gains in the conversion of electricity to hydrogen (and reconversion to electricity) to approach theoretical limits before hydrogen technology is deployed on a global scale.

◆ **Explore advanced concepts in hydrogen production.** Current hydrogen R&D activities have focused much of the available funding for hydrogen production on near-term technologies that can be ready for commercialization to support introduction of hydrogen vehicles soon after a 2015 industry decision-point. As available, funding has also supported advanced renewables-based production approaches such as photoelectrochemical and photobiological pathways. Breakthroughs in these or other advanced production pathways will be necessary to make a true hydrogen economy economically feasible, and the CCTP will be exploring ways to ensure sufficient funding for advanced, post-2015 hydrogen production technologies.

5.3 Renewable Energy and Fuels

Renewable sources of energy include the energy of the sun, the kinetic energy of wind, the thermal

BOX 5-1

RENEWABLE ENERGY AND FUELS TECHNOLOGIES

- Wind Energy
- Solar Photovoltaic Power
- Solar Buildings
- Concentrating Solar Power
- Biochemical Conversion of Biomass
- Thermochemical Conversion of Biomass
- Biomass Residues
- Energy Crops
- Waste-to-Energy
- Photoconversion
- Advanced Hydropower
- Geothermal Energy

energy inside the Earth itself, the kinetic energy of flowing water, and the chemical energy of biomass and waste. These sources of energy, available in one or more forms across the globe, can be converted and/or delivered to end users as electricity, heat, fuels, hydrogen, and useful chemicals and materials. Box 5-1 lists 12 renewable energy technologies, many of which are discussed in *Technology Options for the Near and Long Term*. In the United States in 2003, of the 71.42 quads of net energy supply and disposition (98.22 quads total energy consumption), renewable resources contributed 5.89 quads (8 percent of supply, or 6 percent of the total). Of the renewable energy, 2.78 quads came from hydropower, 2.72 quads from burning biomass (wood and waste), 0.28 quads from geothermal energy, and 0.12 quads from solar and wind energy combined. An additional 0.24 quads of ethanol were produced from corn for transportation (EIA 2005).

The technologies in the suite of renewable energy technologies are in various states of market penetration or readiness. In many cases, industry has the financial incentive to make incremental improvements to commercialized renewable energy technologies, or other policies exist to promote renewable energy development and deployment (research and experimentation tax credit, State renewable portfolio standards). These and other factors are important to consider as CCTP helps to prioritize Federal investments.

Hydropower is well established, but improvements in the technology could increase its efficiency and widen its applicability. Geothermal technologies are established in some areas and applications, but significant improvements are needed to tap broader resources. The installation of wind energy has been rapidly and steadily expanding during the past several years. In the past decade, the global wind energy capacity has increased tenfold—from 3.5 GW in 1994 to almost 59 GW by the end of 2006. Technology improvements will continue to lower the cost of land-based wind energy and will enable access to the immense wind resources in shallow and deep waters of U.S. coastal areas and the Great Lakes near large energy markets. The next generations of solar—with improved performance and lower cost—are in various stages of concept identification, laboratory research, engineering development, and process scale-up. Also, the development of integrated and advanced systems involving solar photovoltaics, concentrating solar power, and solar buildings are in early stages of development; but advances in these technologies are expected to make them competitive with conventional sources in the future.

Biochemical and thermochemical conversion technologies also range broadly in their stages of development, from some that need only to be proved at an industrial scale, to others that need more research, and to others in early stages of scientific exploration. In the general category of photo-conversion, most technical ideas are at the earliest stages of concept development, theoretical modeling, and laboratory experiment. Waste-to-energy accounts for more than 2.5 gigawatts of power production.

The energy-production potential and siting of the various types of renewable energy facilities is dependent on availability of the applicable natural resources. Figures 5-8 through 5-12 show global wind capacity growth and availability of key U.S. renewable resources as estimated by the National Renewable Energy Laboratory (NREL) at the Renewable Resource Data Center (see http://nrel.gov/rredc).

Potential Role of Technology

Renewable energy technologies are generally modular and can be used to help meet the energy needs of a stand-alone application or building, an industrial plant or community, or the larger needs of a national electrical grid or fuel network. Renewable energy technologies can also be used in various

combinations—including hybrids with fossil-fuel-based energy sources and with advanced storage systems—to improve renewable resource availability. Because of this flexibility, technologies and standards to safely and reliably interconnect individual renewable electric technologies, individual loads or buildings, and the electric grid are very important.

In addition, the diversity of renewable energy sources offers a broad array of technology choices that can reduce CO_2 emissions. The generation of electricity from solar, wind, geothermal, or hydropower sources contributes no CO_2 or other GHGs directly to the atmosphere. Increasing the contribution of renewables to the Nation's energy portfolio will directly lower GHG intensity (GHGs emitted per unit of economic activity) in proportion to the amount of carbon-emitting energy sources displaced.

Analogous to crude oil, biomass can be converted to heat, electrical power, fuels, hydrogen, chemicals, and intermediates. Biomass refers to both biomass residues (agricultural wastes such as corn stover and rice hulls, forest residues, pulp and paper wastes, animal wastes, etc.) and to fast-growing "energy crops," chosen specifically for their efficiency in being converted to electricity, fuels, etc. The CO_2 consumed when the biomass is grown essentially offsets the CO_2 released during combustion or processing. Biomass systems actually represent a net sink for GHG emissions when biomass residues are

Global Wind Capacity Growth

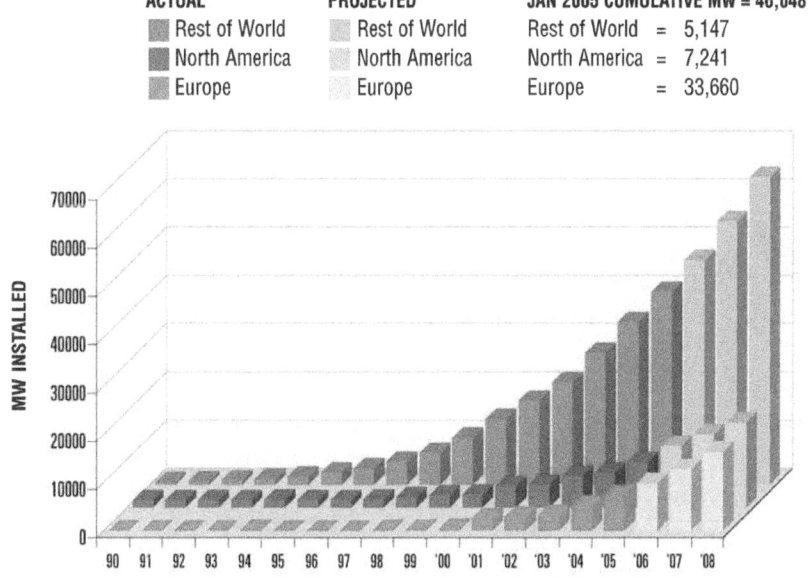

ACTUAL	PROJECTED	JAN 2005 CUMULATIVE MW = 46,048
Rest of World	Rest of World	Rest of World = 5,147
North America	North America	North America = 7,241
Europe	Europe	Europe = 33,660

Figure 5-8. Global Wind Capacity Growth (Source: BTM Consult Aps, March 2003; Windpower Monthly, January 2005; NREL estimate for 2005)

U.S. Biomass Resources

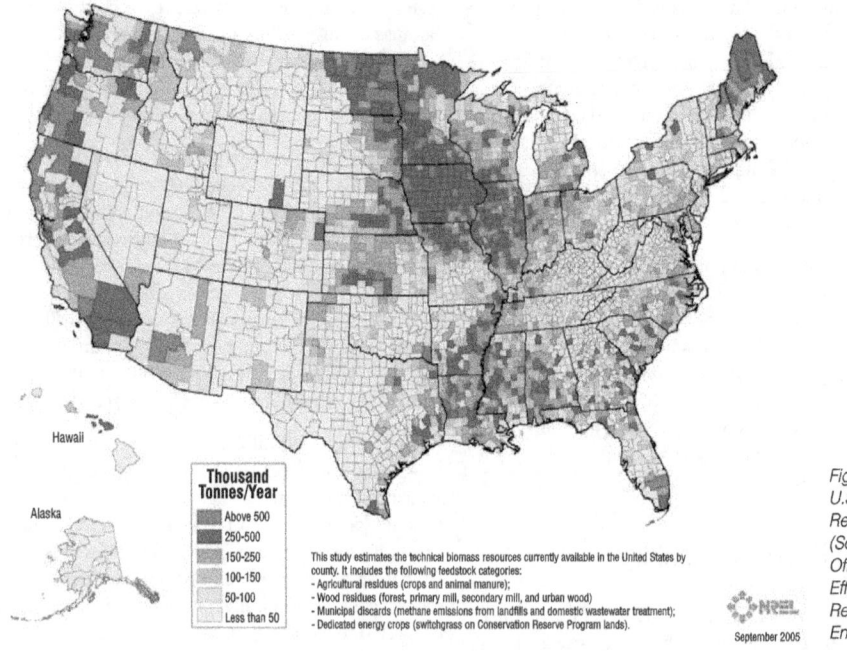

Thousand Tonnes/Year

Above 500
250-500
150-250
100-150
50-100
Less than 50

This study estimates the technical biomass resources currently available in the United States by county. It includes the following feedstock categories:
- Agricultural residues (crops and animal manure);
- Wood residues (forest, primary mill, secondary mill, and urban wood)
- Municipal discards (methane emissions from landfills and domestic wastewater treatment);
- Dedicated energy crops (switchgrass on Conservation Reserve Program lands).

Hawaii

Alaska

September 2005

Figure 5-9. U.S. Biomass Resources (Source: DOE Office of Energy Efficiency and Renewable Energy)

Note: This Biomass Resources map is the "current" resource base (about 500 million dry tons) and does not reflect the potential future resource base as outlined in the "Billion Ton Vision." The map takes into account the resource required for soil conservation. More information can be found at: http://www.nrel.gov/gis/biomass.html.

U.S. Solar Resources

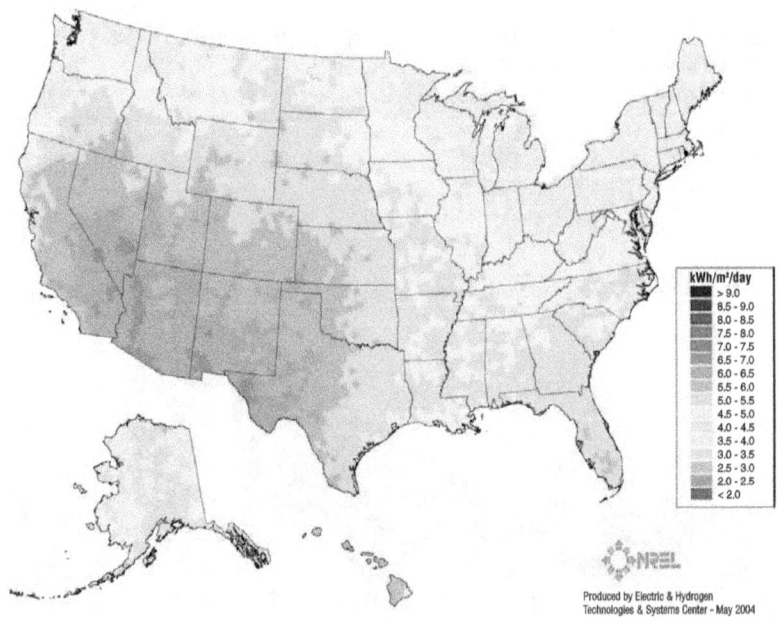

kWh/m²/day

> 9.0
8.5 - 9.0
8.0 - 8.5
7.5 - 8.0
7.0 - 7.5
6.5 - 7.0
6.0 - 6.5
5.5 - 6.0
5.0 - 5.5
4.5 - 5.0
4.0 - 4.5
3.5 - 4.0
3.0 - 3.5
2.5 - 3.0
2.0 - 2.5
< 2.0

Figure 5-10. U.S. Solar Resources (Source: DOE Office of Energy Efficiency and Renewable Energy)

Produced by Electric & Hydrogen Technologies & Systems Center - May 2004

U.S. Wind Resources

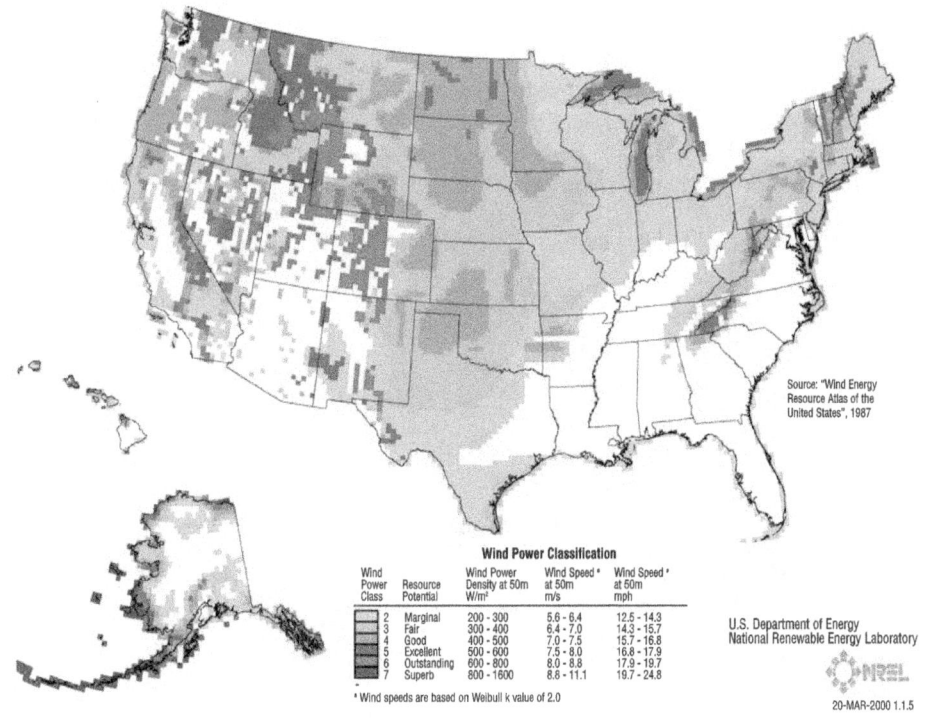

Figure 5-11. U.S. Land-Based Wind Resources
(Source: DOE Office of Energy Efficiency and Renewable Energy)

U.S. Geothermal Resources

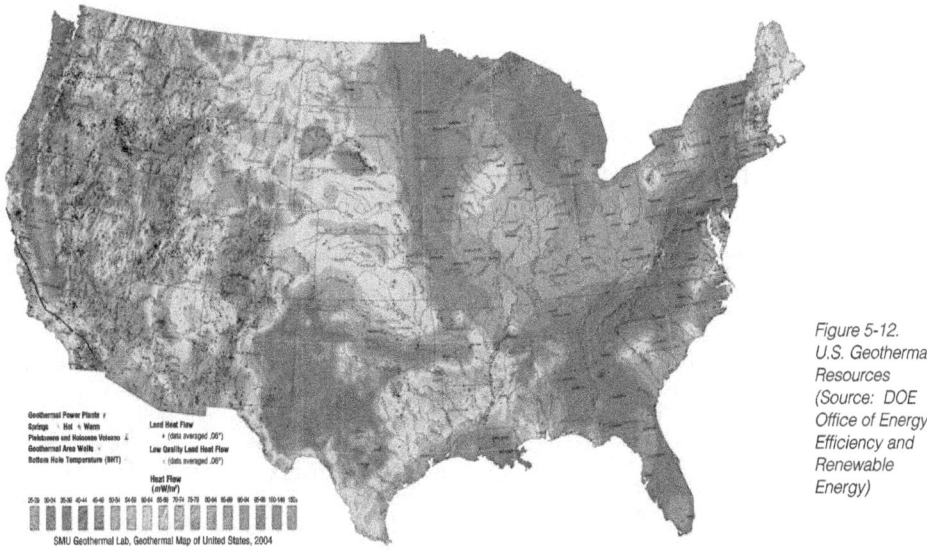

Figure 5-12.
U.S. Geothermal
Resources
(Source: DOE
Office of Energy
Efficiency and
Renewable
Energy)

Bioenergy Cycle

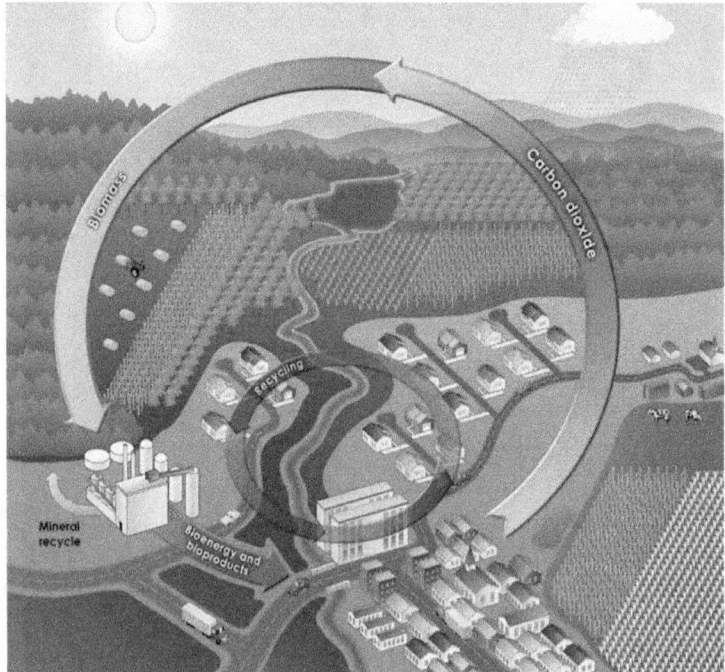

Figure 5-13.
Bioenergy Cycle
Courtesy: DOE/ORNL

used, because this avoids methane emissions that result from landfilling unused biomass (Figures 5-13 and 5-14). Biorefineries of the future could produce value-added chemicals and materials together with fuels and/or power from nonconventional, lower-cost feedstocks (such as agricultural and forest residues and specially grown crops) with no net CO_2 emissions.

Technology Strategy

Given the diversity of the stages of development of the technologies, impacts on different economic sectors, and geographic dispersion of renewable energy sources, it is likely that a portfolio of renewable energy technologies—not just one—will contribute to lowering CO_2 emissions. The composition of this portfolio will change as R&D continues and markets change. Appropriately balancing investments in developing this portfolio will be important to maximizing the effect of renewable energy technologies on GHG emissions in the future.

Transitioning from today's reliance on fossil fuels to a global energy portfolio that includes significant renewable energy sources will require continued improvements in cost and performance of renewable

technologies. This transition would also require shifts in the energy infrastructure to allow a more diverse mix of technologies to be delivered efficiently to consumers in forms they can readily use. For example, changes to the electricity infrastructure are needed to accommodate greatly increased use of renewable electric generation. These changes include additional transmission lines to access those renewable resources that are located far from load centers; grid operating practices and storage to accommodate renewables that are intermittent, such as wind and solar; greater use of renewables in a distributed generation mode; and adapting current fossil generation for biomass co-firing. Fortunately, there already is substantial progress in adapting the electricity infrastructure to enable greater use of renewables generation, and additional changes that would be needed are relatively easy to make in a decade or so. With regard to renewable fuels used for transportation, no significant changes to vehicles and refueling infrastructure are needed until E10 ethanol (90 percent gasoline, 10 percent ethanol) in gasoline blends captures 10 percent of gasoline markets. When 85 percent ethanol/gasoline blends (E85) expand, which is expected if ethanol costs come down further, limited refueling infrastructure modifications will be needed. However, refueling technology needed for E-85 is already well developed, and several

Biomass as Feedstock for a Bioenergy and Bioproducts Industry

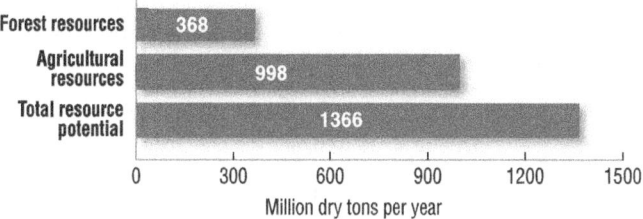

Figure 5-14. Biomass as Feedstock for a Bioenergy and Bioproducts Industry (Source: ORNL 2005)

automobile manufacturers are already selling E-85 vehicles at low or no incremental costs.

In general, as performance continues to improve and costs continue to decline, improved new generations of technologies will replace today's renewable technologies. Combinations of renewable and conventional technologies and systems—and, therefore, integration and interconnection issues—will grow in importance.

The transition from today's energy mix to a state of GHG stabilization can be projected as an interweaving of individual renewable energy technologies with other energy technologies, as well as market developments through the upcoming decades. Today, grid-connected wind energy, geothermal, solar energy, and biopower systems are well established. Demand for these systems is growing in some parts of the world. Solar hot-water technologies are reasonably established, although improvements continue. Markets are growing for small, high-value or remote applications of solar photovoltaics; wind energy; biomass-based CHP; certain types of hydropower; and integrated systems that usually include natural gas or diesel generators. Other technologies and applications today are in various stages of research, development, and demonstration. Possible near-, mid-, and long-term pathways for renewable energy are discussed in the following paragraphs.

In the near term, as system costs continue to decrease, the penetration of off-grid systems could continue to increase rapidly, including integration of renewable systems such as photovoltaics into buildings. As interconnection issues are resolved, the number of grid-connected renewable systems could increase quite rapidly, meeting local energy needs such as uninterruptible power, community power, or peak shaving. Wind energy may expand most rapidly among grid-connected applications, with solar expanding as system costs are reduced. Environment-friendly hydropower systems could be further developed. The use of utility-scale wind technology

is likely to continue to expand onshore and is targeted to become competitive in select offshore locations between 5 and 50 nautical miles from shore and at water depths of 30 meters or less. Small wind turbines are on the verge of operating cost-effectively in most of the rural areas of the United States, and more than 15 million homes have the potential to generate electricity with small wind turbines.[14] With a further maturing of the market, costs will be lowered to compete directly with retail rates for homeowners, farmers, small businesses, and community-based projects.

The biomass near-term outlook includes industry investment to make the production of corn-based ethanol (already produced at nearly 4 billion gallons) more efficient by increasing the quantity of ethanol through residual starch conversion, and conversion of fiber already collected and present at the operating facilities. The inclusion of biochemicals as byproducts will further help to improve the industry's profitability. The Biofuels Initiative launched in FY 2007 will accelerate demonstrations of biorefinery concepts, producing one or more products (bioethanol, bioproducts, electricity, CHP, etc.) from one plant using local waste and residues as the feedstock. Biodiesel use may continue to grow, replacing fossil-fuel-derived diesel fuel. The technology being developed to convert agricultural residues to ethanol is also partially applicable to the conversion of municipal solid waste to ethanol.

A recent assessment of potential biofuel resources concluded that by mid-21st century it would be technically possible to produce enough biomass to displace about one-third of current petroleum consumption. The study made no conclusions about economic feasibility. This level of biofuel production would require economically-competitive technologies to convert cellulosic biomass (rather than just the sugars and starches) into ethanol, and developing cellulosic ethanol technologies is a central aspect of the Biofuels Initiative and the President's Advanced Energy Initiative.

---

14 U.S. Small Wind Turbine Industry Roadmap, NREL Report No. BK-500-31958; DOE/GO-102002-1598, 2002 http://www.nrel.gov/docs/gen/fy02/31958.pdf.

In the mid term, offshore wind energy could begin to expand significantly. Technology development may focus on turbine-support structures suitable for deeper water depths, and reducing turbine system and balance-of-plant costs to offset increased distance from shore, decreased accessibility, and more stringent environmental conditions. Land-based use of wind turbines is also likely to expand for large and small turbines as the costs for these systems continue to decrease. Small turbines may be used to harness wind to provide pumping for farm irrigation, help alleviate water-availability problems, and provide a viable source of clean and renewable hydrogen production.[15]

Reductions in cost could encourage penetration by solar technologies into large-scale markets, first in distributed markets such as commercial buildings and communities, and later in utility-scale systems. Solar systems could also become cost-effective in new construction for commercial buildings and homes. The first geothermal plants using engineered geothermal systems technology could come online, greatly extending access to geothermal resources. Hydropower may benefit from full acceptance of new turbines and operational improvements that enhance environmental performance, lowering barriers to new development.

As a result of the Biofuels Initiative, biorefineries could begin using agricultural and forest residues, and eventually energy crops as primary feedstocks. Assuming success in reducing production costs and expanding the fuels distribution infrastructure, bioethanol and, to a lesser extent, biodiesel could achieve substantial market penetration in the 2030-2040 timeframe. This would be an important step in lowering U.S. dependence on imported petroleum.

In the long term, hydrogen from solar, wind, and possibly geothermal energy could be the backbone of the economy, powering vehicles and stationary fuel cells. Solar technologies could also be providing electricity and heat for residential and commercial buildings, industrial plants, and entire communities in major sections of the country. A major value for solar is that most residential and commercial buildings could generate their own energy on-site. Wind energy could be the lowest-cost option for electricity generation in favorable wind areas for grid power, and offshore systems could become prevalent in many countries by achieving a commercially viable cost by using floating platform technologies. Geothermal systems could be a major source of baseload electricity

for large regions. Biorefineries could be providing a wide range of cost-effective products as rural areas embrace the economic advantages of widespread demand for energy crops. Vehicle fuels could be powered by a combination of hydrogen fuel cells, with some bioethanol and biodiesel in significant markets.

Current Portfolio

The current Federal portfolio of renewable energy supply technologies encompasses the areas described below:

◆ **Wind Energy.** Generating electricity from wind energy focuses on using aerodynamically designed blades to drive generators that produce electric power in proportion to wind speed. Utility-scale turbines can be several megawatts and produce energy at between $0.04-0.06/kWh depending on the wind resource. Smaller turbines (under 100 kW) serve a range of distributed, remote, and stand-alone power applications, producing energy between $0.12-0.17/kWh. While the focus in the last several years has been on low-wind-speed technology R&D for onshore applications, R&D for reducing the cost of offshore systems, based on recent emergence of U.S. land-based wind power development and the assessment of potential national benefits, is also supported. Research activities include wind characteristics and forecasting, aerodynamics, structural dynamics and fatigue, control systems, design and testing of new onshore and offshore prototypes, component and system testing, power systems integration, and standards development. Federal agencies are also collaborating with interested stakeholders on addressing and minimizing siting concerns (i.e., wildlife and acoustics).

Research program goals in this area vary by application. For distributed wind turbines under 100 kW, the goal is to achieve a power production cost of $0.10 to 0.15/kWh in Class 3 winds by 2007. For larger systems greater than 100 kW, the goal is to achieve a power production cost of $0.036/kWh for land-based at sites with average wind speeds of 13 mph (wind Class 4) by 2012, and $0.05/kWh at shallow (depths up to 30 meters) offshore sites with average wind speeds of 15 mph (wind Class 6) by 2014, and $0.05/kWh for transitional (depths up to 60 meters) offshore systems in Class 6 winds by 2016.[16]

---

[15] National Academy of Science, *The Hydrogen Economy: Opportunities, Costs, Barriers, and R&D Needs* http://www4.nationalacademies.org/news.nsf/isbn/0309091632?OpenDocument.

[16] See Section 2.3.1 (CCTP 2005): http://www.climatetechnology.gov/library/2005/tech-options/tor2005-231.pdf.

◆ **Solar Photovoltaic Power.** Generating electricity from solar energy focuses on using semiconductor devices to convert sunlight directly to electricity. A variety of semiconductor materials can be used, varying in conversion efficiency and cost. Today's commercial modules are 11 percent to 13 percent efficient, and grid-tied photovoltaic (PV) systems generate electricity for about $0.18 to 0.23/kWh under ideal siting and financing conditions for commercial (low end of range) and residential (high end of range) systems. Actual levelized cost of energy may be significantly greater in parts of the United States where the solar resource potential is less than ideal. Research activities, conducted with partnerships between the Federal laboratories and the private sector, include the fundamental understanding and optimization of photovoltaic materials, process, and devices; module validation and testing; process research to lower costs and scale up production (Figure 5-15); and technical issues with inverters and batteries.

Research program goals in this area focus primarily on a new initiative—the Solar America Initiative (SAI)—which will accelerate R&D efforts designed to achieve market competitiveness for PV solar electricity by 2015 (i.e., 5 to 10 cents/kWh under ideal siting and financing conditions). The accelerated R&D effort will focus on PV technology pathways that have the greatest potential to lower costs and improve performance. New industry-led R&D partnerships, known as "Technology Pathway Partnerships," will be funded to address the issues of cost, performance, and reliability associated with each technology pathway. Potential partners within the Technology Pathway Partnerships include industry, universities, laboratories, States, and other governmental entities. If the research is successful and if other policies remain in place to promote deployment (production tax credits, State renewable portfolio standards), then by 2015, 5-10 GW of new solar power capacity could be deployed, equivalent to the amount of electricity needed to power 1-2 million homes. This deployment level would result in 10 million metric tons of avoided carbon dioxide emissions in the United States. The interim cost goal is to reduce the 30-year user cost for PV electrical energy to a range of $0.11 to 0.18/kWh under ideal conditions by 2010.[17]

◆ **Solar Heating and Lighting.** Solar heating and lighting technologies are being developed for use in buildings applications that include solar water heating and hybrid solar lighting. Most of the goals for these technologies have been sufficiently developed for use in southern climates and now can be transferred to industry for commercialization.[18]

◆ **Concentrating Solar Power.** Concentrating solar power (CSP) technology utilizes the heat generated by concentrating and absorbing the sun's energy to produce electric power. The concentrated sunlight produces thermal energy at temperatures ranging from 600 degrees F to over 1500 degrees F to run heat, engines, or steam turbines for generating power or producing clean fuels such as hydrogen. The long-term goal is to achieve a power cost of between $0.035/kWh and $0.062/kWh, compared to the cost of between $0.12-0.14/kWh in 2004.[19]

◆ **Biochemical Conversion of Biomass.** Biochemical technology can be used to convert the cellulose and hemicellulose polymers in biomass (agricultural crops and residues, wood residues, trees and forest residues, grasses, and municipal waste) to their building blocks, such as sugars and glycerides. Using either acid hydrolysis (well-established) or enzymatic hydrolysis (being developed), sugars can then be converted to liquid fuels, such as ethanol, chemical intermediates, and other products, such as lactic acid and hydrogen.

Figure 5-15. Research and development activities on solar photovoltaic power, conducted at Federal laboratories and the private sector, help to improve efficiencies, lower production costs, and resolve technical issues.

Courtesy: DOE/NREL, Credit: United Solar Systems Corp.

---

[17] See Section 2.3.2 (CCTP 2005): http://www.climatetechnology.gov/library/2005/tech-options/tor2005-232.pdf.

[18] See Section 2.3.3 (CCTP 2005): http://www.climatetechnology.gov/library/2005/tech-options/tor2005-233.pdf.

[19] See Section 2.3.4 (CCTP 2005): http://www.climatetechnology.gov/library/2005/tech-options/tor2005-234.pdf.

Glycerides can be converted to a bio-based alternative for diesel fuel and other products. Producing multiple products from biomass feedstocks in a biorefinery could ultimately resemble today's oil refinery.

By FY 2007, the goal is to complete a preliminary engineering design package, market analysis, and financial projections for two industrial-scale projects for near-term agricultural pathways (corn wet mill, corn dry mill, oilseed) to produce a minimum of five million gallons of biofuels per year. The intent is to provide proof that the resultant industrial-scale biorefineries could produce and market biofuels at prices competitive with petroleum fuels produced from $50-per-barrel oil. By 2009, develop a conceptual, novel harvesting system and test a wet storage system for agricultural residues. By 2012, the goal is to reduce the estimated cost for producing a mixed, dilute sugar stream suitable for fermentation to ethanol, to $0.096/lb, compared to the cost of $0.15/lb in 2003. If successful, this cost goal would correspond to $1.50 per gallon of ethanol, assuming a cost of $45 per dry ton of corn stover.[20]

◆ **Thermochemical Conversion of Biomass.** Thermochemical technology uses heat to convert biomass into a wide variety of products. Pyrolysis or gasification of biomass produces an oil-rich vapor or syngas, which can be used to generate heat, electricity, liquid fuels, and chemicals. Combustion of biomass (or combinations of biomass and coal) generates steam for electricity production and/or space, water, or process heat, occurring today in the wood products industry and biomass power plants. Analogous to an oil refinery, a biorefinery can use one or more of these methods to convert a variety of biomass feedstocks into multiple products.[21]

◆ **Biomass Residues**. Biomass residues include agricultural residues, wood residues, trees and forest residues, animal wastes, pulp, and paper waste. These must be harvested, stored, and transported on a large scale to be used in a biorefinery. Research activities include improving and adapting the existing harvest collection, densification, storage, transportation, and information technologies to bioenergy supply systems—and developing robust machines for multiple applications.

The mid- to long-term research program goal in this area is to reduce biomass harvesting and storage costs so that the delivered cost will be reduced from $53 per dry ton in 2003 to $45 per dry ton by 2012. The Biofuels Initiative of 2007 proposes to establish three regional biomass development partnerships in conjunction with land grant universities to develop research-specific data on feedstock availability, productivity, markets, and economics by 2008, and an additional partnership by 2009. Engineering designs and techno-economic assessments of integrated wet biomass storage and field preprocessing will be completed by 2008. By 2010, engineering design on multi-crop feedstock depot systems that can receive a variety of feedstocks and preprocess will then be completed.[22]

◆ **Energy Crops.** Energy crops are fast-growing, often genetically improved trees and grasses grown under sustainable conditions to provide feedstocks that can be converted to heat, electricity, fuels such as ethanol, and chemicals and intermediates. Research activities include genetic improvement, pest and disease management, and harvest equipment development to maximize yields and sustainability.

The overall research goal of this program is to advance the concept of energy crops contributing strongly to meet biomass power and biofuels production goals by 2020. Interim goals include: (1) by 2006, to develop feedstock crops with experimentally demonstrated yield potential of 6-8 dry ton/acre/year and accompanying cost-effective, energy-efficient, environmentally sound harvest methods; (2) by 2010, the goal is to identify genes that control growth and characteristics important to conversion processes in few model energy crops and achieve low-cost, "no-touch" harvest/ processing/transport of biomass to the process facility; and (3) by 2020, the goal is to increase yield of useful biomass per acre by a factor of 2 or more compared with year 2000 yields.[23]

◆ **Photoconversion.** Photoconversion processes use solar photons to drive a variety of quantum conversion processes other than solid-state photovoltaics. These processes can produce electrical power or fuels, materials, and chemicals directly from simple renewable substrates such as

---

[20] See Section 2.3.5 (CCTP 2005): http://www.climatetechnology.gov/library/2005/tech-options/tor2005-235.pdf.

[21] See Section 2.3.6 (CCTP 2005): http://www.climatetechnology.gov/library/2005/tech-options/tor2005-236.pdf.

[22] See Section 2.3.7 (CCTP 2005): http://www.climatetechnology.gov/library/2005/tech-options/tor2005-237.pdf.

[23] See Section 2.3.8 (CCTP 2005): http://www.climatetechnology.gov/library/2005/tech-options/tor2005-238.pdf.

water, carbon dioxide, and nitrogen. Photoconversion processes that mimic nature (termed "bio-inspired") can also convert CO_2 into liquid and gaseous fuels. Most of these technologies are at early stages of research where technical feasibility must be demonstrated, but a few (such as dye-sensitized solar cells) are at the developmental level.

The research program in this area is still in an exploratory stage. In the near term, research will focus on applications related to electrical power and high-value fuels and chemicals, where commercial potential may be expected during the next 5 to 10 years. If successful, larger-scale applications of photoconversion technologies may follow in the period from 2010 to 2015, with materials and fuels production beginning in the period 2015 to 2020, and commodity chemicals production in the period from 2020 to 2030.[24]

◆ **Geothermal Energy.** Geothermal sources of energy include hot rock masses, highly pressured hot fluids, hot hydrothermal systems, and shallow warm groundwater. Exploration techniques locate resources to drill; well fields and distribution systems allow the hot fluids to move to the point of use; and utilization systems apply the heat directly or convert it to electricity. Geothermal heat pumps use the shallow earth as a heat source and heat sink for heating and cooling applications. The U.S.-installed capacity for geothermal electrical generation is currently about 2 gigawatts. Government geothermal research and development activities are being transferred to the private sector for commercialization.[25]

◆ **The Green Power Partnership** facilitates purchases of environmentally friendly electricity products generated from renewable energy sources by addressing the market barriers that may be stifling demand. It publishes information about low-cost purchasing strategies, educates partners about features of different green power products, and reduces the transaction costs for organizations interested in making green power purchases.

◆ **The Combined Heat and Power Partnership** provides technical assistance designed to meet CHP project needs along each step of the project development cycle in order to make investments in CHP more attractive. EPA educates industry about the benefits of CHP and project development strategies and provides networking opportunities. EPA also works with State

governments to design air emissions standards and interconnection requirements that recognize the benefits of clean CHP.

◆ **The Renewable Energy Systems and Energy Efficiency Improvements Program** supports biomass/renewable energy related ventures.

◆ **The Renewable Energy and Energy Efficiency Partnership** seeks to accelerate and expand the global market for renewable energy and energy efficient technologies.

Future Research Directions

The current portfolio supports the main components of the technology development strategy and addresses the highest priority current investment opportunities in this technology area. For the future, CCTP seeks to consider a full array of promising technology options. From diverse sources, suggestions for future research have come to CCTP's attention. Some of these, and others, are currently being explored and under consideration for the future R&D portfolio. These include:

◆ **Wind Energy.** Research challenges include developing wind technology that will be economically competitive at low-wind-speed sites without a production tax credit; developing offshore wind technology to take advantage of the immense wind resources in U.S. coastal areas and

Figure 5-16. The Department of Energy is working closely with industry to research and develop new, advanced wind turbine technology.
Courtesy: GE Energy, ©2005, General Electric International, Inc.

[24] See Section 2.3.9 (CCTP 2005): http://www.climatetechnology.gov/library/2005/tech-options/tor2005-239.pdf.

[25] See Section 2.3.11 (CCTP 2005): http://www.climatetechnology.gov/library/2005/tech-options/tor2005-2311.pdf.

the Great Lakes (Figure 5-16); and exploring the role of wind turbines in emerging applications such as electrolytic hydrogen production, water purification, and irrigation.

- ◆ **Solar Photovoltaic Power.** Research would be required to lower the cost of solar electricity further. This can occur through developing "third-generation" materials such as quantum dots and nanostructures for ultra-high efficiencies or lower-cost organic or polymer materials, solving complex integrated processing problems to lower the cost of large-scale production of thin-film polycrystalline devices, optimizing cells and optical systems using concentrated sunlight, and improving the reliability and lowering the cost of inverters and batteries.

- ◆ **Solar Buildings.** Future research could include reducing cost and improving reliability of components and systems, optimizing and integrating solar technologies into building designs, and incorporating solar technologies into building codes and standards.

- ◆ **Concentrating Solar Power.** Future challenges requiring RD&D include reducing cost and improving reliability, demonstrating Stirling engine performance in the field, and developing technology to produce hydrogen from concentrated sunlight and water.

- ◆ **Solar Fuels.** Research could focus on artificial photosynthetic systems that operate at higher rates and efficiencies than natural photosynthetic processes in plants. Such systems might either be photo-biologically-based or electro-photo-chemistry-based and would consume CO_2 and water from the atmosphere and produce H_2, O_2, and (very) small photosynthetic electric currents. Research would include solar-powered photo-catalyzed production of liquid transportation fuels from H_2 and CO_2

- ◆ **Biochemical Conversion of Biomass.** Research is required to gain a better understanding of genomes, proteins, and their functions; the enzymes used for hydrolyzing pretreated biomass into fermentable sugars; the micro-organisms used in fermentation; and new tools of discovery such as bio-informatics, high-throughput screening of biodiversity, directed enzyme development and evolution, and gene shuffling. Research must focus on improving the cost, yield, and equipment reliability for harvesting, collecting, and transporting biomass; pretreating biomass before conversion; lowering the cost of the genetically engineered cellulose enzymes needed to hydrolyze

biomass; developing and improving fermentation organisms; and developing integrated processing applicable to a large, continuous-production commercial facility.

- ◆ **Thermochemical Conversion of Biomass.** Research is needed to improve the production, preparation, and handling of biomass; improve the operational reliability of thermochemical biorefineries; remove contaminants from syngas; and develop cost-competitive catalysts and processes for converting synthetic gases to chemicals, fuels, or electricity. All processes in the entire conversion system must be integrated to maximize efficiency and reduce costs.

- ◆ **Biomass Residues.** Research challenges include developing sustainable agriculture and forest-management systems that provide biomass residues; developing cost-effective drying, densification, and transportation techniques to create more standard feedstock from various residues; developing whole-crop harvest and fractionation systems; and developing methods for pretreatment of residues at harvest locations.

- ◆ **Energy Crops.** Future crop research needs include identifying genes that control growth and characteristics important to conversion processes, developing gene maps, understanding functional genomics in model crops, and applying advanced management systems and enhanced cultural practices to optimize sustainable energy crop production.

- ◆ **Waste-to-Energy.** Waste-to-energy technologies can produce valued energy, with low "net-emissions" of GHGs when the full fuel cycle is considered, by processing and combusting municipal solid waste and doing so in an environmentally acceptable manner with specially designed and equipped modern equipment and pollution controls. Research challenges remain in the areas of better understanding contaminants, pre-cleaning of waste streams, separations and recycling, and overall environmental and safety performance.

- ◆ **Photoconversion.** Photoconversion research requires developing the fundamental scientific understanding of photolytic processes through multidisciplinary approaches involving theory, mechanisms, kinetics, biological pathways and molecular genetics, natural photosynthesis, materials science, catalysts, and catalytic cycles.

- ◆ **Bio-X.** Combines advanced biological processing with emerging advances in other fields such as nanotechnology, computer science, and physics to develop new approaches for renewable technologies.

◆ **Geothermal Energy.** Future research areas that industry may pursue include developing improved methodologies for predicting reservoir performance and lifetime; finding and characterizing underground fracture permeability; developing low-cost innovative drilling technologies; reducing the cost and improving the efficiency of conversion systems; and developing engineered geothermal systems that will allow the use of geothermal areas that are deeper, less permeable, or drier than those currently considered as reserves.

◆ **Advanced Solid State Thermoelectric Devices.** If additional technical progress can be achieved, thermoelectric devices could allow conversion of mid-grade and low-grade heat to electricity at economically attractive efficiencies. To the extent that waste heat might otherwise be discarded, the conversion could add significantly to net energy supply. There could be multiple applications for adding a cogeneration "bottoming cycle" and improving the overall efficiency of any process involving significant amounts of heat, including conversion of solar heat.

Nuclear Fission

Currently, 443 nuclear power plants, operating in 31 nations, generate 17 percent of the world's electricity (Figure 5-1) and provide nearly 7 percent of total world energy (Figure 5-2). Twenty-three new plants are under construction in ten countries. Because they emit no GHGs, today's nuclear power plants avoid the CO_2 emissions associated with combustion of coal or other fossil fuels (Figure 5-17).

During the past 30 years, operators of U.S. nuclear power plants have steadily improved economic performance through reduced costs for maintenance and operations and improved power plant availability, while operating reliably and safely. In addition, science and technology for the safe storage and ultimate disposal of nuclear waste have been advanced. Waste from nuclear energy must be isolated from the environment. High-level nuclear wastes from fission reactors (used fuel assemblies) are stored in contained, reinforced concrete steel-lined pools or in robust dry casks at limited-access reactor sites, until a deep geologic repository is ready to accept and isolate the spent fuel from the environment. Used nuclear fuel contains a substantial quantity of fissionable materials, and advanced technologies may be able to recover energy from this

Figure 5-17. More than 400 successfully operating civilian nuclear power plants around the world avoid billions of tons of CO_2 emissions for the atmosphere.

spent fuel and reduce required repository space and the radiotoxicity of the disposed waste.

While the current application of nuclear energy is the production of electricity, other applications are possible, such as cogeneration of process heat, the generation of hydrogen from water or from methane (with carbon capture or integration with other materials production or manufacturing), and desalination.

Potential Role of Technology

The 103 currently operating U.S. nuclear-reactor units are saving as much as 600 million metric tons of carbon dioxide emissions every year. Through the summer of 2005, 39 of these units have received approval to extend their operating licenses for an additional 20 years; 12 others have applications under review. All of the remaining units most likely will follow suit. Such CO_2 emission mitigation could be increased if new nuclear capacity were to be brought online.

To the extent that the financial risks of new nuclear construction can be addressed and with improvement from new technologies in the longer term, the nuclear option can continue to be an important, growing part of a GHG-emissions-free energy portfolio. Design and demonstration efforts on near-term advanced reactor concepts—in combination with Federal financial risk mitigation tools—will enable power companies to build and operate new reactors that are economical and competitive with other generation technologies, supporting energy security and diversity of supply.

Nuclear Reactors Under Active Construction Worldwide

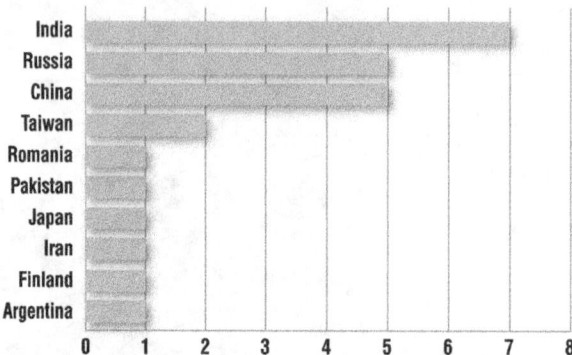

Figure 5-18. Nuclear Reactors Under Active Construction Worldwide (Source: World Nuclear Association[26])

Evolutionary light-water reactors of standardized design (having received U.S. Nuclear Regulatory Commission design certification and having been constructed on schedule in Japan and South Korea) are demonstrated and available now for construction in the United States. Other newer designs should be reviewed and certified over the next several years, making them also available. However, more advanced nuclear energy systems for the longer term have the potential to offer significant advances in the areas of sustainability, proliferation resistance and physical protection, safety, and economics. These advanced nuclear energy systems—described as Generation IV reactors—could replace or add to existing light-water reactor capacity.

Technology Strategy

U.S. leadership is essential to the expansion of nuclear capacity in markets other than Asia and Eastern Europe (Figure 5-18), through deployment of advanced nuclear power plants in the relatively near term. The new Federal regulatory processes for the siting, construction, and operation of new nuclear plants must be demonstrated. In addition, other major obstacles must be addressed, including the initial high capital costs of the first few plants and the business risks resulting from both the costs and the regulatory uncertainty.

In the longer term, advanced nuclear energy systems could serve a vital role in both diversifying the Nation's energy supply and reducing GHG emissions. By successfully addressing the fundamental research and development issues of system concepts that excel in safety, sustainability, cost-effectiveness, and

proliferation resistance, the systems could attract future private-sector sponsorship and ultimate commercialization by the private sector. Advanced nuclear fission-reactor systems aim to extract the full energy potential of the spent nuclear fuel from current fission reactors, while reducing or eliminating the potential for proliferation of nuclear materials and technologies, and reducing both the radiotoxicity and total amount of waste produced.

A key objective of nuclear energy research and development is to enhance the basic technology and, through advanced civilian technology research, chart the way toward the next leap in technology. From these efforts, and those of industry and overseas partners, nuclear energy may continue to fulfill its promise as a safe, advanced, inexpensive, and emission-free approach to providing reliable energy throughout the world.

Current Portfolio

The current Federal portfolio focuses on three areas:

◆ **Research on Nuclear Power Plant Technologies for Near-Term Deployment** is focused on advanced fission reactor designs that are currently available or could be made available with limited additional work to complete design development and deployment in the 2010 time frame.

A Roadmap to Deploy New Nuclear Power Plants in the United States by 2010, issued in October 2001 (DOE 2001), advises DOE on actions and resource requirements needed to put the country on a path to bringing new nuclear power plants on-line in the next decade. The primary purposes

---

[26] See http://www.world-nuclear.org/info/printable_information_papers/reactorsprint.htm.

of the roadmap are to identify the generic and design-specific prerequisites to near-term deployment, to identify those designs that best promise to meet the needs of the marketplace, and to propose recommended actions that would support deployment. These include, but are not limited to, actions to achieve economic competitiveness and timely regulatory approvals.

The Nuclear Power 2010 Program is a joint government/industry cost-shared effort. The program is designed to pave the way for an industry decision to order at least one new nuclear power plant by the end of the decade. Activities under this program support cost-shared demonstration of the Early Site Permit (ESP) and combined Construction and Operating License (COL) processes to reduce licensing uncertainties and minimize the attendant financial risks to the licensee. In addition, the program includes technology research and development to finalize and license a standardized advanced reactor design, which U.S. power-generation companies will find to be more competitive in the deregulated electricity market. The economics and business case for building new nuclear power plants has been evaluated as part of the Nuclear Power 2010 program to identify the necessary financial conditions under which power-generation companies would add new nuclear capacity.

The research program goals in this area are focused on successfully demonstrating the untested regulatory processes for ESP and combined COL, and on the regulatory acceptance (certification) and completion of first-of-a-kind engineering and design. Specific goals include industry decisions to build new nuclear power plants by 2009 with commercial operation in the next decade.[27]

◆ **Research under the Generation IV Nuclear Energy Systems Initiative (Gen IV)** will enhance the viability of advanced nuclear energy systems that offer significant advances in the areas of sustainability, proliferation-resistance, and physical protection, safety, and economics. These newer nuclear energy systems will replace or add to existing light-water reactor capacity and should be available between 2020 and 2030. To develop these advanced reactor systems, DOE manages the Gen IV.

Development of next-generation nuclear energy systems is being pursued by the Generation IV International Forum (GIF) [www.gen-4.org], a

group of 10 leading nuclear nations (Argentina, Brazil, Canada, France, Japan, the Republic of Korea, the Republic of South Africa, Switzerland, the United Kingdom, and the United States) plus the European Atomic Energy Community (Euratom). The GIF has selected six promising technologies as candidates for advanced nuclear energy systems concepts. The Gen IV addresses the fundamental research and development issues necessary to establish the viability of next-generation nuclear energy system concepts. After successfully addressing the viability issues, the systems are highly likely to attract future private-sector sponsorship and ultimate commercialization by the private sector.

The primary focus of these Gen IV systems will be to generate electricity in a safe, economical, and secure manner; other possible benefits include the production of hydrogen, desalinated water, and process heat (Figure 5-19). In particular, making nuclear power sustainable over the long term requires developing and deploying proliferation-resistant fuel recycling technology and fast reactors for breeding new fuel and transmuting higher actinides. The GIF and DOE's Nuclear Energy Research Advisory Committee (NERAC) issued a report on its two-year effort to develop a technology roadmap for future nuclear energy systems (GIF-NERAC 2002). The technology roadmap defines and plans the necessary R&D to support the advanced nuclear energy systems known as Gen IV. DOE also prepared a report to the U.S. Congress regarding how it intends to carry out the results of the Gen IV Roadmap (DOE-NE 2003a).

Goals for next-generation fission energy systems (Gen IV) research are focused on the design of reactors and fuel cycles that are safer, more economically competitive, more resistant to proliferation, produce less waste, and make better use of the energy content in uranium, in accord with the above-mentioned reports and roadmaps.[28]

◆ **The Advanced Fuel Cycle Initiative (AFCI),** under the leadership of DOE, is focused on developing advanced fuel-cycle technologies, which include spent fuel treatment, advanced fuels, and transmutation technologies, for application to current operating commercial reactors and next-generation reactors; and to inform a recommendation by the Secretary of Energy in the 2007-2010 time frame on the need for a second geologic repository.

---

[27] See Section 2.4.2 (CCTP 2005): http://www.climatetechnology.gov/library/2005/tech-options/tor2005-242.pdf.

[28] See Section 2.4.1 (CCTP 2005): http://www.climatetechnology.gov/library/2005/tech-options/tor2005-241.pdf.

Future Nuclear Power Concepts

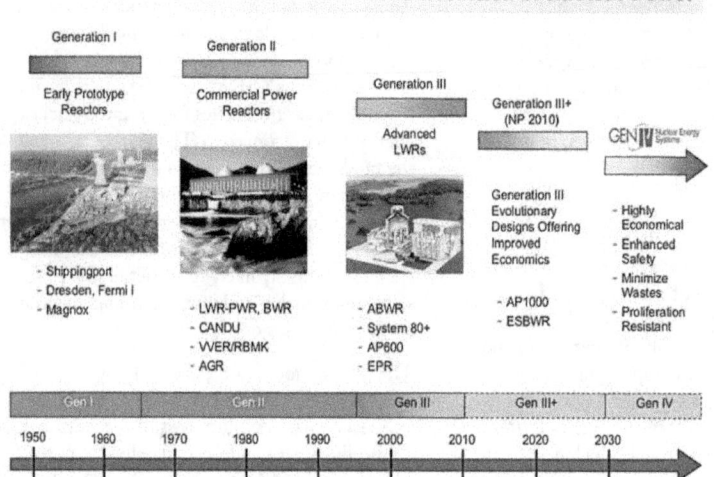

Figure 5-19. Future Nuclear Power Concepts (Source: DOE, Office of Nuclear Energy Internal Document).

The AFCI program will develop technologies to address intermediate and long-term issues associated with spent nuclear fuel. The intermediate-term issues are the reduction of the volume and heat generation of material requiring geologic disposal. The program will develop proliferation-resistant processes and fuels for application to current light-water reactor systems and Gen IV reactor systems to enable the energy value of these materials to be recovered, while destroying significant quantities of plutonium. This work provides the opportunity to optimize use of the Nation's first repository and reduce the technical need for an additional repository. The longer-term issues to be addressed by the AFCI program are the development of fuel-cycle technologies to destroy minor actinides, which would greatly reduce the long-term radiotoxicity and heat load of high-level waste sent to a geologic repository. This will be accomplished through the development of Gen IV fast reactor fuel-cycle technologies and possibly of accelerator-driven systems (DOE-NE 2003b).

Goals for advanced nuclear fuel-cycle research focus on proving design principles of spent-fuel treatment and transmutation technologies, demonstrating the fuel and separation technologies for waste transmutation, and deploying Gen IV advanced fast spectrum reactors that can transmute nuclear waste.[29]

Future Research Directions

The current portfolio supports the main components of the technology development strategy and addresses the highest priority current investment opportunities in this technology area. For the future, CCTP seeks to consider a full array of promising technology options. From diverse sources, suggestions for future research have come to CCTP's attention. Some of these, and others, are currently being explored and under consideration for the future R&D portfolio. These include:

◆ **Reducing the financial risks and costs of advanced nuclear power plants.** Provide for development and demonstration of advanced technologies to reduce construction time for new nuclear power plants and to minimize schedule uncertainties and associated costs for construction.

◆ **Next Generation nuclear plants and advanced fuel cycles.** Develop reactor designs that excel in operational safety, sustainability, and proliferation-resistance, and achieve economy of both capital and operation and maintenance (O&M), including advanced fuels that are more resistant to melting and are operated to minimize wastes. Included in these designs are small, exportable reactors, which can be used by reactor user nations under the Global Nuclear Energy Partnership that will not develop indigenous fuel cycle capabilities. Primary research needs are for development and characterization of fuel fabrication processes; testing of in-reactor fuel performance; power

[29] See Section 2.4.3 (CCTP 2005): http://www.climatetechnology.gov/library/2005/tech-options/tor2005-243.pdf.

conversion-system design and testing, including resolution of uncertainties regarding materials, reliability, and maintainability; and fission-reactor internal design and verification.

◆ **High-temperature materials and heat transfer technology.** Develop and test high-temperature materials (central to future nuclear reactor systems). Low-activation materials that resist corrosion and embrittlement at high temperatures are needed for structural materials within the core and fuel assemblies and for reactor pressure vessel systems. Advanced materials are needed for future heat exchangers, turbine components, radiochemical particulate filters, and transport of heat to hydrogen production plants. Other research needs include the use of alternate working fluids and heat-exchange cycles.

◆ **Closing the fuel cycle.** Investigate proliferation-resistant, closed fuel-cycle concepts that reduce the quantity and heat load of wastes requiring geological emplacement. Compared to other industrial waste, the spent nuclear fuel generated during the production of electricity is relatively small in quantity. However, it is highly radioactive for many thousands of years, and its disposal requires resolution of many political, societal, technical, and regulatory issues. While these issues are being addressed in the license application for the Yucca Mountain repository in Nevada, several countries worldwide have pursued advanced technologies that could treat and transmute spent nuclear fuel from nuclear power plants, such as advanced burner reactors. These technologies have the potential to dramatically reduce the quantity and toxicity of waste requiring geologic disposal. During the past four years, the United States has joined this international effort and found considerable merit in this area of joint advanced research.

◆ **Nuclear computations.** Use the emerging DOE supercomputer system capabilities to combine more rigorous computational models of 3-dimensional radiation transport, thermal transport, fluid flow and more detailed accounting of energy resonances in nuclear cross-section data into powerful new tools for exploring novel reactor concepts. Although the current generation of nuclear computational techniques is adequate for design and evaluation of current nuclear reactors, the computer models use many approximate techniques that ultimately limit the ability to design innovative nuclear fuels and core geometries.

5.5 Fusion Energy

Fusion energy holds the possibility of an almost inexhaustible supply of zero-GHG electricity. Fusion is the power source of the sun and the stars. Lighter elements are "fused" together in the core of the sun, producing heavier elements and prodigious amounts of energy. On Earth, fusion energy has been demonstrated in the laboratory at powers of 5 to 15 million watts, with pulse lengths in the range of 1 to 5 seconds. The goal is for fusion power to eventually be produced at much larger scales and in longer pulse lengths.

Fusion power generation offers a number of advantageous features. The basic sources of fusion fuel, deuterium and tritium, are actually heavy forms of hydrogen. Deuterium is abundantly available because it occurs naturally in water; and tritium can be derived from lithium, a light metal found in the earth's crust. Tritium is radioactive, but the quantities in use at any given time are quite modest and can be safely handled. There are no chemical pollutants or CO_2 emissions from the fusion process. With appropriate advances in materials, the radioactivity of the fusion byproducts would be relatively short-lived, thereby obviating the need for extensive waste management measures.

From a safety perspective, the fusion process poses little radiation risk to anyone outside the facility. Also, since only a small quantity of fuel is in the fusion system at any given time, there is no risk of a critical accident or meltdown, and little after-heat to be managed in the event of an accident. The potential usefulness of fusion systems is great, but significant scientific and technical challenges remain.

Potential Role of Technology

Fusion energy is an attractive option to consider for long-term sustainable energy generation. It would be particularly suited for baseload electricity supply, but could also be used for hydrogen production. With the growth of the world's population expected to occur in cities and megacities, concentrated energy sources (such as fusion energy) that can be located near population centers may be particularly attractive. In addition, the fusion process does not produce GHGs and has attractive inherent safety and environmental characteristics that could help gain public acceptance.

Energy scenarios imposing reasonable constraints on nonsustainable energy sources show that fusion energy could contribute significantly to large-scale electricity production during the second half of the 21st century.

Making fusion energy a part of the future energy solution is among the most ambitious scientific and engineering challenges of our era. The following are some of the major technical questions that need to be answered:

◆ Can burning plasma that shares the characteristic intensity and power of the sun be successfully produced and sustained?

◆ To what extent can models be used to simulate and predict the behavior of the burning, self-sustained fuel required for fusion applications?

◆ How can new materials that can survive the fusion environment (which are needed for fusion power to be commercially viable) be developed?

Answering these questions requires understanding and control of complex and dynamic phenomena occurring across a broad range of temporal and spatial scales. The experiments required for a commercially viable fusion power technology constitute a complex scientific and engineering enterprise.

Technology Strategy

Given the substantial scientific and technological uncertainties that now exist, the U.S. Federal Government will continue to employ a portfolio strategy that explores a variety of magnetic confinement approaches and leads to the most promising commercial fusion concept. Advanced computational modeling will be central to testing the agreement between theory and experiment, simulating experiments that cannot be readily investigated in the laboratory, and exploring innovative designs for fusion plants. To ensure the highest possible scientific return, DOE's Fusion Energy Sciences (FES) program will extensively engage with and leverage other DOE programs and international programs in areas such as magnetic confinement physics, materials science, ion beam physics, and high-energy-density physics. Large-scale experimental facilities may be necessary, and the rewards, risks, and costs of these major facilities will need to be shared through international collaborations. The target physics aspect of inertial fusion is being conducted now through the National Nuclear Security Administration's (NNSA) stockpile stewardship program. The overall FES effort will be organized around a set of four broad goals.

FUSION ENERGY SCIENCES GOAL #1:
Demonstrate with burning plasmas the scientific and technological feasibility of fusion energy. The goal is to demonstrate sustained, self-heated fusion plasma, in which the plasma is maintained at fusion temperatures by the reaction products, a critical step to practical fusion power. The strategy includes the following area of emphasis:

◆ Participate in the international magnetic fusion experiment, ITER (Latin for "the way") project, with the European Union, Japan, Russia, China, South Korea, India, and perhaps others as partners.

FUSION ENERGY SCIENCES GOAL #2:
Develop a fundamental understanding of plasma behavior sufficient to provide a reliable predictive capability for fusion energy systems. Basic research is required in turbulence and transport, nonlinear behavior and overall stability of confined plasmas, interactions of waves and particles in plasmas, the physics occurring at the wall-plasma interface, and the physics of intense ion beam plasmas and high-energy-density plasmas. The strategy includes the following areas of emphasis:

◆ Conduct fusion science research through individual-investigator and research-team experimental, computational, and theoretical investigations

◆ Advance the state-of-the-art computational modeling and simulation of plasma behavior in partnership with the Advanced Scientific Computing Research program in DOE's Office of Science

◆ Support basic plasma science, partly with the National Science Foundation, connecting both experiments and theory with related disciplines such as astrophysics.

FUSION ENERGY SCIENCES GOAL #3:
Determine the most promising approaches and configurations to confining hot plasmas for practical fusion energy systems. The strategy includes experiments and advanced simulation and modeling; innovative magnetic confinement configurations, such as the National Spherical Torus Experiment (NSTX); and a planned compact stellarator experiment, the National Compact Stellarator Experiment (NCSX) at Princeton Plasma Physics Laboratory (PPPL); as well as smaller experiments at multiple sites.

FUSION ENERGY SCIENCES GOAL #4:
Develop the new materials, components, and technologies necessary to make fusion energy a reality. The environment created in a fusion reactor poses great challenges to materials and components. Materials must be able to withstand high fluxes of

high-energy neutrons and endure high temperatures and high thermal gradients, with minimal degradation. The strategy includes the following areas of emphasis:

◆ Design materials at the molecular scale to create new materials that possess the necessary high-performance properties, leveraging investments in fusion energy research with investments in basic materials research

◆ Explore "liquid first-wall" materials to ameliorate first-wall requirements for advanced fusion energy concepts.

Current Portfolio

The current FES program, within DOE's Office of Science, is a program of fundamental research into the nature of fusion plasmas and the means for confining plasma to yield energy. This includes: (1) exploring basic issues in plasma science; (2) developing the scientific basis and computational tools to predict the behavior of magnetically confined plasmas; (3) using the advances in tokamak[30] research to enable the initiation of the burning plasma physics phase of the FES program; (4) exploring innovative confinement options that offer the potential of more attractive fusion energy sources in the long term; (5) developing the cutting-edge technologies that enable fusion facilities to achieve their scientific goals; and (6) advancing the science base for innovative materials to establish the economic feasibility and environmental quality of fusion energy.

The overall effort requires operation of a set of unique and diversified experimental facilities, ranging from smaller-scale university programs to several large national facilities that require extensive collaboration. These facilities provide scientists with the means to test and extend theoretical understanding and computer models, leading ultimately to an improved predictive capability for fusion science.

The two major tokamak experiments, DIII-D at General Atomics and the Alcator C-Mod at MIT, are extensively equipped with sophisticated diagnostics that allow for very detailed measurements in time and spatial dimensions as they continuously push the frontiers of tokamak plasma confinement. They each involve an array of national and international collaborators on the scientific programs.

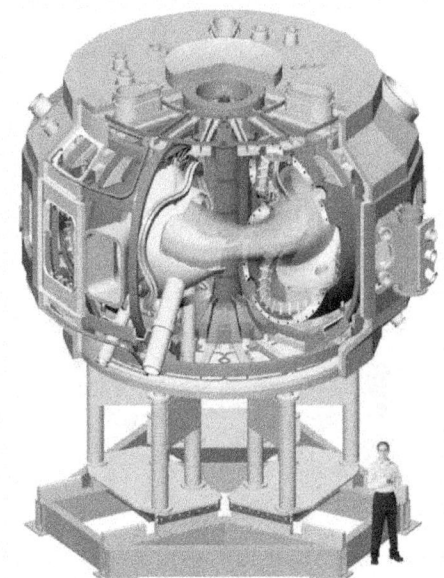

Figure 5-20. Fusion energy affords the possibility of long-term, sustainable energy supply, with little or no emissions of GHGs.
Credit: National Compact Stellarator Experiment, DOE/PPPL

Similarly, the NSTX at PPPL is also a well-diagnosed and highly collaborative experiment on an innovative confinement approach that seems likely to lead to improved understanding of toroidal[31] confinement systems.

An additional innovative concept, the National Compact Stellarator Experiment, is currently being fabricated at PPPL with first operation scheduled for 2009 (Figure 5-20). This machine is a product of new computational capabilities that have optimized the 3-dimensional toroidal magnetic geometry for improved confinement and stability in a compact form.

In addition to these major experiments, there are a larger number of smaller magnetic confinement experiments with more specialized missions. These are generally located at universities and provide an opportunity for student training.

A modest-scale high-energy-density physics program is also underway, with an emphasis on using heavy ion drivers to explore plasma/beam dynamics and warm dense matter with potential applications to future inertial fusion systems. This program also explores innovative approaches to improving inertial fusion such as the fast-ignition experiments. In addition, the FES program benefits from existing experimental programs conducted elsewhere for NNSA's stockpile

---

[30] Tokamak (Acronym created from the Russian words, "TOroidalnaya KAmera ee MAgnitnaya Katushka," or "Toroidal Chamber and Magnetic Coil"): The tokamak is the most common research machine for magnetic confinement fusion today.

[31] Toroidal: in the shape of a torus, or doughnut. Toroidal is a general term that refers to toruses as opposed to other geometries (e.g., tokamaks and stellarators are examples of toroidal devices).

ITER International Magnetic Fusion Experiment

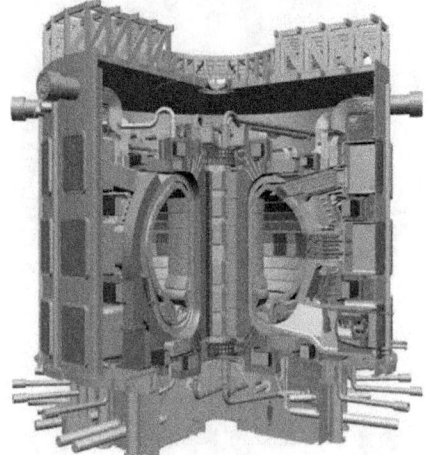

Figure 5-21. *The United States is participating with other countries in the building of ITER, a large-scale fusion process machine, which will provide experimental data to inform future technology directions.*
Credit: ITER

stewardship program and the Department of Defense (DoD). Both the "Z" experiment at Sandia National Laboratories and the OMEGA experiment at the University of Rochester, for example, offer opportunities for improving understanding of high-energy-density physics.

Theory and computing are key parts of the present program, as they provide the intellectual framework for the overall approach to fusion energy, as well as the computer codes, which attempt to systematically rationalize the understanding of fusion plasmas.[32]

Future Research Directions

For the future, CCTP seeks to consider a full array of promising technology options. From diverse sources, suggestions for future research have come to CCTP's attention. Some of these, and others, are currently being explored and under consideration for the future R&D portfolio. These include:

◆ **ITER.** Burning plasmas represent the next major science and technology frontier for fusion research. In the major international effort mentioned above (ITER), the United States, Europe, Japan, China, Russia, the Republic of Korea, and India plan to construct a magnetic

fusion burning plasma science and engineering test facility. The ITER international magnetic fusion experiment is a key part of the U.S. strategy to investigate the underlying science for magnetic confinement fusion energy (MFE), (Figure 5-21). Additional investments in fusion materials, components, and technologies for MFE are contingent upon favorable results from ITER.

◆ **Plasma Confinement Systems.** Prior to the anticipated operation of ITER around 2015, physics research experiments on a wide range of plasma-confinement systems worldwide will continue in preparation for ITER operations. These experiments will include detailed simulations of ITER behavior as well as innovative new ways of operating fusion systems to optimize efficiency. Because of the sophisticated measurement techniques employed on modern fusion experiments, detailed data are already available to validate computer models.[33] Work will also continue on confinement configuration optimization that would allow better understanding or improve the confinement approach for future power systems.

◆ **High-Energy-Density Physics and Related Technology.** In other efforts, the United States is proceeding with high-energy-density physics, the science base for inertial fusion[34], through the development of NNSA's National Ignition Facility (NIF) and other work, including driver, target fabrication, and chamber technologies. The drivers include lasers and pulsed power-driven z-pinches in the NNSA program and heavy ion accelerators. Efforts to explore the understanding and predictability of high-energy-density plasma physics, including the ramifications for energy-producing applications, are also underway.[35] However, any additional investment in the inertial fusion energy approach awaits successful demonstration of ignition and gain in the NIF.

5.6 Summary

This chapter reviews various forms of advanced technology, their potential for reducing emissions from energy supply, and the R&D strategies intended to accelerate their development. Although

---

[32] See Section 2.5.1 (CCTP 2005): http://www.climatetechnology.gov/library/2005/tech-options/tor2005-251.pdf.

[33] For additional information about ITER, see http://www.iter.org/.

[34] Inertial fusion is an alternative approach to creating the physical conditions necessary for light elements to fuse and release binding energy, in which an array of intense x-rays, lasers, or heavy ion beams are used to contain, compress, and heat the fusible materials to fusion enabling conditions.

[35] For additional information about U.S. fusion energy research, see http://www.ofes.fusion.doe.gov/.

uncertainties exist about both the level at which GHG concentrations might need to be stabilized and the nature of the technologies that may come to the fore, the long-term potential of advanced energy supply technologies is estimated to be significant, both in reducing emissions (as shown in Figure 3-19 and highlighted in the figure at the beginning of this chapter) and in reducing the costs for achieving those reductions, as suggested by Figure 3-14. Further, the advances in technology development needed to realize this potential, as modeled in the associated analyses, animate the research and development goals for each energy supply technology area.

As one illustration among the many hypothesized cases analyzed,[36] when GHG emissions were constrained to a high level over the course of the 21st century the lowest-cost arrays of advanced technology in energy supply, when compared to a reference case, resulted in reduced or avoided emissions of roughly between 110 and 210 GtC. This amounted to, roughly, between 20 and 35 percent of all GHG emissions reduced, avoided, captured and stored, or otherwise withdrawn and sequestered, needed to constrain emissions at this level. Similarly, the costs for achieving such emissions reductions, when compared to the reference case, were reduced by

Technologies for Goal #2: Reducing Emissions from Energy Supply

	NEAR-TERM	MID-TERM	LONG-TERM
Fossil Power	• IGCC Commercialization • FutureGen Demonstration • Solid Oxide Fuel Cells • More Efficient, Lower-Cost, Cleaner Coal Plants	• Pre-Combustion Technology for Cleaner Coal-Based Electricity Generation • Zero-Emission Coal Plants (FutureGen) • H_2 Co-Production from Coal/Biomass	• Zero-Emission Fossil Energy
Hydrogen	• Integrated Stationary Fuel Cell System • Codes & Standards • Demonstrations of Renewable Hydrogen Production	• Low-Cost H_2 Storage & Delivery • H_2 Production from Nuclear • H_2 Production from Renewables • Renewable-H_2-Powered Fuel Cell Vehicles	• H_2 & Electric Economy
Renewables	• Lower-Cost Wind Power • Biodiesel, Demos of Cellulosic Ethanol • Photovoltaics on Buildings • Cost-Competitive Solar PV • 1st Generation Biorefinery • Distributed Generation Systems	• Low-Wind-Speed Turbines • Advanced Biorefineries • Cellulosic Biofuels • Community-Scale Solar • Photolytic Water Splitting • Energy Storage Options	• Widespread Renewable Energy • Bio-Engineered Biomass • Bio-inspired Energy & Fuels
Nuclear Fission	• Advanced Fission Reactor and Fuel Cycle Technology • New Fuel Forms and Materials	• GenIV Nuclear Plants • Closed Proliferation-Resistant Fuel Cycles • Minimization of Wastes Requiring Geological Disposal	• Widespread Nuclear Power • Advanced Concepts for Waste Reduction
Fusion Power	• Greater Understanding of Plasmas • Demonstration of Burning Plasmas (ITER) • Identification of Technology Options • Understand Potential of High-Energy-Density Physics Research	• Fusion Pilot Plant Demonstration	• Fusion Power Plants

Figure 5-22. Technologies for Goal #2: Reducing Emissions from Energy Supply (Note: Technologies shown are representations of larger suites. With some overlap, "near-term" envisions significant technology adoption by 10 to 20 years from present, "mid-term" in a following period of 20-40 years, and "long-term" in a following period of 40-60 years. See also List of Acronyms and Abbreviations.)

[36] In Chapter 3, various advanced technology scenarios were analyzed for cases where global emissions of GHGs were hypothetically constrained. Over the course of the 21st century, growth in emissions was assumed to slow, then stop, and eventually reverse in order to ultimately stabilize GHG concentrations in the Earth's atmosphere at levels ranging from 450 to 750 ppm. In each case, technologies competed within the emissions-constrained market, and the results were compared in terms of energy (or other metric), emissions, and costs.

roughly a factor of 3. See Chapter 3 for other cases and other scenarios.

As described in this chapter, CCTP's technology development strategy supports achievements in this range. The overall strategy is summarized schematically in Figure 5-22. Advanced technologies are seen entering the marketplace in the near, mid, and long terms, where the long term is sustained indefinitely. Such a progression, if successfully realized worldwide, would be consistent with attaining the energy supply potential portrayed at the beginning of this chapter.

The timing and pace of technology adoption are uncertain and must be guided by science. In the case of the illustration above, the first GtC per year (1GtC/year) of reduced or avoided emissions, as compared to an unconstrained reference case, would need to be in place and operating between 2040 and 2060. For this to happen, a number of new or advanced energy supply technologies would need to penetrate the market at significant scale before this date. Other cases would suggest faster or slower rates of deployment, depending on assumptions. See Chapter 3 for other cases and other scenarios.

References

Energy Information Administration (EIA). 2006. *Annual Energy Outlook 2006*, DOE/EIA-0383(2006). Washington, DC: U.S. Department of Energy.

Generation IV International Forum (GIF) and the U.S. Department of Energy's Nuclear Energy Research Advisory Committee (NERAC). 2002. *A Technology Roadmap for Generation IV Nuclear Energy Systems*. GIF-002-01, December. www.gen-4.org/Technology/roadmap.htm

International Energy Agency (IEA). 2004. *Key World Energy Statistics 2004*. http://library.iea.org/dbtw-wpd/Textbase/nppdf/free/2004/keyworld2004.pdf

National Research Council (NRC), National Academy of Engineering (NAE). 2004. *The Hydrogen Economy: Opportunities, Costs, Barriers, and R&D Needs*. Washington, DC: The National Academies Press. http://www.nap.edu/books/0309091632/html/

Oak Ridge National Laboratory (ORNL). 2005. *Biomass as Feedstock for a Biorefinery and Bioproducts Industry: The Technical Feasibility of a Billion-Ton Supply*. http://feedstockreview.ornl.gov/pdf/billion_ton_vision.pdf

U.S. Climate Change Technology Program (CCTP). 2003. *Technology Options for the Near and Long Term*. DOE/PI-0002. Washington, DC: U.S. Department of Energy. http://www.climatetechnology.gov/library/2003/tech-options/index.htm Update is at http://www.climatetechnology.gov/library/2005/tech-options/index.htm

U.S. Department of Energy (DOE). 2005. *Hydrogen Fuel Cells and infrastructure Technologies Multi-Year Research, Development and Demonstration Plan, 2003-2010*. Washington, DC: U.S. Department of Energy. http://www.eere.energy.gov/hydrogenandfuelcells/mypp/

U.S. Department of Energy (DOE). 2004. *Hydrogen Posture Plan, an Integrated Research, Development, and Demonstration Plan*. Washington, DC: U.S. Department of Energy. http://www.eere.energy.gov/hydrogenandfuelcells/pdfs/hydrogen_posture_plan.pdf

U.S. Department of Energy (DOE), Office of Nuclear Energy Science and Technology (NE). 2001. *A Roadmap to Deploy New Nuclear Power Plants in the United States by 2010*. Washington, DC: U.S. Department of Energy. http://nuclear.gov/nerac/ntdroadmapvolume1.pdf and http://nuclear.gov/nerac/NTDRoadmapVolII.PDF

U.S. Department of Energy, (DOE), Office of Nuclear Energy Science and Technology (NE). 2003a. *The U.S. Generation IV Implementation Strategy*. Washington, DC: U.S. Department of Energy. http://nuclear.gov/reports/Gen-IV_Implementation_Plan_9-9-03.pdf

U.S. Department of Energy, (DOE), Office of Nuclear Energy Science and Technology (NE). 2003b. Report to Congress on Advanced Fuel Cycle Initiative: the Future Path for Advanced Spent Fuel Treatment and Transmutation Research. Washington, DC: U.S. Department of Energy. http://nuclear.gov/reports/AFCI_CongRpt2003.pdf

Capturing and Sequestering Carbon Dioxide

echnologies and improved management systems for carbon dioxide (CO_2) capture, storage, and sequestration can potentially reduce CO_2 emissions significantly and help slow the growth in atmospheric CO_2 concentrations. The relative significance of this potential is suggested by Figure 3-19 and highlighted in the figure at right, which draws insights from one set of scenarios analyses that explored various ways to reduce emissions through a suite of these kinds of technologies.

Energy supply technologies incorporating carbon capture and storage were found capable of contributing significantly to future near-zero or very low emissions energy supply. When combined with other sequestration technologies capable of capturing CO_2 from the atmosphere, reduced, avoided, or sequestered global carbon emissions, compared to a reference case, and depending upon assumptions, ranged from low amounts up to nearly 300 gigatons of carbon (GtC) over the course of the 21ˢᵗ century. Although bracketed by a number of uncertainties, this range suggests both the potential role for advanced technology and a long-term goal for contributions from this area in the future global economy.

The three main focus areas for R&D related to carbon cycle management include: (1) the capture of CO_2 emissions from large point sources, such as coal-based power plants, oil refineries, and industrial processes, coupled with storage in geologic formations or other storage media; (2) enhanced carbon uptake and storage by terrestrial biotic systems—terrestrial sequestration; and (3) improved understanding of the potential for ocean storage and sequestration methodologies.[1]

If current world energy production and consumption patterns persist into the foreseeable future, fossil fuels will remain the mainstay of global energy production well into the 21ˢᵗ century. The Energy Information Administration (EIA) projects that by 2025, about 88 percent of global energy demand will be met by fossil fuels, because fossil fuels will likely continue to yield competitive advantages relative to other alternatives (EIA 2004a). In the United States, the use of fossil fuels in the electric power industry accounted for 39

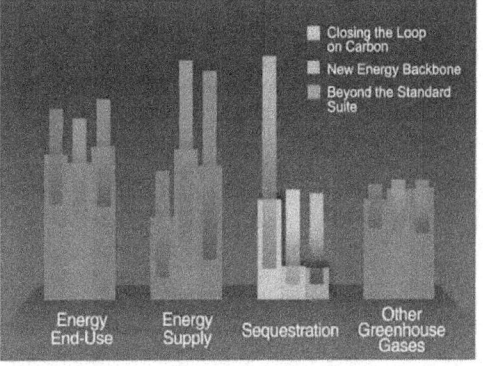

Sequestration
Potential Contributions to Emissions Reduction

Potential contributions of Carbon Capture, Storage, and Sequestration to cumulative GHG emissions reductions to 2100, across a range of uncertainties, for three advanced technology scenarios. See Chapter 3 for details.

percent of total energy-related CO_2 emissions in 2003, and this share is expected to slightly increase to 41 percent in 2025. In 2025, coal is projected to account for 50 percent of U.S. electricity generation and for an estimated 81 percent of electricity-generated CO_2 emissions. Natural gas is projected to account for 24 percent of electricity generation and about 15 percent of electricity-related CO_2 emissions in 2025 (EIA 2005).

Many scenarios of the future suggest that world coal markets will continue to grow steadily over the course of the 21ˢᵗ century, in the absence of CO_2 emissions restrictions. While increased energy efficiency, and use of renewable and nuclear energy afford good

[1] In this Plan, the three approaches are collectively referred to as "capturing and sequestering carbon dioxide" or "capturing and sequestering carbon."

opportunities for reducing CO_2 emissions, fossil fuel reserves are abundant and economical, making their continued use an attractive option. In various advanced technology scenarios where CO_2 capture and storage technology were assumed to become a cost-competitive technology strategy, fossil-based energy continued to supply a large portion of total electricity consumed into the future (e.g., various studies estimated a 55-70 percent share), even under high carbon management requirements.

Human activities related to land conversion and agricultural practices have also contributed to the buildup of carbon dioxide to the atmosphere. During the past 150 years, land use and land-use changes were responsible for one-third of all human emissions of CO_2 (IPCC 2000). Over the next 100 years, global land-use change and deforestation are likely to account for at least 10 percent of overall human-caused CO_2 emissions. The dominant drivers of current and past land-use-related emissions of CO_2 are the conversion of forest and grassland to crop and pastureland and the depletion of soil carbon through agricultural and other land-management practices (IPCC 2000). Past CO_2 emissions from land-use activities are potentially reversible, and improved land-management practices can actually restore depleted carbon stocks. Therefore, there are potentially large opportunities to increase terrestrial carbon sequestration.

The potential storage and sequestration capacity for CO_2 in various "sinks" is large. Some estimates indicate that about 83 to 131 gigatons of carbon (GtC) could be sequestered in forests and agricultural soils by 2050 (IPCC 2001b), while others estimate geologic storage capacities within a broad range of 300 to 3,200 GtC (IEA 1994a, 1994b, 2000). The ocean represents the largest potential sink for anthropogenic CO_2. Analysis indicates that the ocean is currently absorbing passively some 7.3 Gt of excess atmospheric CO_2 per year (Sabine et al. 2004), partially offsetting the impact on atmospheric concentrations of CO_2 from annual anthropogenic emissions of CO_2 of about 25 Gt per year. The potential storage capacity of the ocean is largely unknown, although some researchers estimate that it might hold thousands of GtC or greater (Herzog 2001, Smith and Sandwell 1997, Hoffert et al. 2002).

There are potential ancillary benefits associated with carbon capture, storage, and sequestration. Many land-management practices that sequester carbon can improve water quality, reduce soil erosion, and benefit wildlife. The injection of CO_2 into geologic structures can be beneficially used to enhance recovery of oil from depleted oil reservoirs and the

recovery of methane from unmineable coal seams.

Carbon capture, storage, and sequestration technologies have become a high-priority R&D focus under CCTP because they hold the potential to reduce CO_2 emissions from point sources, as well as from the atmosphere, and to enable continued use of coal and other fossil fuels well into the future. Near-term R&D opportunities include optimizing carbon sequestration and management technologies and practices in terrestrial systems, and accelerating the development of technologies for capturing and geologically storing CO_2 for enhanced oil recovery (EOR). Longer-term R&D opportunities include further development of other types of geologic storage and terrestrial sequestration options, as well as furthering the understanding of both the role oceans might play in storing carbon and the potential consequences of using the oceans for carbon sequestration.

In 2005, the Intergovernmental Panel on Climate Change (IPCC) released its *Special Report on Carbon Dioxide Capture and Storage* (IPCC 2005). While this report is not focused on future R&D options, it serves as an authoritative reference on the state-of-the-art methods in CO_2 capture and storage.

The remaining sections in this chapter summarize the current and potential future research activities and challenges associated with developing carbon sequestration technology. In each section, the description of the current R&D activities includes a hyperlink to the CCTP report, *Technology Options in the Near and Long Term* (CCTP 2005).

6.1 Carbon Capture

Point source CO_2 emissions from power plants vary depending on the combustion fuel, technology, and operational use. Concentrating and capturing CO_2 from flue gas is a technological challenge. Flue gas from conventional coal-fired power plants contains 10 to 12 percent of CO_2 by volume, and flue gas from integrated gasification combined cycle (IGCC) plants contains between 5 and 15 percent CO_2. For a combined cycle gas turbine system, the CO_2 concentration is about 3 percent. The CO_2 in flue gases must be concentrated to greater than 90 percent for most storage, conversion, or reuse applications. Thus, R&D programs are targeted at capture systems that can produce a concentrated and pressurized stream of CO_2 at relatively low cost.

Potential Role of Technology

Large CO_2 point sources, such as power plants, oil refineries, cement plants, and other industrial facilities are considered the most viable sites for CO_2 capture. The current technology for CO_2 capture uses a class of chemical absorbents called amines that remove CO_2 from the gas stream and produce byproduct food-grade CO_2 often used in carbonated soft drinks and other foods. However, the current absorbent process is costly and energy intensive, increasing the cost of a coal-fired plant by 50 to 80 percent (Davison et al. 2001) and energy reductions on the order of 30 percent of the net power generation rate (DOE 1999). Thus, several R&D opportunities are being pursued to reduce CO_2 capture costs and lessen the energy reductions in power generation, or the "net energy penalty."

Technology Strategy

Realizing the possibilities for point source CO_2 capture employs a research portfolio that covers a wide range of technology areas, including post-combustion capture, oxy-fuel combustion, and pre-combustion decarbonization. R&D investments in technologies that use pure oxygen during combustion, pre-combustion de-carbonization technologies, regenerable sorbents, advanced membranes, and hydrate formation can potentially reduce costs, as well as the net energy penalty. After component performance evaluations are completed, the next short-term step would be to conduct pilot scale and slip stream (i.e., diversion of a small stream from the total emissions of an existing plant) level testing of the most promising capture technologies. Larger or full-scale tests might be appropriate within the next few decades to demonstrate and have a suite of capture technologies available for deployment. Fully integrated capture and storage system demonstration (i.e., FutureGen) helps to enable commercial deployment to mitigate the financial and technical performance risks associated with any new technology that must maintain a high availability, such as required by the power generation sector.

Current Portfolio

The metrics and goals for CO_2 capture research are focused on reducing the cost and energy penalty, because analysis shows that CO_2 capture drives the cost of sequestration systems. Similarly, the goals and

metrics for carbon storage, measurement, and monitoring are focused on ensuring permanence and safety. All three research areas work toward the overarching program goal of 90 percent CO_2 capture, with 99 percent storage permanence at less than 20 percent increase in the cost of energy services by 2007, and less than 10 percent by 2012. A large-scale demonstration (i.e., FutureGen) would still be necessary.

Across the current Federal portfolio, agency activities are focused on a wide range of technical issues.[2]

◆ For new construction or re-powering of existing coal-fired power plants, there are pre-combustion decarbonization technology options that provide a pure stream of CO_2 as well as hydrogen at relatively low incremental cost. The most promising option, and the primary focus of current R&D, is **gasification**, in which the hydrocarbon is partially oxidized, causing it to break up into hydrogen (H_2), carbon monoxide (CO), and CO_2, and possibly some methane and other light hydrocarbons. The CO can be reacted with water to form H_2 and CO_2, and the CO_2 and H_2 can be separated. The H_2 used in a combustion turbine or fuel cell, and the CO_2 can be stored.

◆ New technologies to reduce the capital and energy penalty costs for **post-combustion capture** are also currently under development and include regenerable sorbents, advanced membranes, and novel concepts. One such novel concept, forming CO_2 hydrates to facilitate capture, could be especially attractive for advanced coal conversion systems like the IGCC. A challenge for post-combustion capture is the large amount of gas that must be processed per unit of CO_2 captured. This is especially true for combustion turbines where the concentration of CO_2 in the flue gas can be as low as 3 percent. One area of research is developing gas/liquid contactors where CO_2 gas is chemically absorbed into a liquid, and the resulting mixture is then separated.

◆ **Oxygen-fired combustion** is also being researched for large CO_2 point sources to determine if CO_2 can be recovered at reasonable cost. In oxygen-fired combustion, oxygen, instead of air, is used in combustion of petroleum coke, coal, or biomass fuels. Oxygen-fired combustion may also be implemented in power systems in which gaseous fuels are combusted with oxygen in the presence of recycled water to produce a steam/CO_2 turbine drive gas. Water is condensed

---

2 See Section 3.1.1 (CCTP 2005): http://www.climatetechnology.gov/library/2005/tech-options/tor2005-311.pdf.

WEYBURN II CO_2 STORAGE PROJECT

DOE is participating in this commercial-scale project that is using CO_2 for EOR. CO_2 is being supplied to the oil field in southern Saskatchewan, Canada, via a 320 kilometer pipeline from a North Dakota coal gasification facility. The goal is to determine the performance and undertake a thorough risk assessment of CO_2 storage in conjunction with its use in EOR. The project will include extensive above and below ground CO_2 monitoring.

from the steam/CO_2 exiting the turbine, leaving sequesterable CO_2. Current R&D investments are focusing on both pulverized coal and circulating fluidized bed designs, and both new plant and retrofit applications. Flue gas can be recycled to control operational characteristics such as thermal flow and flame temperature. Reducing recycle gas in the combustion process results in a higher flame temperature and potentially higher operating efficiency, but this can create other operational challenges. Oxygen is generally supplied via air separation, but "chemical looping" options that extract oxygen from minerals, which are subsequently recirculated and regenerated, are being considered. In addition, there is research underway on low-cost oxygen separation technologies, such as oxygen transport membranes.

◆ A number of collaborative efforts are currently underway that will contribute to this strategy. **Regional Carbon Sequestration Partnerships** have been organized within the United States, and include networks of state agencies, universities, and private companies focused on determining suitable approaches for capturing and storing CO_2. Four Canadian Provinces are also participating in the effort. The Partnerships are developing a framework to identify, validate, and potentially test the carbon capture and storage technologies best suited for each geographic region and its point sources. During Phase II, beginning in 2005, the Partnerships will pursue technologies for small-scale sequestration validation testing.

◆ The DOE Carbon Sequestration Program is participating **in collaborations with**

international partners in developing new capture and storage technologies. Among these are a cooperative agreement with Canada (Weyburn Project) (Box 6-1) and the Sleipner North Sea Project. The Carbon Sequestration Leadership Forum (CSLF) (Box 6-3) is an international collaborative effort to focus international attention on the development of carbon capture and storage technologies.

Future Research Directions

The current portfolio supports the main components of the technology development strategy and addresses the highest priority current investment opportunities in this technology area. For the future, CCTP seeks to consider a full array of promising technology options. From diverse sources, suggestions for future research have come to CCTP's attention. Some of these, and others, are currently being explored and under consideration for the future R&D portfolio. These include:

◆ Reduce the costs for sorbents, reducing regeneration energy requirements, and increasing sorbent life.

◆ Increase understanding of the CO_2 purity requirements to ensure that CO_2 transportation and storage operations are not compromised. In CO_2 transportation, small quantities of SO_2 can lead to two-phase flow and pipeline pressure loss. The presence of water and other minute contaminants might promote acid formation and lead to pipeline and wellbore integrity problems. The history of transporting CO_2 in pipelines that contain substantial amounts of SO_X and NO_X is limited. These components can also impact the integrity of reservoir caprock.

◆ Develop pre-and post-combustion CO_2 capture technologies that reduce the economic impacts of contaminants in a gas stream. For example, the corrosive nature of some of the contaminants can complicate CO_2 separation processes. Too much nitrogen in the CO_2 can significantly increase the cost of compression prior to geologic storage.

◆ Develop pre- and post-combustion CO_2 capture technologies that enable storage of criteria pollutants (SO_X, NO_X, H_2S) with the CO_2. In this area, the criteria pollutants are not separated from the CO_2 stream, but rather stored along with the CO_2.

◆ Continue to improve the cost-effectiveness of CO_2 separation membranes. Performance is improved by more cost-effective designs and materials with increased selectivity to CO_2 (increased CO_2 concentration per single membrane pass), increased throughput (increased flow rate per single membrane pass), and improved chemical stability (a measure of how well the membrane resists chemical reaction with its environment).

◆ Continue to lower the costs of oxygen used by coal-fueled power plants with separation technologies such as oxygen transport membranes. Success in this area could reduce the costs of oxy-combustion technologies (e.g., circulating fluidized bed designs), as well as gasification technologies.

◆ Develop an integrated modeling framework for evaluating alternative CO_2 capture technologies for existing and advanced electric power plants.

◆ Pursue innovative, potentially high-payoff concepts that build on current approaches or that offer entirely new pathways. This would encompass areas such as advanced materials, and chemical and biological processes. Examples include ionization of CO_2, using CO_2 solvents, novel microporous metal organic frameworks (MOFs) suitable for CO_2 separation (Box 6-2), and metabolic engineering to create strains of microbes that feed off CO_2 and produce useful chemical byproducts.

◆ Continue system integration and advancements of classical MEA-based systems for near-term CO_2 availability.

BOX 6-3

CARBON SEQUESTRATION LEADERSHIP FORUM (CSLF)

Established by the State Department and DOE in February 2003, the CSLF coordinates data gathering, R&D, and joint projects to advance the development and deployment of geologic carbon sequestration technologies worldwide. The CSLF is a particularly attractive mechanism for achieving international cooperation for larger field tests. See http://fossil.energy.gov/programs/sequestration/cslf

6.2 Geologic Storage

Different types of geologic formations can store CO_2, including depleted oil reservoirs, depleted gas reservoirs, unmineable coal seams, saline formations, shale formations with high organic content, and others. Such formations have provided natural storage for crude oil, natural gas, brine, and CO_2 over millions of years. Each type of formation has its own mechanism for storing CO_2 and a resultant set of research priorities and opportunities. Many power plants and other large point sources of CO_2 emissions are located near geologic formations that are amenable to CO_2 storage. For example, DOE, along with private and public sector partners, is conducting research on the suitability of geologic formations at the Mountaineer Plant in West Virginia.

BOX 6-2

METAL ORGANIC FRAMEWORKS

Scientists have recently developed improved capabilities to synthesize a class of chemical compounds called metal organic frameworks (MOFs), and "tune" their macromolecular properties. Through a project funded by the DOE Sequestration Program, a team of researchers is measuring the CO_2 adsorption isotherms of a set of MOFs to develop a better understanding of what MOF characteristics affect CO_2 adsorption. There have been some early promising results. For example, one particular MOF exhibited a CO_2 sorption capacity that was significantly better than commercially available zeolite sorbents. The increased storage capacity can lower the size and cost of a CO_2 capture system.

MOF 177, Yaghi et. al Nature 427, 523-527 (2004)

Potential Role of Technology

Geologic formations offer an attractive option for carbon storage. The formations are found throughout the United States, and there is extensive knowledge about many of them from the experience of exploration and operation of oil and gas production (Box 6-4). Opportunities exist in the near-term to combine CO_2 storage with EOR and enhanced coal-bed methane (ECBM) recovery using injected CO_2. In 2000, 34 million tons of CO_2, roughly equivalent to annual emissions from 6 million cars, were injected as part of EOR activities in the United States.

Coal-bed methane has been one of the fastest growing sources of domestic natural gas supply. Pilot projects have demonstrated the value of CO_2 ECBM recovery as a way to increase production of this resource.

In the long-term, CO_2 storage in saline and depleted gas formations is being explored. One project is currently in commercial operation, where one million tons of CO_2 per year are being injected in a saline formation at the Sleipner natural gas production field in the North Sea. The Frio Brine Pilot experiment near Houston, Texas, is the first U.S. field test to investigate the ability of saline formations to store greenhouse gases (GHGs). In October 2004, 1,600 tons of CO_2 were injected into a mile-deep well. Extensive methods were used to characterize the formation and monitor the movement of the CO_2. The site is representative of a very large volume of the subsurface from coastal Alabama to Mexico and will provide experience useful in planning CO_2 storage in high-permeability sediments worldwide.

The overall estimated capacity of geologic formations appears to be large enough to store decades to centuries worth of CO_2 emissions, although the CO_2 storage potential of geologic reservoirs depends on many factors that are, as yet, poorly understood. For example, characteristics of reservoir integrity, volume, porosity, permeability, and pressure vary widely even within the same reservoir, making it difficult to establish a reservoir's storage potential with certainty. Assessments of storage capacity could help to better understand the potential of geologic formations for CO_2 storage.

Technology Strategy

Potential CO_2 sources and sinks vary widely across the United States, and the challenge is to understand the economic, health, safety, and environmental implications of potential large-scale geologic storage projects. The geologic storage program was initiated in 1997 and initially focused on smaller projects. However, field testing is the next step to verify the results of smaller-scale R&D, and the program is taking on larger projects, as knowledge grows and opportunities become available.

In the near-term, activities will focus on addressing important carbon storage-related issues consistent with the Carbon Sequestration Technology Roadmap and Program Plan (DOE 2005). Among these activities are developing an understanding of the behavior of CO_2 when stored in geologic formations. Long-term activities could include understanding and reducing potential health, safety, environmental, and economic risks associated with geologic sequestration.

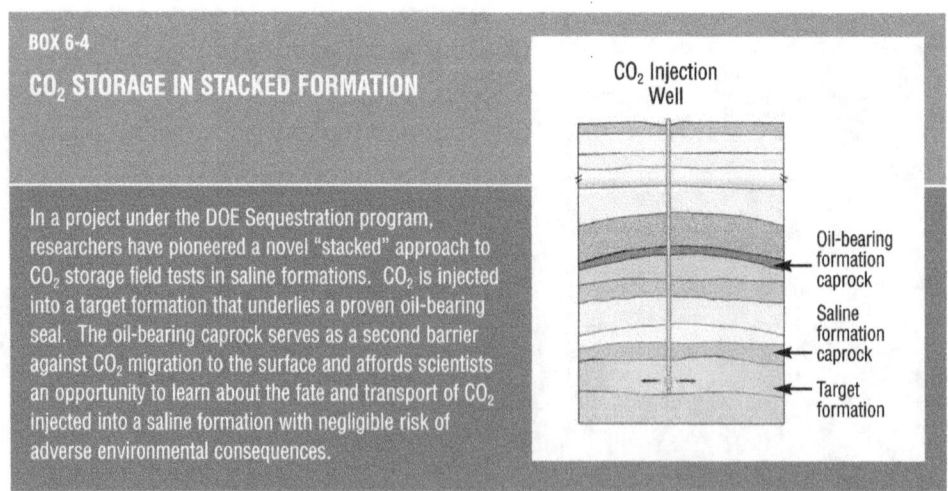

BOX 6-4

CO_2 STORAGE IN STACKED FORMATION

In a project under the DOE Sequestration program, researchers have pioneered a novel "stacked" approach to CO_2 storage field tests in saline formations. CO_2 is injected into a target formation that underlies a proven oil-bearing seal. The oil-bearing caprock serves as a second barrier against CO_2 migration to the surface and affords scientists an opportunity to learn about the fate and transport of CO_2 injected into a saline formation with negligible risk of adverse environmental consequences.

CO₂ Injection Well

Oil-bearing formation caprock

Saline formation caprock

Target formation

Courtesy of DOE/NETL

Regional domestic partnerships and international cooperation are viewed as key to deploying carbon storage technologies. Field validation activities test the large-scale viability of point-source capture and storage systems and demonstrate to interested parties the potential of these systems.

Current Portfolio

The goal of geologic storage R&D portfolio is to advance technologies that would enable development of domestic CO_2 underground storage repositories capable of accepting around one billion tons of CO_2 per year. Toward this goal, there is a need to demonstrate that CO_2 storage underground is safe and environmentally acceptable, and an acceptable GHG mitigation approach. Another need is to demonstrate an effective business model for CO_2 EOR and ECBM, where significantly more CO_2 is stored for the long-term than under current practices.

The Federal portfolio for geologic storage activities includes several major thrusts designed to move technologies from early R&D to deployment.[3]

Core RD&D focuses on understanding the behavior of CO_2 when stored in geologic formations. For example, studies are being conducted to determine the extent to which CO_2 moves within the geologic formation, and what physical and chemical changes occur to the formation when CO_2 is injected. This information is needed to ensure that CO_2 storage will not impair the geologic integrity of an underground formation and that CO_2 storage is secure and environmentally acceptable. There are three major research thrusts:

◆ **Knowledge Base and Technology for CO_2 Storage Reservoirs.** These activities seek to increase the knowledge base and technology options. The petroleum industry has built significant experience over the past few decades on how to inject carbon dioxide into oil reservoirs for EOR. Many of the issues related to injection technologies and gas compression have already been solved. Because oil and gas reservoirs have been able to store gases and other hydrocarbons for geologically significant periods of time (hundreds of thousands to millions of years), they likely have caprocks that will be good seals for CO_2 as well. Furthermore, CO_2 can potentially enhance oil and gas production, which can help mitigate carbon storage costs. However, because the petroleum industry understandably has been

focused on resource recovery and not on CO_2 storage, it has not developed procedures to maximize the amount of CO_2 that is stored or to track the CO_2 once it is has been injected to ensure that it remains in the ground. In addition, most well-developed oil fields, by definition, contain many wells that have pierced the caprock for the field, creating potential leakage pathways for CO_2. Research is currently underway to develop technologies to locate abandoned wells, to track the movement of CO_2 in the ground, and to ensure long-term storage, as well as to optimize costs, assess performance, and reduce uncertainties in capacity estimates.

Another attractive option is carbon storage in deep, unmineable coal seams. Not only do these formations have high potential for adsorbing CO_2 on coal surfaces, but the injected CO_2 can displace adsorbed methane, thus producing a valuable byproduct and decreasing the overall storage cost. One potential barrier is the tendency of coal to swell in volume when adsorbing CO_2. This can cause a sharp drop in permeability, thereby impeding the flow of CO_2 and the recovery of methane. Laboratory research, modeling, and field studies are currently being implemented and proposed to gain a better understanding of the processes behind coal swelling and determine if it will be a significant barrier to sequestration in coal seams.

Another option is the use of large saline formations for CO_2 storage, a relatively new concept. About two-thirds of the United States is underlain by deep saline formations that have significant sequestration potential. Since the water in the saline formations is typically not suitable for irrigation or consumption, many opportunities exist for CO_2 to be injected without adverse impacts. The storage capacity of saline formations is enhanced because of the ability of CO_2 to dissolve in the aqueous phase. But, there are uncertainties associated with the heterogeneous reactions that may occur between CO_2, brine, and minerals in the surrounding strata, especially with respect to reaction kinetics. For example, saline formations contain minerals that could react with injected CO_2 to form solid carbonates, which would eliminate potential migration out of the reservoir. On the negative side, the carbonates could plug the formation in the immediate vicinity of the injection well. Researchers are looking into multiphase behavior of CO_2 in saline aquifers and the volume, fate, and

---

[3] See Section 3.1.2 (CCTP 2005): http://www.climatetechnology.gov/library/2005/tech-options/tor2005-312.pdf.

transport of the stored CO_2. New technologies and techniques are being developed to reduce cost and inefficiency due to leaks and to better define the geology of the saline aquifers. A recent review article addresses the technological challenges of sequestering carbon dioxide in saline formations and coal seams (White et al. 2003).[4]

◆ **Measurement and Monitoring.** These activities are described more fully in Chapter 8. An important R&D need is to develop a comprehensive monitoring and modeling capability that not only focuses on technical issues, but also can help ensure that geologic storage of CO_2 is safe. Long-term geologic storage issues, such as leakage of CO_2 through old well bores, faults, seals, or diffusion out of the formation, need to be addressed. Many tools exist or are being developed for monitoring geologic storage of CO_2, including well testing and pressure monitoring; tracers and chemical sampling; surface and borehole seismic monitoring; and electromagnetic/geomechanical meters, such as tiltmeters. However, the spatial and temporal resolution of these methods may not be sufficient for performance confirmation and leak detection.

◆ **Health, Safety, and Environmental Risk Assessment.** Assessing the risks of CO_2 release from geologic storage sites is fundamentally different from assessing risks associated with hazardous materials, for which best practice manuals are often available. In some cases, geologic storage sites may exist near populated areas. Although CO_2 is not toxic or flammable, it can cause suffocation if present at high concentrations. Therefore, the mechanism for potential leaks must be better understood. The

assessment of risks includes identifying potential subsurface leakage modes, the likelihood of an actual leak, leak rate over time, and the long-term implications for safe carbon storage. Diagnostic options need to be developed for assessing leakage potential on a quantitative basis.

Two activities cited in Section cited in Section 6.1.3 will continue to play an important role in encouraging the deployment of technologies developed under the core RD&D program. The Regional Partnerships Program[5] is building a nationwide network of Federal, State, and private sector partnerships to determine the most suitable technologies, regulations, and infrastructure for future point source carbon capture, storage, and geologic sequestration in different areas of the country. The Carbon Sequestration Leadership Forum is facilitating the development and worldwide deployment of technologies for separation, capture, transportation, and long-term storage of CO_2.

In addition, the FutureGen project (Box 6-5) is expected to be the world's first coal-fueled prototype power plant that will incorporate geological storage. It will provide a way to demonstrate some of the key technologies developed with Federal support, and demonstrate to the public and regulators the viability of large-scale carbon storage.

Future Research Directions

The current portfolio supports the main components of the technology development strategy and addresses the highest priority current investment opportunities in this technology area. For the future, CCTP seeks to consider a full array of promising technology options. From diverse sources, suggestions for future

---

BOX 6-5

FUTUREGEN

FutureGen is a public-private initiative to build the world's first integrated carbon capture/storage and hydrogen production power plant. When in operation, the prototype will be the cleanest fossil fuel power plant in the world. An industrial consortium representing the U.S. coal and power industry will work closely with DOE to implement this project. Other countries, including India and South Korea, have recently agreed to participate in the Program. See http://www.netl.doe.gov/coalpower/sequestration/futureGen/main.html.

Courtesy of DOE/NETL

---

4   See Section 3.1.2 (CCTP 2005): http://www.climatetechnology.gov/library/2005/tech-options/tor2005-312.pdf.

5   For more information on the Regional Partnerships Program, see http://fossil.energy.gov/programs/sequestration/partnerships.

research have come to CCTP's attention. Some of these, and others, are currently being explored and under consideration for the future R&D portfolio. These include:

- Defining the factors that determine the optimum conditions for sequestration in geological formations, such as depleting oil and gas reservoirs, saline formations, and coal seams, as well as unconventional hydrocarbon bearing formations.

- Developing the ability to predict and optimize CO_2 storage capacity and resource recovery.

- New storage engineering practices that maximize pore volume utilization and accelerate capillary, solubility, and mineral trapping for long-term storage.

- Developing the ability to continuously track the fate and transport of injected CO_2 in different formations. Areas of R&D include geophysical arrays that provide real-time, low-cost, and high-resolution data; and surface and near-surface monitoring techniques such as surface CO_2 flux detectors, injecting tracers in soil gas, and measuring changes in shallow aquifer chemistry for CO_2 leakage.

- Developing models to simulate the migration of CO_2 throughout the subsurface and the effects of injection on the integrity of caprock structures.

- Developing advanced subsurface imaging and alteration of fluid-rock interactions.

- Understanding geochemical reactions (Box 6-6) and harnessing them to enhance containment.

- Developing injection practices that preserve cap integrity, and practices to mitigate leakage to the atmosphere. These practices would include new materials and methods for sealing wells.

- Developing an understanding of CO_2 reactions and movement in shales and other unconventional hydrocarbon-bearing formations that will permit the economic recovery of these hydrocarbons.

- Developing technologies to conduct underground (in-situ) liquefaction and gasification of solid hydrocarbon deposits, such as oil shale and coal.

- Developing cost-effective systems to integrate energy conversion with carbon capture, geologic storage, and subsurface conversion of CO_2 into benign materials or useful byproducts (e.g., through biogeochemical processes that can create methane or carbonates).

BOX 6-6

COAL SWELLING

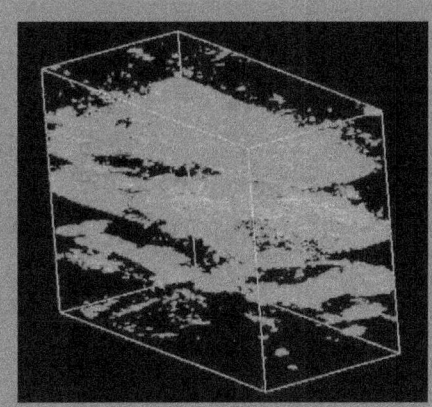

Computed tomography (CT) scans were taken to image the inside of a block of coal before and after the introduction of high pressure CO_2. Comparison of the two scans showed where CO_2-induced swelling took place in the coal, shown in green. The swelling phenomena is a research challenge, since it can adversely impact the economics of enhanced coal bed methane recovery by reducing the flow of CO_2 into the coal.

Courtesy of DOE/NETL

- Developing improved methods and data for estimating the overall costs of geologic sequestration, including capture, compression, and transportation.

- Improving the understanding of the key elements for effective risk management of geologic storage. Technical objectives are to (1) characterize available storage formations in terms of size and location, (2) characterize leakage rates in order to establish risk management approaches and policies, and (3) conduct large-scale testing on representative formations.

- Reducing the cost of geologic sequestration.

- Improving CO_2 transport systems to provide for public acceptance and regulatory approval, including providing for early leak detection and warning, preventing major pipeline failures, and linking to a national pipeline infrastructure.

- Pursuing breakthrough concepts to reach long-term program goals. Breakthrough concepts are revolutionary and transformational approaches with potential for low-cost, permanence, and large global capacity. For example, some of the lowest cost estimates for capture/sequestration options are for systems where flue gas components from coal-fueled plants are not scrubbed but rather stored in geologic formations with CO_2. This eliminates the need for costly flue gas cleanup systems, but the potential effects of this option are unknown. Technological innovations could come from concepts associated with areas not normally related to traditional energy R&D fields.

In the long-term, CO_2 capture can be integrated with geologic storage and/or conversion. Many CO_2 conversion reactions are attractive, but too slow for economic chemical processes. Use of impurities in captured CO_2 (e.g., SO_X and NO_X) or additives could possibly enhance geologic storage and provide an opportunity to combine CO_2 emissions reduction with criteria pollutant emissions reduction.

Field tests are the next step to verify R&D results. It is possible that additional tests will eventually be carried out through the Regional Partnerships Program based on analysis of CO_2 sources and sinks by participants to determine the highest benefit projects.

6.3 Terrestrial Sequestration

Terrestrial sequestration can play a significant role in addressing the increase of CO_2 in the atmosphere. A wide range of technologies and practices, including tree planting, forest management, and conservation tillage practices are available to increase the sequestration of carbon in plants and soils. Terrestrial sequestration activities can provide a positive force for improving landscape-level land management and provide significant additional benefits to society, such as improvements in wildlife and fisheries habitat, enhanced soil productivity, reduction in soil erosion, and improved water quality. Terrestrial sequestration represents a set of technically and commercially viable technologies that have the capability to reduce the rate of CO_2 increase in the atmosphere. Given the size and productivity of the U.S. land base, terrestrial sequestration has distinct economic and environmental advantages. Globally, the potential for terrestrial sequestration is also significant, due in part to low-cost opportunities to reduce ongoing emissions from current land-use practices and land conversion

and to enhance carbon stocks via afforestation, forest restoration, and improved forest and agricultural management.

Terrestrial sequestration technologies refer broadly to equipment, processes, decision tools, management systems and practices, and techniques that can enhance carbon stocks in soils, biomass, and wood products, while reducing CO_2 concentrations in the atmosphere. Extensions of terrestrial sequestration can use sustainably generated biomass to displace fossil fuels. Examples of terrestrial sequestration technologies include conservation tillage, conservation set-asides, cover crops, buffer strips, biomass energy crops, active forest management, active wildlife habitat management, low-impact harvesting, precision use of advanced information technologies, genetically improved stock, wood products life-cycle management, and advanced bioproducts.

Potential Role of Technology

Increasing terrestrial carbon stocks is attractive because it can potentially offset a major fraction of emissions and serve as a bridge over an interim period, allowing for development of other low-CO_2 or CO_2-free technologies. Carbon stock management technologies and practices that enhance soil and forest carbon sinks need to be maintained once the carbon stock reaches higher levels. Although the benefits can be temporarily reversed by fire, plowing of cropland soils, and other disturbances, the potential improvements in carbon stocks are of such magnitude that they can play a significant overall role in addressing the increase in atmospheric CO_2 emissions from the United States and globally throughout the 21st century.

Other opportunities described in this section can provide benefits essentially indefinitely. For example, changes in crop management practices can reduce annual emissions of trace GHGs; sustainable biomass energy systems can displace fossil fuels and provide indefinite net CO_2 emissions reductions; and enhanced forest management and conversion to durable wood products provide a mechanism to allow forests to continually sequester carbon.

Estimates of the global potential for terrestrial sequestration activities remain uncertain. Such estimates are generally of the technical potential (i.e., the biophysical potential of managed ecosystems to sequester carbon), and disregard market and policy considerations. The IPCC (IPCC 2001c) estimates such technical potential of biological mitigation options (i.e., forest, agricultural, and other land-management activities) to be on the order of 100 GtC cumulative by 2050, at costs ranging from about $0.10 to about $20/t carbon in tropical countries, and from $20/t carbon to $100/t in non-tropical countries. Technical potential estimates for the United States range widely, depending on assumptions about biophysical sequestration rates per hectare, the land area available for different activities, and other factors.

Widely cited estimates of U.S. technical potential for carbon sequestration include about 55–164 teragrams of carbon (TgC) per year for potential sequestration on croplands (Lal et al. 1998); 29–110 TgC per year on grazing lands (Follett et al. 2001); 210 TgC per year on forest land (Joyce and Birdsey 2000); and 91–152 TgC per year on dedicated bioenergy croplands (Tuskan and Walsh 2001). In addition, dedicated bioenergy crops would substitute for fossil fuels, leading to an estimated 450 TgC reduction of CO_2 emissions (Tuskan and Walsh 2001).

These estimates generally represent technical potential that does not reflect barriers to implementation, competition across land uses and sectors, or landowner response to public policies and economic incentives. A recent study of cropland (Eve et al. 2002) indicates a potential of about 66 TgC per year on croplands, toward the lower end of the Lal et al. (1998) range. With regard to bioenergy, a recent DOE/USDA analysis estimates that U.S. forest and agricultural lands could sustainably supply up to 1,300 Tg of biomass/year for bioenergy, similar to the findings of Tuskan and Walsh, but without major shifts in land use or food or fiber production (Perlack et al. 2005). Such a quantity of biomass could displace over 30 percent of current U.S. petroleum consumption.

Technology Strategy

Realizing the opportunities to sequester carbon in terrestrial systems will require managing resources in new ways that integrate crosscutting technologies and practices. A balanced portfolio is needed that supports basic science, technological development, emerging technology demonstrations, innovative partnerships with the private sector, and techniques and metrics for measuring success.

An array of actual and potential technologies can be found in the short-, mid-, and long-terms. In the short-term, some technologies and practices being

routinely used can be expanded to increase carbon sequestration. In addition, improvements to many current systems are needed to enable them to enhance above- and below-ground carbon stocks, and manage wood products pools. In the mid to long-term, research can focus on options that take advantage of entirely new technologies and practices.

In the near- and long-term, the R&D portfolio is based on the following:

◆ Design, develop and demonstrate carbon management strategies consistent with economic and environmental goals for terrestrial ecosystems.

◆ Improve the understanding of the relationship of carbon management and ecosystem goods and services.

◆ Determine how terrestrial systems' capacities can be manipulated to enhance carbon sequestration by increasing pool sizes, areal extent, rates of carbon accumulation, and/or longevity of carbon storage in pools.

◆ Analyze the relationship between natural resource and agricultural policy, and terrestrial sequestration technologies and identifying ways to maximize synergies and avoid potential conflicts between the two.

◆ Analyze the relationship between energy policy and terrestrial sequestration technologies to enhance understanding of the potential carbon benefits associated with different biofuels.

◆ Evaluate existing and new market-based adoption and diffusion strategies for terrestrial sequestration technologies.

◆ Optimize management practices and techniques, informed by analyses of and accounting for all associated GHG emissions and removals, including, to the extent practicable, associated climate-related impacts (such as changes in albedo or surface roughness).

◆ Improve methods of measuring changes in carbon pools and verifying sequestration rates.

◆ Develop and analyze incentives for implementation.

Current Portfolio

Much of the research currently underway that could have applications for increasing terrestrial carbon

Terrestrial Sequestration: Woody Crops

Figure 6-1. Growing short rotation woody crops, shown above, provides opportunities to sequester carbon in the soil and biomass feedstocks.

Courtesy of DOE/NREL, Credit: Warren Gretz

sequestration is being undertaken for multiple reasons, often unrelated to climate change. Significant investments are being made in developing sustainable natural resource management systems that provide economic and environmental benefits. In particular, advances have been made in increasing forest productivity, developing effective and environmentally sound uses of crop fertilizers, enhancing soil quality, and in producing biomass feedstocks (Figure 6-1).

Across the current Federal portfolio of terrestrial sequestration-related RD&D, multi-agency activities are focused on a wide range of issues, including the following:

♦ Cropland management practices can increase the amount of carbon stored in agricultural soils by increasing plant biomass inputs or reducing the rate of loss of soil organic matter to the atmosphere as CO_2. Precision agriculture is a form of site-specific management that can be adapted for improving soil carbon sequestration through a customized carbon sequestering management plan. The goals of this activity are to quantify the carbon sequestration potential of each technology and management practice for various crop production systems, climates, and soils; for various crop production systems, soil types, and geographical areas develop the combinations of

practices that optimize soil carbon sequestration, crop production, and profits; develop decision support tools for farmers, other land managers, and policy makers that provide guidance for land-management decisions. For example, create databases that answer questions about how changing from one land-use practice to another will affect carbon sequestration, production, and profits.[6]

♦ Conversion of marginal croplands to other less-intensive land uses to conservation reserve and buffer areas. The goals of this activity are to quantify the carbon sequestration potential of cropland conservation programs for various climates and soils; develop the combination of practices (e.g., plant species, siting, establishment practices) that optimize carbon sequestration and minimize production losses for various types of cropland conservation practices; and develop decision support tools for farmers, other land managers, and policy makers to inform cropland conservation policies and the relative costs and benefits of different cropland conservation approaches, both in terms of carbon sequestration and production.[7]

♦ Evaluation of advanced forest and wood products management that may offer significant carbon sequestration opportunities. The goals and milestones of this activity are to increase energy efficiency of forest operations; develop and apply models to better understand the economics of achieving certain GHG mitigation goals through improved forest management; sensors/monitors and information management systems; advanced fertilizers, technologies, and application strategies to improve fertilizer efficiency and reduce nitrogen fertilizer inputs; integrated management strategies and systems to increase nutrient and water use efficiency, increase CO_2 uptake and sequestration and reduce emissions; and wood product management and substitution strategies. The milestones are to have initial systems models and prototype operation on major plantation types in place by 2007 and to deploy first-generation integrated system models and technology by 2010.[8]

♦ Grazing management to increase amount of carbon in soils. The goals of this activity are to construct quantitative models that describe site-specific interactions among grazing systems, vegetation, soil and climate, and the effects on

GHG dynamics; and to develop decision support tools to inform the relative costs and benefits of different grassland management scenarios for carbon sequestration and other conservation benefits.[9]

◆ Restoration of degraded rangelands using low-cost, reliable technologies. The goals of this activity are to develop low-cost, reliable technologies for the restoration of vegetation on degraded arid and semi-arid rangelands; improve decision support for the application of low-cost technologies, such as fire, to control invasive species and to reduce GHG emissions from mesic rangelands; and to develop seed production technology to produce low-cost seeds for reestablishing desired rangeland species. Currently costs are high and seed supply is limited for many cultivars.[10]

◆ Wetland restoration and management for carbon sequestration and GHG offsets. The goals of this activity are to evaluate various management practices on restored wetlands; delineate and quantify carbon stocks in U.S. wetlands by region and type; develop and demonstrate integrated management strategies for wetland carbon sequestration; and identify wetland areas most likely to be impacted by climate change and prioritize areas for protection.[11]

◆ Use of biotechnology for modifying the chemical composition of plants and micro-organisms to enhance carbon sequestration (Box 6-7). The goals of this activity are to identify the traits needed in plants and micro-organisms to increase soil carbon sequestration capacity; determine the feasibility of using biotechnology to modify the traits of plants and micro-organisms that can affect soil carbon sequestration; develop systems for monitoring non-target environmental affects associated with plant modifications; develop methods to incorporate genetically modified plant and micro-organisms into cropland and conservation reserve and buffers systems.[12]

BOX 6-7

PHYSIOLOGICAL MECHANISMS OF GROWTH, RESPONSE, AND ADAPTATION IN FOREST TREES

Enhancing the natural capacity of terrestrial ecosystems to store carbon is a viable strategy for stabilizing rising CO_2 concentrations in the atmosphere. However, gains in improving the sequestration potential of croplands, grasslands, and forest lands could be enhanced by major scientific advancements in understanding the processes that control the initial uptake, ultimate chemical forms, and subsequent carbon transfer in plants and soils.

Research carried out by the USDA and DOE is underway to determine the mechanisms that control the quantity and quality of carbon allocated to stems, branches, leaves, and roots of trees as a means of understanding the biological processes that underlie carbon sequestration in trees and soils; understanding controlling genetic mechanisms; and selecting, testing, and demonstrating useful genotypes. Research is focused on several species, including hybrid poplar, willow, and loblolly pine. The studies are designed to determine the interaction of physiological and biogeochemical processes and water and nutrient management on carbon fixation, allocation, storage, and dynamics in forest systems. Field and laboratory studies are being used to quantify and understand carbon dynamics, both above and below ground. Forest researchers hope that these and similar studies will provide the scientific foundation for managing forest systems to enhance carbon sequestration, and improve environmental quality and productivity.

Courtesy of DOE/Office of Science

---

9 See Section 3.2.1.4 (CCTP 2005): http://www.climatetechnology.gov/library/2005/tech-options/tor2005-3214.pdf.

10 See Section 3.2.1.5 (CCTP 2005): http://www.climatetechnology.gov/library/2005/tech-options/tor2005-3215.pdf.

11 See Section 3.2.1.6 (CCTP 2005): http://www.climatetechnology.gov/library/2005/tech-options/tor2005-3216.pdf.

12 See Section 3.2.2.1 (CCTP 2005): http://www.climatetechnology.gov/library/2005/tech-options/tor2005-3221.pdf.

◆ Development of terrestrial sensors, measurements, and modeling. The goals of this activity are to develop a new generation of sensors, probes, and other instruments to measure soil carbon, GHG flux in situ across a wide variety of agricultural ecosystems.[13]

◆ Measuring, monitoring, and verification for forests. The goals of this activity are to develop technologies for remote sensing data collection and analysis, in situ instrumentation and monitoring systems, and other measuring and monitoring technologies.[14]

◆ USDA is providing incentives and supporting voluntary actions by private landowners to reduce GHG emissions and increase carbon sequestration through the portfolio of conservation programs administered by the Department. USDA's actions include financial incentives, technical assistance, demonstrations, pilot programs, education, and capacity building, along with measurements to assess the success of these efforts.

Future Research Directions

The current portfolio supports the main components of the technology development strategy and addresses the highest priority current investment opportunities in this technology area. For the future, CCTP seeks to consider a full array of promising technology options. From diverse sources, suggestions for future research have come to CCTP's attention. Some of these, and others, are currently being explored and under consideration for the future R&D portfolio. These iclude:

◆ Quantifying the carbon sequestration potential for management practices and techniques across all major land uses, including cropland, forests, grasslands, rangelands, and wetlands; across cultivation and management systems; and across regions.

◆ Designing, developing, and testing management systems to increase carbon sequestration, maintain storage, and minimize net GHG emissions while meeting economic (i.e., forest and agricultural production) and environmental goals. Using a systems approach across sectors and gases will improve the understanding of how technologies are configured to work in a synergistic manner. An example of this approach is in the production of biofuel crops that enhance carbon sequestration, and reduce nitrogen releases to the atmosphere.

◆ Developing bioenergy and additional durable uses of bio-based products and improve management of residues and wood products.

◆ Improving biomass supply technologies (harvesting, handling, onsite separation and processing, transportation) to reduce costs and impacts; and enhancing techniques that improve yields, transport, and efficiency of conversion to fuels.

◆ Exploring the use of trees and other vegetative cover in urban environments to both sequester carbon and reduce the urban heat island effect

◆ Evaluating terrestrial carbon stock vulnerabilities and stability.

◆ Improving the understanding of the implications of potential sequestration options on the emissions of other GHGs through comprehensive accounting of all GHG emissions and sinks as land-based carbon sequestration technologies are implemented.

◆ Improving the performance of technologies and practices to provide additional benefits, including improvements in wildlife habitat; water and air quality; soil characteristics such as stability, water infiltration and retention; and nutrient retention.

◆ Enhancing sequestration potential through the use of advanced technologies, including bio-based and biotechnology techniques to enhance seed stock qualities, precision water and nutrient application, land management using geographic information system and other tools, and alternative tillage, harvest and fertilizing (e.g., char-based fertilizer) techniques.

◆ Developing novel alternative technologies such as high-lignin trees for combustion and low-lignin trees to reduce paper processing costs and improved digestibility of fodder and forage.

◆ Researching biotechnology (genomics, genetics, proteomics), related to biological and ecological processes affecting carbon allocation, storage, and system capacity. Improved understanding of the functional genomics of high-potential biomass crops can increase yields and provide a more effective basis for increasing the conversion efficiency of biomass of fuels, chemicals, and other bioproducts.

◆ Improving observation and quantification of deforestation, and cost and benefit analysis of options to reduce deforestation.

---

[13] See Section 3.2.3.1 (CCTP 2005): http://www.climatetechnology.gov/library/2005/tech-options/tor2005-3231.pdf.

[14] See Section 3.2.3.2 (CCTP 2005): http://www.climatetechnology.gov/library/2005/tech-options/tor2005-3232.pdf.

6.4 Ocean Sequestration

Because of the large CO_2 storage capacity of the ocean, increasing the carbon uptake and storage of carbon in the oceans cannot be ignored (Figure 6-2). Indeed, the ocean is currently playing an important role in consuming significant amounts of anthropogenic CO_2 via passive air-sea exchange, biological uptake, and ocean mixing (e.g., Sabine et al. 2004). This natural rate of CO_2 uptake (about 7.3Gt CO_2/yr), however, is not keeping pace with the rate of current anthropogenic emissions. Also, there are consequences. Ocean acidification that is accompanying the air-sea flux, for example, could have undesirable environmental consequences, if allowed to continue (e.g., Caldeira and Wickett 2003, Feely et al. 2004, Orr et al. 2005).

Figure 6-2. *While the oceans play an important role in taking up large amounts of CO_2, there is a need for a better understanding of the potential role of ocean sequestration as a mitigation strategy and its environmental consequences.*
Credit: IStockphoto

To understand the additional role the ocean could play in mitigating the effects CO_2 emissions on atmospheric concentrations, several issues must be addressed, including the capacity of the ocean to sequester CO_2, its effectiveness at reducing atmospheric CO_2 concentration levels, the depth and form (e.g., molecular or chemically bound, gas, liquid, or solid) for introduction of the carbon, and the potential for adverse environmental consequences. Ocean storage has not yet been deployed or thoroughly tested, but there have been small-scale field experiments and 25 years of theoretical, laboratory, and modeling studies of intentional ocean storage of CO_2. Nevertheless, there is still much that is unknown and more needs to be learned about the potential environmental consequences to ocean ecosystems and natural biogeochemical cycles.

Although there are a variety of potential ocean carbon sequestration options (see Future Research Directions), two strategies have received the most attention: (1) direct injection of a relatively pure stream of CO_2 into the ocean's deep interior, and (2) iron fertilization to stimulate the growth of nutrient-constrained biota and enhance the ocean's natural biological pump. It is generally thought that direct injection of CO_2 may be technically feasible and effectively isolate CO_2 from the atmosphere for at least several centuries. The primary concerns relate to possible adverse environmental effects. In contrast, the technical feasibility and effectiveness of ocean fertilization remain open to question. Further, whereas direct injection approaches seek to minimize ecosystem impacts, ocean fertilization depends upon the manipulation of ecosystem function over large areas of the ocean's surface.

Over the period of centuries, it is estimated, the oceans will passively take up about 70 percent of global fossil carbon emissions, as CO_2 diffuses into the ocean, is transported across the ocean thermocline, and mixed into deep ocean waters (IPCC 2001a). Direct injection of captured CO_2 would seek to augment to this natural CO_2 flux to the deep sea and, thus, more rapidly slow or reverse the increase in atmospheric CO_2 concentrations. The potential for the ocean to absorb CO_2 over the long-term is large relative to that which would be generated by fossil-fuel resources. But several factors may affect the capacity and desirability of direct injection. Unless consumed by biological or chemical processes, excess CO_2 placed in the deep sea will eventually, via diffusion and ocean circulation, interact with the atmosphere, adding some part of the injected CO_2 to the atmospheric burden. For example, injection of about 8,000 Gt CO_2 to the deep ocean will eventually produce atmospheric CO_2 concentrations of about 750 ppm, even in the absence of additional CO_2 release to the atmosphere. Experiments and models have shown that high concentrations of CO_2 depress ocean pH (i.e., acidification), and thus may harm marine organisms and biogeochemical processes (e.g., Portner et al. 2004, TRS 2005). The true scope and magnitude of such effects could be the subject of further study. Alternatives to direct injection and fertilization have been proposed for CO_2 mitigation strategies. While they may avoid the preceding concerns, they may have environmental, capacity, and cost limitations of their own (see Future Research Directions).

Potential Role of Technology

Ocean sequestration offers the potential to significantly reduce the level of CO_2 concentrations in the atmosphere. There are many technological options envisioned for accomplishing this. Under the direct injection approach, for example, CO_2 could be captured from large point sources, (e.g., fossil-fired power plants, industrial processes, etc.), and then pressurized to liquid form (a supercritical liquid) and injected at depths of 2,000 to 3,000 meters below the surface. Once there, because its density as a liquid is greater than that of sea water, it would be expected to remain for centuries. However, this option has yet to be tested or deployed in a continuous mode at industrial concentrations.

Technology Strategy

The key to any successful technology strategy in this area is to assess adequately (a) the potential of ocean-based options as mitigation strategies; (b) the potential adverse impacts on the ocean biosphere; and the (c) potential effectiveness as evaluated against specific R&D criteria. This includes a research portfolio that seeks to determine, via experimentation and computer simulations, the potential for storing anthropogenic CO_2 in the world's oceans while minimizing negative environmental consequences.

Various studies based on models and ocean observations indicate that the isolation of CO_2 from the atmosphere generally increases with the depth of injection. In the near-term, the key research questions that are related to direct injection involve evaluating the impact of added CO_2 and/or nutrients on marine ecosystems and the biogeochemical cycles to which they contribute. This is being investigated through both observations and modeling of marine organisms and ecosystems, as is now being funded by DOE and the National Science Foundation (NSF), among others. In the long-term, R&D activities could focus on improving an understanding of the effects of elevated concentrations of CO_2 on marine organisms and ecosystems.

Another potential area of study is the effectiveness and environmental and ecological consequences of iron fertilization. Alternative ocean CO_2 mitigation strategies (see "Future Research Directions") pose a different set of environmental and efficacy concerns that need to be evaluated should the effects of direct injection prove to be unacceptable.

Current Portfolio

Ongoing research activities target ocean carbon sequestration using direct injection and iron fertilization. These activities are summarized below:

◆ **Direct Injection.** Currently, the technology exists for the direct injection of CO_2. Previous laboratory experiments concentrated on establishing an understanding of the processes that occur when CO_2 comes into contact with high pressure seawater. As a result, a much better understanding of the influence of CO_2 hydrates (or clathrates, "solids" in which gas molecules are held in place) on the dissolution processes exists. Additional research conducted by DOE's Oak Ridge National Laboratory simulated a negatively buoyant clathrate. In addition, the Monterey Bay Aquarium Research Institute demonstrated that CO_2 clathrates tend to be negatively buoyant at depths below 3,000 meters. This property of clathrates would presumably reduce the potential ecological impact of CO_2 on the shallow layers of the ocean, where most marine life occurs. It would also increase the length of time that injected CO_2 would remain in the ocean, thus enhancing the effectiveness of CO_2 sequestration by injection. The goal of this R&D activity is to demonstrate that CO_2 direct injection is safe and environmentally acceptable.[15]

Future Research Directions

The current portfolio supports the main components of the technology development strategy and addresses the highest priority current investment opportunities in this technology area. For the future, CCTP seeks to consider a full array of promising technology options. From diverse sources, suggestions for future research have come to CCTP's attention. Some of these, and others, are currently being explored and under consideration for the future R&D portfolio. These include:

---

[15] See Section 3.3.1 (CCTP 2005): http://www.climatetechnology.gov/library/2005/tech-options/tor2005-331.pdf.

◆ **Direct Injection.** R&D related to direct injection involves improving our understanding of the long-term effects of elevated concentration of CO_2 on marine organisms and ecosystems, as well as mitigation strategies. This could include both in situ and laboratory experiments combined with a program of process modeling aimed at a predictive capability for both biological and physico-chemical parameters.

◆ **Iron Fertilization.** There are a multitude of R&D opportunities regarding the effectiveness and environmental consequences of ocean fertilization. One question is whether iron enrichment increases the downward transport of carbon from the surface waters to the deep sea. This would help in predicting whether fertilization is an effective carbon sequestration mechanism. Other important questions could be explored: What are the long-term ecological consequences of iron enrichment on surface water community structure, and on mid-water and benthic processes? How can carbon export best be verified?

◆ **Enhanced Chemical CO_2 Uptake.** The uptake of CO_2 by an aqueous solution can be enhanced by the addition of OH- and/or CO_3- ions. Thus, Kheshgi (1995) pointed out that this could be done on a large scale by adding lime (CaO or CaOH) to the ocean to facilitate its abiotic CO_2 uptake from the atmosphere via the reaction: $Ca(OH)_2 + 2CO_2 \rightarrow Ca^{2+} + 2HCO_3^-$. Importantly, this form of CO_2 mitigation would (1) avoid the need for point-source CO_2 capture, separation, and purification (unlike direct injection, but similar to ocean fertilization); (2) prevent increased ocean acidity because the added CO_2 is neutralized to calcium bicarbonate dissolved in seawater; and (3) permanently store the added carbon in an ionic form that is already abundant in the ocean and not easily degassed back to the atmosphere. The concerns with this approach include the cost and carbon intensity of producing lime from the calcination of limestone, its transport to and dispersal in the ocean, and the environmental consequences of doing so.

◆ **Enhanced Carbonate Weathering.** CO_2 in power plant flue gases or other industrial gas emissions streams can be brought in contact with calcium carbonate and water, as in weathering processes, and a spontaneous chemical reaction takes place [$CO_2 + H_2O + CaCO_3 \rightarrow Ca^{2+} + 2(HCO_3^-)$]. The resulting dissolved calcium bicarbonate ions can be injected into the ocean (Rau and Caldeira 1999, 2002). This would avoid the need for molecular CO_2 capture and purification and would convert most of the CO_2 to relatively benign, ionic species. Modeling studies showed that such carbon storage would be effective for thousands of years and with less impact to ocean pH than directly injecting a comparable quantity of carbon as molecular CO_2 in the ocean (Caldeira and Rau 2000). Initial cost estimates have shown that for treatment of coastal CO_2 point sources this form of CO_2 mitigation would be less expensive than more conventional molecular CO_2 capture and geologic storage (Sarv and Downs 2002, Rau et al. 2004). However, the true cost, capacity, effectiveness, and environmental impact of this approach need further evaluation.

◆ **Ocean Burial of Crop Residue.** It has been suggested that organic waste from agriculture be actively buried on the ocean floor, thus enhancing the natural air-to-land-to-ocean carbon sink represented by plant production, soil formation, soil erosion, and river transport to the sea (Metzger and Benford 2001). This approach would prevent some if not most of the oxidation of residue biomass on land and thus eliminate the resulting flux of CO_2 back to the atmosphere. Ocean sites with existing permanent anoxia (e.g. offshore from major river deltas) could be used to slow or avoid oxidation of the biomass once on the ocean floor prior to its permanent burial by natural sedimentation. Concerns to be more thoroughly addressed include (1) the cost of collecting, bundling, transporting, and sinking the residue; (2) the consequences to the fertility of the remaining cropland; and (3) the ultimate impacts to the marine environment.

◆ **Ocean Disposal of CO_2 Emulsions.** Golomb et al. (2001, 2004) have shown that CO_2 can form a dense emulsion when combined $CaCO_3$ (e.g., limestone) particles under pressure. Such emulsions could be formed prior to or during ocean CO_2 injection, with the resulting CO_2-rich mass sinking to and stored on the ocean floor. Studies suggest that at deep ocean temperatures and pressures, the CO_2 might be sequestered indefinitely by this approach. The method and cost of (1) initial CO_2 capture and purification, (2) limestone/carbonate preparation, and (3) transporting reactants to ocean sites, as well as the marine environmental consequences of this approach are among the issues that remain to be addressed in detail.

◆ **Other Methods.** The preceding list of CO_2 mitigation options involving the ocean may not be exhaustive, and any future research portfolio should be open to the possibility of new approaches or mix of approaches.

In summary, the ocean is currently playing an important role in mitigating significant amounts of anthropogenic CO_2 via passive air-to-sea transfer. The chemical impacts accompanying this flux, including ocean acidification, may have serious environmental consequences. Any scheme that introduces additional molecular CO_2 (unreacted or uncombined) to the ocean will contribute to these impacts. There are alternative, potentially promising ways for ocean carbon addition that lessen or avoid these impacts. However, such approaches are likely to be attended by other unresolved issues of their own, and the economic and environmental costs and benefits of such schemes could be the subject of further research. All options for safely using the ocean's potential for carbon uptake need to be seriously and carefully considered.

6.5 Summary

The development of the technical, economic, and environmental feasibility and acceptability of CO_2 sequestration strategies has important implications for meeting the needs for food, fiber, and energy while minimizing GHG emissions. As the current energy infrastructure evolves around fossil fuels, the viability of sequestration could provide many options for a future of near-net-zero GHG emissions. Carbon sequestration has the potential to reduce the cost of stabilizing GHG concentrations in the atmosphere, conceivably at lower costs than other alternatives, if successful, and further support domestic and global economic growth. If carbon sequestration were to prove technically and economically viable, fossil fuels could continue to play an important role as a primary energy supply.

This chapter reviews various forms of advanced technology, their potential for reducing emissions by capturing, storing, and sequestering carbon dioxide, and the R&D strategies intended to accelerate the development of these technologies. Although uncertainties exist about both the level at which GHG

concentrations might need to be stabilized and the nature of the technologies that may come to the fore, the long-term potential of advanced technologies to capture, store, and sequester carbon dioxide is estimated to be significant, both in reducing emissions (as shown in the figure at the beginning of this chapter) and in reducing the costs for achieving those reductions, as suggested by Figure 3-14. Further, the advances in technology development needed to realize this potential, as modeled in the associated analyses, animate the R&D goals for each carbon dioxide capture and sequestration technology area.

As one illustration among the many hypothetical cases analyzed,[16] GHG emissions were constrained to a high level over the course of the 21st century in such a way that a stabilized GHG concentration levels could ultimately be attained. The lowest-cost arrays of advanced technology in capturing, storing, and sequestering carbon dioxide, when compared to a reference case, resulted in reduced or avoided emissions of between 10 and 110 GtC over 100 years. The breadth of this range is due to a large degree of uncertainty at this point in time in the cost and viability of some sequestration technologies. For perspective, these quantities amounted to, roughly, between 2 and 20 percent of all GHG emissions reduced, avoided, captured and stored, or otherwise withdrawn and sequestered needed to attain this level over the same period. Similarly, the costs for achieving such emissions reductions, when compared to the reference case, were reduced by roughly a factor of 3. See Chapter 3 for other cases and other scenarios.

As described in this chapter, CCTP's technology development strategy supports achievements in this range. The overall strategy is summarized schematically in Figure 6-3. Advanced technologies are seen entering the marketplace in the near-, mid-, and long-terms, where the long-term is sustained indefinitely. Such a progression, if successfully realized worldwide, would be consistent with attaining the potential for carbon dioxide capture and sequestration portrayed at the beginning of this chapter.

The timing and pace of technology adoption are uncertain and must be guided by science and supported by appropriate policies (see Approach 7, Chapters 2 and 10). In the case of the illustration above, the first GtC per year (1GtC/year) of reduced

---

[16] In Chapter 3, various advanced technology scenarios were analyzed for cases where global emissions of GHGs were hypothetically constrained. Over the course of the 21st century, growth in emissions was assumed to slow, then stop, and eventually reverse in order to ultimately stabilize GHG concentrations in the Earth's atmosphere at levels ranging from 450 to 750 ppm. In each case, technologies competed within the emissions-constrained market, and the results were compared in terms of energy (or other metric), emissions, and costs.

or avoided emissions, as compared to an unconstrained reference case, would need to be in place and operating, roughly, as early as 2040. For this to happen, a number of new or advanced technologies to capture, store, and sequester carbon dioxide would need to penetrate the market at significant scale before this date. Other cases would suggest faster or slower rates of deployment. See Chapter 3 for other cases and other scenarios.

Throughout Chapter 6, the discussions of the current activities in each area support the main components

of this approach to technology development. The activities outlined in the current portfolio sections address the highest-priority investment opportunities for this point in time. Beyond these activities, the chapter identifies promising directions for future research, identified in part by the technical working group and assessments and inputs from non-Federal experts. CCTP remains open to a full array of promising technology options as current work is completed and changes in the overall portfolio are considered.

Technologies for Goal #3: CO$_2$ Capture, Storage, and Sequestration

	NEAR-TERM	MID-TERM	LONG-TERM
Carbon Capture	• CSLF and CSRP • Post Combustion Capture • Pre-Combustion Technologies • Oxy-Fuel Combustion • Oxygen Separation Technologies	• Capability to Capture Most CO$_2$ Emissions • Novel Capture Technologies • Low-Cost Oxygen • Biomass Coupled with CCS	• Novel In-Situ CO$_2$ Conversion • Capture CO$_2$ Directly from Atmosphere
Geologic	• Reservoir Characterization • Safety, Health, and Environmental Risk Assessment • Understand Underground CO$_2$ Reactions & Microbial Processes • Enhanced Hydrocarbon Recovery • Enhanced Coal-Bed Methane • Large-Scale Demonstration • CO$_2$ Transport Network Design	• Geologic Storage Proven Safe • Well Sealing Techniques Demonstrated • Mineralization: Solid Carbonates • Reliable and Accurate Inventory Monitoring • Well-Established CO$_2$ Transport Infrastructure	• Sufficient CO$_2$ Storage Capacity • Track Record of Successful CO$_2$ Storage Experience
Terrestrial	• Reforestation • Soils Conservation • Vegetation in Urban Settings	• Soils Uptake & Land Use • Inter-relationship among CO$_2$, CH$_4$ & N$_2$O • Sequestration Decision Support Tools • M&M Tools to Validate Terrestrial Sequestration • Bio-Based & Recycled Products	• Biological Sequestration • Large-Scale Sequestration • Minimal Deforestation • Carbon & CO$_2$ Based Products & Materials
Ocean	• Effective Dilution of Direct Injected CO$_2$	• Ocean CO$_2$ Biological Impacts Addressed • Carbonate Dissolution / Alkalinity Addition	• Safe Long-Term Ocean Storage

Figure 6-3. Technologies for Goal #3: CO$_2$ Capture, Storage, and Sequestration
(Note: Technologies shown are representations of larger suites. With some overlap, "near-term" envisions significant technology adoption by 10–20 years from present, "mid-term" in a following period of 20–40 years, and "long-term" in a following period of 40–60 years. See also List of Acronyms and Abbreviations.)

6.6 References

Boyd, P.W., A.J. Watson, C.S. Law, E.R. Abraham, T. Trull, R. Murdoch, D.C.E. Bakker, A.R. Bowie, K.O. Buessler, H. Chang, M.A. Charette, P. Croot, K. Downing, R.D. Frew, M. Gall, M. Hadfield, J.A. Hall, M. Harvey, G. Jameson, J. La Roche, M.I. Liddicoat, R. Ling, M. Maldonado, R.M. McKay, S.D. Nodder, S. Pickmere, R. Pridmore, S. Rintoul, K. Safi, P. Sutton, R. Strzepek, K. Tanneberger, S.M. Turner, A. Waite, and J. Zeldis. 2000. A mesoscale phytoplankton bloom in the polar southern ocean stimulated by iron fertilization. *Nature* 407:695-702.

Caldeira K, Rau, GH. 2000. Accelerating carbonate dissolution to sequester carbon dioxide in the ocean: geochemical implications. *Geophysical Research Letters* 27 (2): 225-228.

Caldeira K, Wickett ME. 2003. Anthropogenic carbon and ocean pH. *Nature* 425 (6956): 365-365.

CCTP (See U.S. Climate Change Technology Program)

Coale K.H., K.S. Johnson, S.E. Fitzwater, R.M. Gordon, S. Tanner, F.P. Chavez, L. Ferioli, C. Sakamoto, P. Rogers, F. Millero, P. Steinberg, P. Nightingale, D. Cooper, W.P. Cochran, M.R. Landry, J. Constantinou, R. Rollwagen, A. Trasvina, and R. Kudela. 1996. A massive phytoplankton bloom induced by an ecosystem-scale iron fertilization experiment in the equatorial Pacific Ocean. Nature 383:495-501.

Coale, K.H., K.S. Johnson, S.E. Fitzwater, S.P.G. Blain, T.P. Stanton, and T.L. Coley. 1998. IronEx-I, an in situ iron-enrichment experiment: experimental design, implementation and results. Deep-Sea Research, Part II: *Topical Studies in Oceanography* 45:6 919-945.

Davison, Jr., P. Freund, and A. Smith. 2001. *Putting carbon back into the ground.* Technical Report. Paris: International Energy Agency (IEA) Greenhouse Gas R&D Programme. http://www.ieagreen.org.uk/putcback.pdf

DOE (See U.S. Department of Energy)

Energy Information Administration (EIA). 2004. *International energy outlook 2004.* Washington, D.C: U.S. Department of Energy. http://www.eia.doe.gov/oiaf/ieo/pdf/0484(2004).pdf

Energy Information Administration (EIA). 2005. *Annual energy outlook 2005.* Washington, DC: U.S. Department of Energy. http://www.eia.doe.gov/oiaf/aeo/pdf/0383(2005).pdf

Eve, M.D., M. Sperow, K. Paustian, and R. Follett. 2002. National-scale estimation of changes in soil carbon stocks on agricultural lands. *Environmental Pollution* 116: 431-438.

Feely R.A., C.L. Sabine, K. Lee, W. Berelson, J. Kleypas, V.J. Fabry, and F.J. Millero. 2004. Impact of anthropogenic CO_2 on the $CaCO_3$ system in the oceans. *Science* 305 (5682): 362-366.

Follett R.F., J.M. Kimble, and R. Lal. 2001. The potential of US grazing lands to sequester carbon and mitigate the greenhouse effect. New York: Lewis Publishers.

Golomb, D., and A. Angelopoulos. 2001. A benign form of CO_2 sequestration in the ocean. Greenhouse gas control technologies, *Proceedings of the 5th International Conference on Greenhouse Gas Control Technologies.* Victoria, Australia: CSIRO Publishing: Collingwood.

Golomb, D, E. Barry, D. Ryan, C. Lawton, and P. Swett. 2004. Limestone-particle-stabilized macroemulsion of liquid and supercritical carbon dioxide in water for ocean sequestration. *Environmental Science and Technology* 38: 4445-4450

Herzog H. 2001. What future for carbon capture and sequestration? *Environmental Science and Technology* 35(7):148A-153A.

Hoffert, et al. 2002. Advanced technology paths to global climate stability: energy for a greenhouse planet. *Science* 298:981-7.

Intergovernmental Panel on Climate Change (IPCC). 2000. *Land use, land-use change, and forestry.* Cambridge, UK: Cambridge University Press.

Intergovernmental Panel on Climate Change (IPCC). 2001a. Chapter 3: the carbon cycle and atmospheric carbon dioxide. In *Climate change 2001: Working Group I: the scientific basis*. Cambridge, UK: Cambridge University Press. http://www.grida.no/climate/ipcc_tar/wg1/095.htm

Intergovernmental Panel on Climate Change (IPCC). 2001b. *Summary for policymakers to climate change 2001: synthesis report of the IPCC third assessment report*. Cambridge, UK: Cambridge University Press. http://www.ipcc.ch/pub/un/syreng/spm.pdf

Intergovernmental Panel on Climate Change (IPCC). 2001c. Chapter 4: technological and economic potential of options to enhance, maintain, and manage biological carbon reservoirs and geo-engineering. In *Climate change 2001: mitigation: contribution of working group iii to the third assessment report*. Cambridge, UK: Cambridge University Press. http://www.grida.no/climate/ipcc_tar/wg3/155.htm

Intergovernmental Panel on Climate Change (IPCC). 2005. *Special Report on Carbon Dioxide Capture and Storage*. Cambridge, UK: Cambridge University Press.

International Energy Agency (IEA) Greenhouse Gas Programme (GHG). 1994a. *Carbon dioxide utilisation*. Cheltenham, UK.

International Energy Agency (IEA) Greenhouse Gas Programme (GHG). 1994b. *Carbon dioxide disposal from power stations*. Cheltenham, UK.

International Energy Agency (IEA) Greenhouse Gas Programme (GHG). 2000. *Barriers to overcome in implementation of CO_2 capture and storage*. Report PH3/22. Cheltenham, UK.

Joyce, L.A., and R. Birdsey, eds. 2000. The impact of climate change on American's forests: a technical document supporting the 2000 USDA Forest Service RPA assessment, 133. Gen. Tech. Rep. RMRS-GTR-59. Fort Collins, Colorado: U.S. Department of Agriculture. http://www.fs.fed.us/rm/pubs/rmrs_gtr059.pdf

Kheshgi Hs. 1995. Sequestering atmospheric carbon-dioxide by increasing ocean alkalinity. *Energy 20* (9): 915-922.

Lal et al. 1998. The potential of U.S. cropland to sequester carbon and mitigate the greenhouse effect. Chelsea, MI: Ann Arbor Press.

Martin, J.H., K.H. Coale, K.S. Johnson, S.E. Fitzwater, R.M. Gordon, and 39 others. 1994. Testing the iron hypothesis in ecosystems of the equatorial Pacific Ocean. *Nature* 371.

Metzger, R.A., and G. Benford. 2001. Sequestering of atmospheric carbon through permanent disposal of crop residue. Climatic Change 49 (1-2): 11-19.

Orr, J.C., V.J. Fabry, O. Aumont, L. Bopp, S.C. Doney, R.A. Feely, A. Gnanadesikan, N. Gruber, A. Ishida, F. Joos, R.M. Key, K. Lindsay, E. Maier-Reimer, R. Matear, P. Monfray, A. Mouchet, R.G. Najjar, G.K. Plattner, K.B. Rodgers, C.L. Sabine, J. L. Sarmiento, R. Schlitzer, R.D. Slater, I.J. Totterdell, M.F. Weirig, Y. Yamanaka, and A. Yool. 2005. Anthropogenic ocean acidification over the twenty-first century and its impact on calcifying organisms. *Nature* 437 (7059): 681-686.

Perlack, R.D., L.L. Wright, A. Turhollow, R.L. Graham, B. Stokes, and D. Erbach. 2005. *Biomass as feedstock for a bioenergy and bioproducts industry: the technical feasibility of a billion-ton annual supply*. ORNL/TM-2005/66. Oak Ridge, TN: Oak Ridge National Laboratory. http://feedstockreview.ornl.gov/pdf/billion_ton_vision.pdf

Portner, H.O., M. Langenbuch, and A. Reipschlager. 2004. Biological impact of elevated ocean CO_2 concentrations: Lessons from animal physiology and earth history. *Journal of Oceanography* 60: 705-718.

Post, W.M., C. Izaurralde, J.D. Jastrow, B.A. McCarl, J.E. Amonette, V.L. Bailey, P.M. Jardine, T.O. West, and J. Zhou. 2004. Enhancement of carbon sequestration in US soils. *Bioscience* 54:895-908.

G.H. Rau, and K. Caldeira. 1999. Enhanced carbonate dissolution: a means of sequestering waste CO_2 as ocean bicarbonate. *Energy Conversion and Management* 40 (17): 1803-1813.

G.H. Rau, and K. Caldeira. 2002. Minimizing effects of CO_2 storage in oceans. *Science* 295 (5553): 275-276.

G.H. Rau, K.G. Knauss, W.H. Langer, and K. Caldeira. 2004. CO_2 mitigation via accelerated limestone weathering. American Chemical Society, Division of Fuel Chemistry Preprints 49 (1): 376.

Sabine C.L., R.A. Feely, N. Gruber, R.M. Key, K. Lee, J.L. Bullister, R. Wanninkhof, C.S. Wong, D.W.R. Wallace, B. Tilbrook, F.J. Millero, T.H. Peng, A. Kozyr, T. Ono, and A.F. Rios. 2004. The oceanic sink for anthropogenic CO_2. *Science* 305 (5682): 367-371.

Sarv H., and W. Downs. 2002. CO_2 capture and sequestration using a novel limestone lagoon scrubber - A white paper. Alliance, OH: McDermott Technology, Inc.

Smith, W.H.F., and D.T. Sandwell. 1997. "Global seafloor topography from satellite altimetry and ship depth soundings." *Science* 277:1957-62.

The Royal Society. 2005. "Ocean acidification due to increasing carbon dioxide". The Royal Society, London. ISBN 0 85403 617 2 http://www.royalsoc.ac.uk/document.asp?id=3249

Tuskan, G.A., and M.E. Walsh. 2001. "Short-rotation woody crop systems, atmospheric carbon dioxide and carbon management: a U.S. case study." *The Forestry Chronicle* 77:259-264.

U.S. Climate Change Technology Program (CCTP). 2005. *Technology options for the near and long term*. Washington, DC: U.S. Department of Energy. http://www.climatetechnology.gov/library/2005/tech-options/index.htm

U.S. Department of Energy (DOE). 1999. *Carbon sequestration research and development*. DOE/SC/FE-1. Washington, DC: U.S. Department of Energy. http://fossil.energy.gov/programs/sequestration/publications/1999_rdreport/index.html

U.S. Department of Energy (DOE). 2005. *Carbon sequestration technology roadmap and program plan 2005*. Washington, DC: Office of Fossil Energy. http://fossil.energy.gov/programs/sequestration/publications/programplans/2005/sequestration_roadmap_2005.pdf

U.S. Department of Energy (DOE), National Energy Technology Laboratory (NETL). 2004. A sea floor gravity survey of the Sleipner Field to monitor CO_2 migration. Project Facts. http://www.netl.doe.gov/publications/factsheets/project/Proj247.pdf

White, C.M., B.R. Strazisar, E.J. Granite, J.S. Hoffman, and H.W. Pennline. 2003. "Separation and capture of CO_2 from large stationary sources and sequestration in geological formations—coalbeds and deep saline aquifers." *Journal of Air & Waste Management* 53:645-715.

Reducing Emissions of Non-CO$_2$ Greenhouse Gases

everal gases other than carbon dioxide (CO$_2$) are known to have greenhouse gas (GHG) warming effects. When concentrated in the Earth's atmosphere, these "non-CO$_2$" GHGs can contribute to climate change. The more significant of these are methane (CH$_4$), which can arise from natural gas production, transportation and distribution systems, bio-degradation of waste in landfills, coal mining, and agricultural production; nitrous oxide

(N$_2$O) from industrial and agricultural activities; and certain fluorine-containing substances, such as hydrofluorocarbons (HFCs), perfluorocarbons (PFCs), and sulfur hexafluoride (SF$_6$) from industrial sources (Box 7-1).

The Intergovernmental Panel on Climate Change's (IPCC) *Third Assessment Report* (IPCC 2001) states that "well-mixed" non-CO$_2$ gases, including methane, nitrous oxide, chlorofluorocarbons, and other gases with high global warming potentials (GWPs) may be responsible for as much as 40 percent of the estimated increase in radiative climate forcing between the years 1750 and 2000.[1] In addition, emissions of black carbon (soot), organic carbon and other aerosols, as well as tropospheric ozone and ozone precursors, have important effects on the Earth's overall energy balance.

Developing technologies for commercial readiness that can reduce emissions of these non-CO$_2$ GHGs

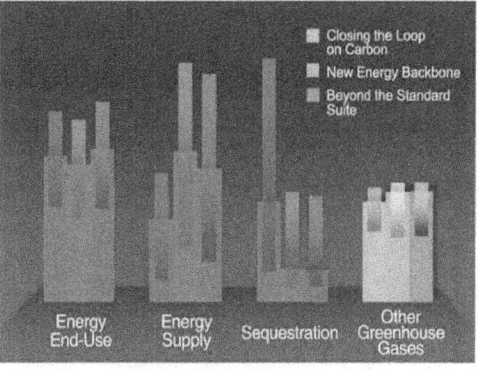

Legend:
- Closing the Loop on Carbon
- New Energy Backbone
- Beyond the Standard Suite

Categories: Energy End-Use, Energy Supply, Sequestration, Other Greenhouse Gases

**Other Greenhouse Gases
Potential Contributions to Emissions Reduction**

Potential contributions of Other Greenhouse Gases to cumulative GHG emissions reductions to 2100, across a range of uncertainties, for three advanced technology scenarios. See Chapter 3 for details.

BOX 7-1

WHAT ARE THE "OTHER" GHGs?

The term "non-CO$_2$ GHGs" covers a broad category of gases and aerosols, but usually refers to methane, nitrous oxide, and the high global warming potential (GWP) gases hydrofluorocarbons (HFCs), perfluorocarbons (PFCs), and sulfur hexafluoride (SF$_6$). Tropospheric ozone, tropospheric ozone precursors, and black carbon (soot) also have important climatic effects. Of these, only ozone is a GHG. Chlorofluorocarbons (CFCs) and other related chemicals contribute to both global warming and stratospheric ozone depletion. Because these chemicals are already being phased out under the Montreal Protocol, they are not addressed in this plan. To streamline terminology for purposes of readability, and unless otherwise noted, the terms "non-CO$_2$ GHGs" and "other GHGs", include methane, nitrous oxide, high-GWP gases, tropospheric ozone, tropospheric ozone precursors, and black and organic carbon aerosols.

---

[1] The radiative forcing due to increases in the well-mixed GHGs between the years 1750 and 2000 is estimated to be 2.43 Wm-2: 1.46 Wm-2 from CO$_2$; 0.48 Wm-2 from CH$_4$; 0.34 Wm-2 from the halocarbons (CFC and HCFC); and 0.15 Wm-2 from N$_2$O.

is an important component of a comprehensive strategy to address concerns about climate change. A recent modeling study (Placet et al. 2004) showed that there is a considerable amount of uncertainty about future rate of growth of non-CO_2 emissions, but most models project that emissions will increase over time in the absence of constraints (see Chapter 3). One set of scenarios that included a wide range of advanced technologies[2] for reducing emissions of non-CO_2 gases showed that emissions could potentially be reduced by a range of between 125 and 160 gigatons (Gt) of carbon-equivalent emissions (cumulatively) over a 100-year planning horizon, as shown in Figure 3-19 and highlighted on the figure above. Although bracketed by a range of uncertainties, this figure suggests both the potential role for advanced technology and a long-term goal for contributions from other GHGs in the future global economy.

In the context of global warming, emissions of the

non-CO_2 GHGs are usually converted to a common and roughly comparable measure of the "equivalent CO_2 emissions." This conversion is performed based on physical emissions, weighted by each gas' global warming potential (GWP). The GWP is the relative ability of a gas to trap heat in the atmosphere over a given timeframe, compared to the CO_2 reference gas (per unit weight). GWP values allow for a comparison of the impacts of emissions and reductions of different gases, although they typically have an uncertainty of ±35 percent (EPA 2005). The choice of timeframe is significant and can change relative GWPs by orders of magnitude. All non-CO_2 gases are compared to CO_2, which has a GWP of one. The GWPs of other GHGs, using a 100-year time horizon, range from 23 for methane to 22,200 for SF_6, as shown in Box 7-2.

Non-CO_2 gases have different GWPs due to differences in atmospheric lifetimes and effectiveness in trapping heat. Methane and some HFCs have relatively short atmospheric lifetimes as compared to other non-CO_2 gases. Thus, emissions reductions among these gases manifest themselves as lower atmospheric concentrations in a matter of a few decades. PFCs and SF_6, in contrast, can remain in the atmosphere for thousands of years. Emissions of these GHGs essentially become permanent additions to the Earth's atmosphere, with concomitant increases in the atmosphere's ability to capture and retain radiant heat. Finally, tropospheric ozone and black carbon aerosols (soot) are very short-lived in the atmosphere (i.e., remaining airborne for a period of days to weeks) and therefore do not become well-mixed in the atmosphere. Primarily for this reason, GWP metrics have not been assigned to these gases and aerosols, but they are nonetheless recognized as significant contributors to climate change.

There is a strong record of successful collaboration between industry and government to reduce emissions of non-CO_2 gases, and these partnerships provide a solid foundation from which to pursue additional technological developments and more substantial future emission reductions. Some highlights of the current activities include:

◆ Industry and the U.S. Environmental Protection Agency (EPA) have developed nine successful public/private partnerships to reduce emissions of methane and high-GWP gases.[3] These programs have led to substantial emission reductions; with U.S. methane emissions in 2003 10 percent below

BOX 7-2

GLOBAL WARMING POTENTIALS OF SELECTED GHGs (100-Year Time Horizon)

GAS	GWP
Carbon dioxide (CO_2)	1
Methane (CH_4)	23
Nitrous oxide (N_2O)	296
Hydrofluorocarbons:	
HFC-23	12000
HFC-125	3400
HFC-134a	1300
HFC-143a	4300
HFC-152a	120
HFC-227ea	3500
HFC-236fa	9400
HFC-43-10mee	1500
Fully Fluorinated Species:	
CF_4	5700
C_2F_6	11900
C_4F_{10}	8600
C_6F_{14}	9000
SF_6	22200

(Source: IPCC 2001)

---

[2] The technologies discussed in this chapter were included in this set of scenarios.

[3] The Landfill Methane Outreach Program, Natural Gas STAR Program, AgSTAR Program, Coalbed Methane Outreach Program, SF_6 Emission Reduction Partnership for Electric Power Systems, Voluntary Aluminum Industrial Partnership, SF_6 Emission Reduction Partnership for the Magnesium Industry, PFC Reduction/Climate Partnership with the Semiconductor Industry, and HCFC-22 Partnership Program.

BOX 7-3

Methane to Markets

The United States is collaborating with 17 countries (Argentina, Australia, Brazil, Canada, China, Colombia, Ecuador, Germany, India, Italy, Japan, Mexico, Nigeria, Russia, South Korea, Ukraine, and the United Kingdom) and over 220 organizations from the private sector, financial community, and other governmental and non-governmental institutions to undertake activities to capture and use methane at landfills, coal mines, oil and gas systems, and agricultural operations.

The United States is committing up to $53 million over the next five years to facilitate the development and implementation of methane projects in developing countries and countries with economies in transition. EPA plays a lead role in the partnership and coordinates efforts with several other departments, including the Departments of State and Energy, the U.S. Trade and Development Agency and the U.S. Agency for International Development. See http://www.methanetomarkets.org.

1990 levels and emissions of many sources of high-GWP gases also declining (EPA 2005). They also provide excellent forums for transferring technical information in an efficient and cost-effective manner. The partnership programs host or participate in annual technical conferences with the respective industries. Public-private partnerships help facilitate effective use of the technologies that are or will soon become available.

◆ The Federal Government is currently addressing agricultural sources of methane and nitrous oxide through a combination of voluntary partnerships and research, development, and demonstration (RD&D) efforts. Cooperative efforts between Government and the agriculture industry are needed to evaluate and develop technologies for lowering N_2O emissions from soils and methane emissions from livestock enteric fermentation.

◆ The U.S. Department of Energy (DOE) and EPA have teamed to co-fund the development of the first ventilation air methane (VAM) project in the United States utilizing a thermal flow reversal reactor to oxidize mine ventilation air, which contains low concentrations of methane. The process generates thermal energy that can have many uses. EPA is also working cooperatively with Natural Resources Canada (NRCan) to deploy a similar technology developed by NRCan's CANMET Energy Technology Centre (CETC).

◆ An international network of those involved in research on non-CO_2 GHGs has been formed by the International Energy Agency (IEA) Greenhouse Gas R&D Programme, EPA, and the European Commission Directorate General Environment. The experts involved in this network cover emissions, abatement options, and systems modeling for policy advice. The network provides an international forum for identification of needed research, as well as creating opportunities for international deployment of non-CO_2 emission reduction technologies.

◆ An international analytical effort has been undertaken by the Stanford Energy Modeling Forum (EMF) to better characterize the role of non-CO_2 mitigation in addressing climate change.[4] This multi-year effort has led to the development of data on the cost and performance of currently available and near-to-market technologies to reduce non-CO_2 emissions. In addition, the 19 international modeling teams participating in the project have incorporated data on non-CO_2 gases into their economic and integrated assessment models and are improving the capabilities needed to analyze comprehensive climate strategies focusing on both CO_2 and non-CO_2 options.

◆ Established in November 2004, the Methane to Markets Partnership (Box 7-3) is a new global initiative to advance international cooperation on the recovery and use of methane as a valuable clean energy source. The partnership will increase energy security, enhance economic growth, improve air quality, improve industrial safety, and reduce GHG emissions throughout the world. Methane to Markets has the potential to reduce net methane emissions by up to 50 million metric

---

[4] Results from this study, EMF 21, are to be published in a special issue of the Energy Journal in 2005. See http://www.stanford.edu/group/EMF/research/index.htm.

tons of carbon equivalent annually by 2015 and continue at that level or higher in the future.

These partnerships and others that are discussed in this chapter demonstrate the potential for significant near-term emission reductions from currently available technologies. In addition, longer-term analyses have identified the potential for current and future technologies to lead to even more significant emission reductions. Historically, non-CO_2 gases were either not included or were treated in a cursory manner in climate change modeling and scenario studies. This situation is changing, however, and many modelers are incorporating the non-CO_2 gases into their models and are developing the capability to assess the role of the non-CO_2 gases in addressing climate change. Studies published to date indicate that substantial mitigation of future increases in radiative forcing could be achieved by reducing emissions of these other GHGs. It is possible that such reductions could contribute as much as one-half of the abatement levels needed to stay within a total radiative forcing gain that would be consistent with commonly discussed stabilization ranges of CO_2 concentrations.[5]

Achieving significant reductions in the emissions of the non-CO_2 gases is possible, taking into account the current achievements in reducing emissions as well as the results of detailed analyses of the technical and economic potential to reduce emissions from particular sources and sectors. Based on the information presented in this chapter, it is possible to achieve CH_4 emissions reductions of 40 to 60 percent by 2050, and 45 to 70 percent by 2100. Emissions of N_2O can be reduced by 25 to 30 percent by 2050, and 50 percent by 2100 (DeAngelo et al. forthcoming, Delhotal et al. forthcoming). In addition, it is possible to reduce emissions of high-GWP gases by 55 to 75 percent by 2050, and 60 to 80 percent by 2100 (Schaefer 2006).

There are a number of potentially fruitful areas for technologies to mitigate growth in emissions of non-CO_2 GHGs and strong promise that over time emissions could be reduced substantially. The strategy for addressing non-CO_2 GHGs has two key elements. First, it focuses on the key emission sources of these GHGs and identifies specific mitigation options and research needs by gas, sector, and source. Given the diversity of emission sources, a generalized technology approach is not practical. Second, the strategy emphasizes both the expedited development

Target Areas for Reducing Emissions of Non-CO_2 GHGs (2000 Emissions in Tg CO_2 Equivalent)

TARGET AREA	U.S. EMISSIONS	% OF TOTAL U.S. NON-CO_2	GLOBAL EMISSIONS	% OF GLOBAL NON-CO_2
CH_4 Emissions from Energy and Waste	371	34	2836	31
CH_4 and N_2O Emissions from Agriculture	444	41	5428	60
Emissions of High Global Warming Potential (GWP) Gases	139	13	368	4
N_2O Emissions from Combustion and Industrial Sources	98	9	390	4
Emissions of Tropospheric Ozone Precursors and Black Carbon	N/A*			

* Emissions estimates exist but they cannot be converted into CO_2 equivalent units.
Sources: EPA 2005, 2004

Table 7-1. Target Areas for Reducing Emissions of Non- CO_2 GHGs(2000 Emissions in Tg CO_2 Equivalent) [6]

---

[5] U.S. Climate Change Science Program, Prospectus for Synthesis and Assessment Product 2.1. http://www.climatescience.gov/Library/sap/default.htm.

[6] For this chapter, the GWP-weighted emissions of methane (estimated at 21) are presented in terms of equivalent emissions of carbon dioxide (CO_2), using units of teragrams of carbon dioxide equivalents (Tg CO_2 equivalent). To convert the emission estimates included in this chapter to giga-tonnes of carbon (GtC), multiply the emissions estimate by .000272. For example, 200 Tg CO_2 equivalent X (.000272) = .054 GtC.

U.S. and Global Methane (CH₄) Emissions from Energy and Waste (2000 Emissions in Tg CO₂ Equivalent)

SOURCE	U.S. EMISSIONS	% OF TOTAL U.S. NON-CO₂ GHG EMISSIONS	GLOBAL EMISSIONS	% OF GLOBAL NON-CO₂ GHG EMISSIONS
Landfills	130.7	12	814	9
Coal Mining	56.2	5	439	5
Natural Gas and Oil Systems	149.7	14	1013	11
Wastewater Treatment	34	3	569	6
Total	371	34	2836	31

* Emissions estimates exist but they cannot be converted into CO₂ equivalent units. Sources: EPA 2005, 2004

Table 7-2. U.S. and Global Methane (CH₄) Emissions from Energy and Waste (2000 Emissions in Tg CO₂ Equivalent)

and deployment of near-term and close-to-market technologies and expanded R&D into longer-term opportunities leading to large-scale emission reductions. By stressing both near- and long-term options, the strategy offers maximum climate protection in the near term and a roadmap to achieve dramatic gains in later years.

The discussion of the key emission sources of other GHGs is organized around five broad categories—or "target areas"—listed in Table 7-1. Following the table, each target area is discussed in subsequent technology sections. Each of these technology sections includes a sub-section describing the current portfolio. The technology descriptions include a link to the CCTP Technology Options for the Near and Long Term (CCTP 2003).

7.1 Methane Emissions from Energy and Waste

In 2000, methane emissions from the energy and waste sectors accounted for 31 percent of global non-CO₂ GHG emissions (Table 7-2), and nearly 50 percent of global methane emissions. The major emission sources in these sectors include coal mining, natural gas and oil systems, landfills, and wastewater treatment. As Table 7-2 shows, among the energy and waste-related methane emission sources, oil and gas systems, and landfills are the largest emission sources, accounting for 9 and 11 percent, respectively, of global non-CO₂ emissions.

The energy and waste sectors present some of the most promising and cost-effective near-term reduction opportunities. Reducing methane emissions, the primary component of natural gas, can be cost-effective in many cases due to the market value of the recovered gas. Efforts in the United States to voluntarily encourage these economically attractive opportunities have already been successful by focusing on the deployment of available, cost-effective technologies. As Table 7-3 shows, emissions from the key sources in the United States have declined in absolute terms by about 16 percent since 1990, equal to about 65 teragrams of carbon dioxide equivalent (Tg CO₂ equivalent).

Despite this success, significant opportunities remain for further emission reductions through the expanded deployment of currently available technologies and the development of promising new technologies. These longer-term technologies could lead to substantial additional methane reductions in the future. The remainder of this section discusses these technical opportunities for the three major emission sources in this category: landfills, oil and gas systems, and coal mines.

Landfills

Methane emissions from landfills result from the decomposition of organic material (yard waste, food waste, etc.) by bacteria in an anaerobic environment. Emission levels are affected by site-specific factors such as waste composition, moisture, and landfill size. Landfills are the second largest anthropogenic methane emission source in the United States, releasing an estimated 131 Tg CO₂ equivalent to the atmosphere in 2003 (EPA 2005). Globally, landfills

Change in U.S. Methane (CH$_4$) Emissions from Energy and Waste

SOURCE	1990 EMISSIONS	2000 EMISSIONS	% CHANGE
Landfills	172	130.7	- 24
Coal Mining	82	56.2	- 32
Natural Gas & Oil	148	149.7	+1
Total	402	337	- 16

Source: EPA 2005.

Table 7-3. Change in U.S. Methane (CH$_4$) Emissions from Energy and Waste

are also a significant emission source, accounting for an estimated 814 Tg CO$_2$ equivalent in 2000 or almost 10 percent of global non-CO$_2$ emissions (Table 7-2). The majority of emissions currently come from developed countries, where sanitary landfills facilitate the anaerobic decomposition of waste. Emissions from developing countries, however, are expected to increase as solid waste will be increasingly diverted to managed landfills as a means of improving overall waste management. By 2020, three regions are projected to each account for more than 10 percent of global methane emissions from landfills: Africa (16 percent), Latin America (13 percent) and Southeast Asia (12 percent) (EPA 2004).

Potential Role of Technology

The principal approach to reduce methane emissions from landfills involves the collection and combustion (through use for energy or flaring) of landfill gas (LFG). LFG utilization technologies can be divided into two main categories: electricity generation and direct gas use. About 75 percent of the projects in the United States involve electricity generation, using reciprocating engines or combustion turbines. Direct use technologies account for about 25 percent of total projects, but their implementation has grown in recent years. Some of these technologies use LFG directly as a medium-Btu fuel, while others require the gas to be upgraded and delivered to a natural gas pipeline.

Technology Strategy

Additional CH$_4$ emission reductions at landfills can be achieved through RD&D efforts focused on improvements in LFG collection efficiency, gas

utilization technologies, and alternatives to existing solid waste management practices. In the near term, RD&D efforts focused on improving collection efficiency and demonstrating promising emerging gas use technologies can yield significant benefits. These approaches could increase emission reductions from the waste currently contained in landfills, which will emit CH$_4$ for 30 or more years. Longer-term reductions will result from research on advanced utilization technologies and development of solid waste management alternatives, such as bioreactor landfills.

Current Portfolio

The current Federal portfolio focuses on three areas:

◆ Research and development of anaerobic and aerobic bioreactor landfills that more quickly stabilize the readily decomposable organic constituents of the waste stream through enhanced microbiological processes. The goal is to have three to five commercial full-scale anaerobic and aerobic bioreactor landfill demonstration units operational by the close of 2006 plus increased market penetration 2007–2012. An additional goal is to further evaluate environmental and public-health impacts, and design and operational issues.[7]

◆ R&D of emerging technologies that facilitate the conversion of LFG to readily usable forms, such as compressed natural gas/liquefied natural gas, and methanol/ethanol. Near-term goals to convert landfill gas to alternative uses include verifying performance of LNG conversion technology application on landfill gas and converted vehicle

---

[7] See Section 4.1.1 (CCTP 2005): http://www.climatetechnology.gov/library/2005/tech-options/tor2005-411.pdf.

performance, development of additional commercially available LNG vehicles (e.g., solid waste collection trucks), and development of distribution/fueling infrastructure. Mid-term goals target research on cost-effective separation technology applications for pipeline quality gas production (Figure 7-1) and to evaluate and demonstrate technologies for producing commercial carbon dioxide.[8]

◆ R&D on improving LFG collection efficiency and enhancing electricity production from LFG through new and improved electricity generation technologies (fuel cells, microturbines, Organic Rankine Cycle, and Stirling-Cycle engines).[9]

Future Research Directions

The current portfolio supports the main components of the technology development strategy and addresses the highest priority current investment opportunities in this technology area. For the future, CCTP seeks to consider a full array of promising technology options. From diverse sources, suggestions for future research have come to CCTP's attention. Some of these, and others, are currently being explored and under consideration for the future R&D portfolio.

Future applied research efforts in the near term could focus efforts to improve LFG collection efficiencies, including research on the design, construction, and operational effectiveness of horizontal wells and other new gas collection systems. Research could also be targeted on the development of additional economical gas utilization technologies and optimizing methane oxidation by cover soils or other advanced cover materials. Development and deployment of near-term technologies to recover LFG from current waste disposal sites could reduce emissions by 50 percent (Delhotal et al. forthcoming).

Over the long term, emissions could theoretically be eliminated through the commercialization and deployment of advanced waste processing and treatment systems such as integrated systems approaches for waste management that could reduce the magnitude of landfill waste and nearly eliminate new landfill waste, such as:

◆ Source-separation of the solid waste stream into processing categories (recyclables, organics, inerts, etc.) for complete recycling and reuse. This could include (1) designing products to tag and identify waste for recycling; (2) facilitating the decomposition of organics through mechanical

Figure 7-1. Capturing and marketing methane emissions from energy and waste systems can be an economically attractive means for reducing GHG emissions.
Courtesy: EPA

biological treatment, followed by rapid and controlled aerobic composting of drier feedstocks, and anaerobic decomposition of wet organics in digesters along with enhanced methane gas recovery; and (3) alternatively, using engineered bacteria that process/break down organic waste without producing methane.

◆ Centralized or distributed waste management systems that include on-site conversion of waste to hydrogen, other fuels, or electricity. These systems would include pyrolosis, whereby waste is significantly reduced in volume to a glass-like cullet, and gasification, whereby waste is converted to a liquid fuel.

◆ Potential technology options include sort/weight recognition technology; tagging and tracking technology; large and small-scale waste conversion to fuels, power, and products; and genetically engineered bacteria.

Coal Mines

Coal mines are a significant methane emission source in the United States and worldwide, accounting for

---

[8] See Section 4.1.2 (CCTP 2005): http://www.climatetechnology.gov/library/2005/tech-options/tor2005-412.pdf.

[9] See Section 4.1.3 (CCTP 2005): http://www.climatetechnology.gov/library/2005/tech-options/tor2005-413.pdf.

about 10 percent of total anthropogenic methane emissions (EPA 2004). Methane trapped in coal deposits and in the surrounding strata is released during normal mining operations in both underground and surface mines. In addition, handling of the coal after mining (e.g., through storage, processing, and transportation) results in methane emissions. Underground mines are the largest source of coal mine methane (CMM) emissions.

Emissions of CMM in the United States in 2000 were 56 Tg CO_2 equivalent and are projected to increase to 70 Tg CO_2 equivalent by 2010 (EPA 2005). Worldwide emissions of methane from the coal industry are estimated to be 432 Tg CO_2 equivalent and are expected to rise to 495 Tg CO_2 equivalent by the year 2010 as coal production increases (EPA 2004). Globally, almost all CMM emissions come from the major coal producing countries and regions of China; India; the United States; the Confederation of Independent States; Australia; Central, Eastern, and Western Europe; the United Kingdom; and Southern Africa.

Underground mines present the greatest opportunities for reducing emissions; however, emission reductions are also possible at surface mines. Emissions from both underground and surface mines vary, depending on the technology used to mine the coal, the rate of coal production, the technologies employed to remove the methane from the mines, and the local geological conditions.

Potential Role of Technology

Upstream and downstream technologies are integral to reducing methane emissions from coal mines. The most important upstream technological contributions are in the recovery of methane from mine degasification operations and in the oxidation of low-concentration methane in mine ventilation air. Degasification systems are used to remove methane from the coal seams to provide for a safe working environment. These systems generally consist of boreholes drilled into the coal seams and adjacent strata, with in-mine and surface gathering systems used to extract and collect methane. CMM can be recovered in advance of mining or after mining has occurred, and recovery may consist of surface wells, in-mine boreholes, or some combination of the two.

From a technical viewpoint, the most appropriate drainage technology depends on the surface topography, subsurface geology, reservoir characteristics, mine layout, and mine operations. Degasification technologies are used around the world and are commonplace in most of the aforementioned countries. Surface gob wells are used to extract methane after mining has occurred, and in-mine horizontal boreholes are standard at many gassy mines. However, advanced degasification employing long-hole in-mine directional drilling has only been successful in a limited number of countries, including the United States, Australia, China, Japan, United Kingdom, Germany, and Mexico; it is currently being tested in Ukraine. Only the United States and Australia have had success with pre-mine drainage using surface wells. Although gas drainage is practiced primarily at underground mines, drainage is also occurring at surface mines in some countries, including the United States, Australia, and Kazakhstan. Horizontal boreholes can be drilled into the coal seam ahead of mining and the methane extracted.

In a number of countries, commercially applied technologies have led to large reductions in CMM emissions through use of the captured methane. These technologies have included the use of CMM as fuel for power generation (primarily internal combustion engines), injection into the natural gas pipeline system and local gas distribution networks, boiler fuel for use at the mine, local heating needs, thermal drying of coal, vehicle fuel, and as a manufacturing feedstock (e.g., methanol, carbon black, and dimethyl ether production). Technology advances in gas processing over the past decade have also resulted in projects to upgrade the quality of CMM and liquefy the gas, which in turn provide more end-use options and improve access to markets.

Although considerable effort is still directed at improving methane drainage recovery efficiencies and broadening the application of end-use technologies, attention is also focused on the capture and use of coal mine ventilation air methane (VAM). Mine ventilation air generally contains less than 1 percent methane in accordance with regulatory standards. The low concentration greatly limits possible uses of the methane. However, VAM is the largest source of underground methane emissions, and presents a significant opportunity to further mitigate GHG emissions from coal mines if capture and use technologies can be successfully applied. Worldwide VAM emissions in 2000 were 238 Tg CO_2 equivalent and are expected to increase to 282 Tg CO_2 equivalent by 2010 and 308 Tg CO_2 equivalent by 2020. Emissions of VAM in the United States in 2000 were about 37 Tg CO_2 equivalent and are anticipated to rise slightly to 40 Tg CO_2 equivalent by 2010 and remain steady thereafter (EPA 2003a).

Technology Strategy

RD&D efforts aimed at emerging methane reduction technologies for coal mines could target VAM and advanced coalbed methane drilling techniques. The development of technologies to use VAM will enable overall emission reductions at underground mines to reach 90 percent, as compared to the current technical recovery limit of 30 to 50 percent (EPA 1999). The most promising approach for recovering VAM emissions is through commercialization of technologies that convert the low-concentration (typically under 1 percent) methane directly into heat using thermal or catalytic flow reversal reaction processes. The heat can then be employed for power production or other heating. Demonstration projects in Australia, Canada, and the United Kingdom have shown that these technologies can be technically viable. The world's first commercial unit is expected to be operative in Australia in 2006, generating enough thermal energy to supply a 6-MW steam turbine. Future efforts will need to focus on continued testing and commercial deployment of VAM combined with market development support to ensure that it is seen by industry as an energy resource, rather than being vented to the atmosphere.

The other potentially important approach to reduce emissions is the development of advanced drilling technologies. Over the 1990s, advances in steerable motors and stimulation techniques have increased the ability to recover a higher percentage of the total methane in coal seams. This methane, much of which is high quality, may then find a viable market. The most promising technologies include in-mine and surface directional drilling systems, which may enable fewer wells to produce more gas, and advanced stimulation techniques, such as nitrogen injection, that increase the recovery efficiency of surface wells. There is also considerable interest in CO_2 injection; however, this is currently not an option for mine degasification. Injecting the CO_2 into the coal seam renders the coal seams unmineable due to the hazard of releasing too much CO_2 into the mine workings. Although it is difficult to characterize the potential for enhanced gas drainage, these technologies have been shown to obtain drainage efficiencies of 70 to 90 percent (EPA 1999). Future RD&D activities will need to focus on the continued testing and commercial deployment of directional drilling and use of other gases in coalbed methane recovery. In addition, market development support will be needed to ensure that increased drained emissions are put to productive use, rather than vented to the atmosphere.

Figure 7-2. Capturing and using methane from coal mine ventilation air represents an economic opportunity to reduce GHG emissions.

Courtesy: DOE

Current Portfolio

The current Federal portfolio focuses on two areas:

◆ Research on advances in **coal mine ventilation air systems** is focused on use of VAM in flow reversal reactors; concentrators to increase the methane concentration to levels that will support oxidation; lean fuel turbines; and as combustion air in small-scale reciprocating engines or large-scale mine-mouth power plants, or as co-combustion medium with waste coal (Figure 7-2). The goal of coal mine ventilation air systems' research, development, demonstration, and deployment (RDD&D) program is market penetration by 2005–2010, ultimately leading by the end of the program to the majority of ventilation air methane emissions mitigated.[10]

◆ Research on advances in **CMM recovery systems** is focused on improving mine drainage system technology through improved directional drilling technologies, in-mine hydraulic fracturing techniques, development of nitrogen and inert gas injection techniques and improved drilling technologies.[11]

◆ The **Coalbed Methane Outreach Program** (CMOP) is working to demonstrate technologies that can eliminate the remaining emissions from degasification systems, and is addressing methane

---

[10] See Section 4.1.4 (CCTP 2005): http://www.climatetechnology.gov/library/2005/tech-options/tor2005-414.pdf.

[11] See Section 4.1.5 (CCTP 2005): http://www.climatetechnology.gov/library/2005/tech-options/tor2005-415.pdf.

emissions in mine ventilation air. Due to enhanced market opportunities for natural gas and power, further refinement of technical options for the capture and utilization of mine methane, a growing reliance on methane degasification in the Western United States, and CMOP's anticipated success in reducing ventilation air methane over the next few years.

Future Research Directions

◆ The current portfolio supports the main components of the technology development strategy and addresses the highest priority current investment opportunities in this technology area. For the future, CCTP seeks to consider a full array of promising technology options. RD&D efforts will be focused on achieving full commercialization and deployment of VAM and advanced coalbed methane drilling techniques. These technologies alone could reduce emissions from underground mining operations by 90 percent (EPA 2003a).

◆ RD&D efforts will be focused on developing new, fully automated mining systems that will almost eliminate methane emissions. Since underground mining represents about 83 percent of U.S. coal mine methane emissions, this would represent the potential for a 75 percent reduction in overall U.S. methane emissions from this source.

Natural Gas and Petroleum Systems

Methane emissions from the oil and gas industry accounted for approximately 11 percent of global non-CO_2 emissions in 2000 (EPA 2004). Russia and the United States accounted for over 30 percent of global methane emissions from oil and gas systems. Emissions occur throughout the production, processing, transmission, and distribution systems and are generally process related. Normal operations, routine maintenance, and system upsets are the primary contributors. Emissions vary greatly from facility to facility and are largely a function of operation and maintenance procedures and equipment. However, over 90 percent of methane emissions from oil and gas systems are associated with natural gas rather than oil-related operations (EPA 2005, 2004).

As demand for oil and gas increases, global methane emissions are projected to increase by more than 72 percent between 1990 and 2020 (EPA 2004). However, in many developed countries there is

increasing concern about the contribution of oil and gas facilities to deteriorating local air quality, particularly emissions of non-methane volatile organic compounds (NMVOCs). Measures designed to mitigate NMVOC emissions, such as efforts to reduce leaks and venting, have the ancillary benefit of reducing methane emissions. In addition, as economies in many Eastern European countries undergo restructuring, efforts are underway to modernize gas and oil facilities. For example, Germany expects to reduce emissions from the former East German system through upgrades and maintenance. Russia also plans to focus on opportunities to reduce emissions from its oil and gas system as part of modernization.

Potential Role of Technology

Reducing methane emissions from the petroleum and natural gas industries necessitates both procedural and technology improvements. Methane emission reduction strategies generally fall into one of three categories: (1) technologies or equipment upgrades that reduce or eliminate equipment venting or fugitive emissions, (2) improvements in management practices and operational procedures, or (3) enhanced management practices that take advantage of improved technology. Each of these technologies and management practices requires a change from business as usual in the schedule and conduct of daily operations. To date, over 90 emission reduction opportunities have been identified by corporate partners in EPA's Natural Gas STAR Program. In many cases, these actions are cost-effective and widely applicable across industry sectors.

Technology Strategy

Despite the current availability of cost-effective methane emission reduction opportunities in the natural gas and petroleum industry, RDD&D efforts could have an important impact on future methane emissions. Both in the near and long terms, RDD&D efforts could focus on increasing market penetration of current emission reduction technologies, improving leak detection and measurement technologies, and developing advanced end-use technologies.

◆ **Current Emission Reduction Technologies** – Perhaps the greatest environmental benefits would be associated with an enhanced demonstration and deployment effort focused on currently available emission reduction technologies. In 2000, deployment of these technologies in the United States reduced emissions by 15 Tg CO_2

equivalent, approximately 12 percent of total industry emissions (EPA 2005). An enhanced effort would encourage additional technology penetration and emissions reductions.

♦ **Leak Detection and Measurement** – Additional benefits could be realized through improvements in and deployment of leak detection and measurement technologies. Although potential industry-wide emission reductions are difficult to quantify, improved identification and quantification of methane losses and leaks would promote mitigation activities. New technologies will allow for quick, relatively inexpensive detection of leaks that are cost-effective to repair. Some of the emerging leak detection and measurement technologies include the Hi-Flow™ Sampler and hand-held optimal imaging cameras that can visualize methane leaks (e.g., Image Multi-Spectral Sensor [IMSS] camera).

♦ **Advancing End-Use Technologies** – Research aimed at advancing fuel cell and microturbine technologies could reduce emissions at remote well sites by enabling remote power generation at these locations. For example, power generated from the lower-quality gas can be used to support instrument air systems and eliminate the need for gas-driven pneumatic devices and pumps.

Current Portfolio

The current Federal R&D portfolio primarily focuses on leak detection measurement and monitoring technologies for natural gas systems. Advanced leak detection and measurement technologies enable quick and cost-effective detection and quantification of fugitive methane leaks (Figure 7-3). Natural gas systems' RDD&D goals related to measurement and monitoring technologies are focused on completing of the development and deployment of advanced measurement technologies like the Hi-Flow™ and on advancing the development of imaging technology for methane leak measurement and facilitate demonstration and deployment.[12]

Future Research Directions

The current portfolio supports the main components of the technology development strategy and addresses the highest priority current investment opportunities in this technology area. For the future, CCTP seeks to consider a full array of promising technology options. From diverse sources, suggestions for future research have come to CCTP's attention. Some of

these, and others, are currently being explored and under consideration for the future R&D portfolio.

Pipelines carrying natural gas as well as facilities where natural gas is liquefied are a source of fugitive emissions of methane. Advances in materials, seals, and valve technology could eliminate or reduce these emissions at the source. Possible research may include:

♦ Development of more accurate and cost-effective leak detection and measurement equipment, which could be effective in reducing fugitive and vented emissions from gas production, processing, transmission, and distribution operations.

♦ Long-term research to explore revolutionary equipment designs. This might focus on "smart equipment," such as smart pipes or seals, that could alert operators to leaks or self-repairing pipelines made of material that can regenerate and automatically seal leaks. Development of additional technologies could enable emission reductions of 50 percent by the middle of the century.

Enhanced leak-detection and measurement efforts can yield significant methane emission reductions. Demonstration of improved technologies has indicated that emissions at compressor stations and gas-processing plants can be reduced cost effectively by as much as 80 to 90 percent. More importantly, an enhanced demonstration and deployment effort focused on currently available emission reduction technologies would encourage additional technology

Figure 7-3. Advances in leak detection and measurement systems for natural gas pipelines are enabling significant reductions in methane emissions.
Courtesy: ITT Corporation

---

[12] See Section 4.1.6 (CCTP 2005): http://www.climatetechnology.gov/library/2005/tech-options/tor2005-416.pdf.

U.S. and Global CH_4 and N_2O Emissions from Agriculture (2000 Emissions in Tg CO_2 Equivalent)

SOURCE	U.S. EMISSIONS	% OF TOTAL U.S. NON-CO_2 GHG EMISSIONS	GLOBAL EMISSIONS	% OF GLOBAL NON-CO_2 GHG EMISSIONS
N_2O Emissions from Agriculture	282	26%	2875	32%
Enteric Methane Emissions	116	11%	1712	19%
Methane Emissions from Manure	38	3%	199	2%
Methane Emissions from Rice Production	8	< 1%	643	7%
Total	443	40%	5429	60%

Sources: EPA 2005, 2004.

Table 7-4. U.S. and Global CH_4 and N_2O Emissions from Agriculture (2000 Emissions in Tg CO_2 Equivalent)

penetration. In the United States alone, this effort could reduce emissions by an estimated 37 Tg CO_2 equivalent in 2010.

7.2 Methane and Nitrous Oxide Emissions from Agriculture

Over 40 percent of total U.S. non-CO_2 GHGs come from methane (CH_4) and nitrous oxide (N_2O) emissions from agriculture (EPA 2005). Globally, agricultural sources of methane and nitrous oxide contribute an estimated 5,428 Tg CO_2 equivalent, nearly 60 percent of global non-CO_2 emissions (EPA 2004). These emissions result from natural biological processes inherent to crop and livestock production and cannot be realistically eliminated, although they can be reduced. For example, emissions of oxides of nitrogen (NO_X) can likely be decreased by 15 to 35 percent through programs that improve crop nitrogen use efficiency, through plant fertilizer technology, precision agriculture, and plant genetics (DeAngelo 2006). Table 7-4 shows N_2O and methane emissions from agricultural sources (Tg CO_2 equivalent).

Key research efforts have focused on the largest agriculture GHG emission sources:

◆ Nitrous oxide emissions from agricultural soil management.

◆ Methane and nitrous oxide emissions from manure management.

◆ Methane emissions from livestock enteric fermentation.

Advanced Agricultural Systems for Nitrous Oxide Emissions Reductions

Low efficiency of nitrogen use in agriculture is primarily caused by large nitrogen losses due to leaching and gaseous emissions (ammonia, nitrous oxide, nitric oxide, and nitrogen). In general, N_2O emissions from mineral and organic nitrogen can be decreased by nutrient and water management practices that optimize a crop's natural ability to compete with processes that result in plant-available nitrogen being lost from the soil-plant system.

Potential Role of Technology

Key technologies in the area of nutrient management can be applicable to N_2O mitigation. They focus on the following areas:

◆ **Precision agriculture** – targeted application of fertilizers, water, and pesticides.

◆ **Cropping system models** – tools to assist farmer management decisions.

◆ **Control release fertilizers and pesticides** – delivery of nutrients and chemicals to match crop demand and timing of pest infestation.

◆ **Soil microbial processes** – use of biological and

chemical methods, such as liming, to manipulate microbial processes to increase efficiency of nutrient uptake, suppress N_2O emissions, and reduce leaching.

◆ **Agricultural best management practices** – limiting N-gas emissions, soil erosion, and leaching.

◆ **Soil conservation practices** – utilizing buffers and conservation reserves.

◆ **Livestock manure utilization** – development of mechanisms to more effectively use livestock manure in crop production.

◆ **Plant breeding** – to increase nutrient use efficiency and decrease demand for pesticides.

Technology Strategy

Technologies and practices that increase the overall nitrogen efficiency while maintaining crop yields represent viable options to decrease N_2O emissions. Focused RDD&D efforts are needed in a number of areas to develop new technologies and expanded deployment of commercially available technologies and management practices (Figure 7-4).

◆ Further development of precision agriculture technologies to meet the fertilizer and energy reduction goals could lead to increased adoption of these technologies and improved performance.

◆ "Smart materials" for prescription release of nutrients and chemicals for major crops currently require modest breakthroughs in materials technology to reach fruition.

◆ Soil microbial processes could also be manipulated to increase N-use efficiency; however, further development is needed to ensure full efficacy and avoid the introduction of environmental risks.

◆ First-generation integrated system models, technology, and supporting education and extension infrastructure need to be implemented, and research on using these techniques to improve management expanded.

◆ Genetically designed major crop plants could utilize fertilizer more efficiently.

◆ Increased extension efforts are needed to fully utilize best management practices.

◆ Basic research on process controls and field monitoring programs are needed to ensure that theoretical understanding exists as technology

Figure 7-4. *Technologies and farming practices, such as precision agriculture and no-till planting, can increase the overall nitrogen efficiency while maintaining crop yields, resulting in reduced nitrous oxide emissions.*
Photo: Tim McCabe, USDA Natural Resources Conservation Service

evolves and that changes in management practices to mitigate GHG emissions actually function as theorized.

◆ Accurate measurement technologies and protocols are needed for assessment and verification.

Current Portfolio

Although many mitigation options for N_2O emissions can be readily identified, their implementation has not been carried out on a large scale. Other than programs to limit nitrogen losses, programs that directly address the issue of N_2O emissions from agricultural soil management are very limited. The current Federal portfolio focuses on N_2O emissions from agricultural soil management; precision agriculture; understanding and manipulation of soil microbial processes; expert system management; and the development of inexpensive, robust measurement and monitoring technologies. Research for reductions in N_2O emissions focus on improved production efficiencies and reduced energy consumption by developing and deploying precision agriculture technologies, sensors/monitors and information-management systems, and smart materials for prescription release utilized in major crops. An additional goal is to improve fertilizer efficiency and reduce nitrogen inputs by developing advanced fertilizers and technologies, methods of manipulating soil microbial processes, and genetically designed major crop plants.[13]

---

[13] See Section 4.2.1 (CCTP 2005): http://www.climatetechnology.gov/library/2005/tech-options/tor2005-421.pdf.

Future Research Directions

The current portfolio supports the main components of the technology development strategy and addresses the highest priority current investment opportunities in this technology area. For the future, CCTP seeks to consider a full array of promising technology options. The current portfolio supports the main components of the technology development strategy and addresses the highest priority current investment opportunities in this technology area. For the future, CCTP seeks to consider a full array of promising technology options. From diverse sources, suggestions for future research have come to CCTP's attention. Some of these, and others, are currently being explored and under consideration for the future R&D portfolio.

In general, an improved understanding of the interaction and interrelationship among methane, carbon dioxide, and nitrous oxide emissions in agricultural environments is needed. This should involve a systems approach across gases and agricultural systems to synergize related technologies. Other possible further research activities include:

♦ Precision agriculture in general requires advances in rapid, low-cost, and accurate soil nutrient and physical property characterization; real-time characterization of crop water need; real-time crop yield and quality characterization; real-time insect and pest infestation characterization; autonomous control systems; and integrated physiological model and massive data/information management systems.

♦ Improved understanding of specific soil microbial processes is required to support development of methods for manipulation of these processes and to identify how manipulation impacts GHG emissions.

♦ To continue to improve systems management, models that represent an accurate understanding of plant physiology must be coupled with soil process models, including decomposition, nutrient cycling, gaseous diffusion, water flow, and storage on a mass balance basis, to understand how ecosystems respond to environmental and management change.

Other options could include improved utilization of the nitrogen in manure on croplands/pasturelands to offset use of synthetic nitrogen and decrease the quantity of nitrogen excreted from livestock by better matching the intake of nitrogen (e.g., protein) with

the actual dietary requirements of the animals. A large portion of the N_2O emissions from soils comes from livestock waste directly deposited on pastures, and this has significant mitigation potential both in the United States and globally.

Wide-scale implementation of these technologies and improved management systems in the United States could lead to reductions in nitrous oxide emissions from agriculture of 15 to 35 percent. In some developing countries, where greater inefficiencies are identified and where potential use of nitrogen is likely to increase greatly in the future as the demand for more crop and pasture production increases, the potential is even greater.

Methane and Nitrous Oxide Emissions from Livestock and Poultry Manure Management

Globally, nitrous oxide and methane emissions from livestock and poultry manure management totaled approximately 400 Tg CO_2 equivalent in 2000 (EPA 2004). Livestock and poultry manure has the potential to produce significant quantities of methane and nitrous oxide, depending on the waste management practices. When manure is stored or treated in systems that promote anaerobic conditions, such as lagoons and tanks, the decomposition of the biodegradable fraction of the waste tends to produce methane. When manure is handled as a solid, such as in stacks or deposits on pastures, the biodegradable fraction tends to decompose aerobically, greatly reducing methane emissions; however, this practice increases emissions of nitrous oxide, which has a greater global warming potential. Practices are needed that minimize both GHGs simultaneously.

Potential Role of Technology

Methane reduction and other environmental benefits can be achieved by utilizing a variety of technologies and processes. Aeration processes, such as aerobic digestion, auto-heated aerobic digestion, and composting, remove and stabilize some pollutant constituents from the waste stream. These technologies facilitate the aerobic decomposition of waste and prevent methane emissions. Anaerobic digestion systems, in contrast, encourage methane generation, and the collection and transfer of manure-generated off-gases to energy-producing combustion devices (such as engine generators, boilers, or odor control flares). Solids separation processes remove some pollutant constituents from the waste stream through gravity, mechanical, or chemical methods. These processes create a second waste stream that must be managed using techniques different from those already in use to manage liquids or slurries. Separation processes offer the opportunity to stabilize solids aerobically (i.e., to control odor and vermin propagation).

Technology Strategy

Methane collection from anaerobic digestion systems plays an important role in reducing emissions from livestock manure management (Figure 7-5). In addition, these systems can provide additional odor-control and energy benefits by collecting and producing electricity from the combustion of methane-using devices, such as engine generators and boilers. Although the use of commercial farm-scale anaerobic digesters has increased over the past five years due to private sector activities, significant opportunity remains. Currently there are only 12 companies that provide proven commercial-scale anaerobic digestion systems and gas utilization options for farm applications in the United States. As of 2003, an estimated 40 anaerobic digester systems, which produce about 1 million kWh/year, were in use at commercial swine and dairy farms in the United States (EPA 2003b).

Expanded technology research and extension efforts could include commercial-scale demonstration projects and evaluation of emerging technologies to determine their effectiveness in reducing emissions, overall environmental benefits, and cost-effectiveness. For example, a number of emerging anaerobic digester systems adopted from the sewage industry are currently under evaluation for farm-scale applications. In addition, it is important to encourage research on odor and nitrogen emission control and ensure that it is coordinated with research on methane production and emission technology development.

Current Portfolio

Methane reduction and other environmental benefits can be achieved by utilizing a variety of technologies and processes, including aeration processes to remove and stabilize some pollutant constituents from the waste stream; anaerobic digestion systems that collect and transfer manure-generated off-gases to energy-producing combustion devises (such as engine generators, boilers, or odor control flares); and solids separation processes to remove some pollutant constituents from the waste stream. The goals of this

Figure 7-5. Methane collection from anaerobic digestion systems plays an important role in reducing emissions from livestock manure management.

Courtesy: EPA

research activity are to reduce costs and improve biological efficiencies of methane and nitrous oxide emissions by developing new types of digesters; developing separation processes for solid and liquid fractions; and on developing, applying, and evaluating process performance of aeration systems for manure waste streams. The current Federal portfolio focuses these technologies.[14]

Future Research Directions

The current portfolio supports the main components of the technology development strategy and addresses the highest priority current investment opportunities in this technology area. For the future, CCTP seeks to consider a full array of promising technology options. From diverse sources, suggestions for future research have come to CCTP's attention. Some of these, and others, are currently being explored and under consideration for the future R&D portfolio.

⬧ Future research could address technologies to reduce carbon in waste lagoons by solids separation and increase aeration of lagoon waste systems. Additional research could facilitate the shift from anaerobic lagoons to solid waste management systems.

⬧ Future research can lead to improved separation processes that remove solids from liquids for improved waste management and stabilization development of new types of digestors with reduced costs and improved biological efficiencies, development of centralized anaerobic digestion

---

[14] See Section 4.2.2 (CCTP 2005): http://www.climatetechnology.gov/library/2005/tech-options/tor2005-422.pdf.

systems for multiple farm operations, and development of aeration processes and pollution control methods for manure waste streams.

Expanded extension efforts to the livestock, agricultural, energy, and regulatory communities in a number of key livestock-producing states (for example, by expanding the activities currently conducted through the AgSTAR Program[15]), could lead to additional emissions reductions in the United States. In addition, research that utilizes new technological developments in analytical instrumentation and molecular biology related to a commercial farm's operational ability would be useful. If such activities were undertaken globally, the emission reductions could be substantial.

Methane Emissions from Livestock Enteric Fermentation

Methane emissions from enteric fermentation are the second largest global agricultural GHG source, contributing an estimated 1712 Tg CO_2 of emissions in 2000 (EPA 2004). Methane emissions occur through microbial fermentation in the digestive system of livestock. The amount of methane emitted depends primarily on the animal's digestive system, and the amount and type of feed. Ruminant livestock such as dairy cattle, beef cattle, and buffalo emit the most methane per animal, while non-ruminant livestock such as swine, horses, and mules emit less. Because methane emissions represent an economic loss to the farmer—where feed is converted to methane rather than to product output—viable mitigation options can entail efficiency improvements to reduce methane emissions per unit of beef or milk.

Potential Role of Technology

Reductions in this energy loss can be achieved through increased nutritional efficiency. The goal of much livestock nutrition research has been to enhance production efficiency in order to indirectly reduce methane per unit of product through breed improvements, increased feeding efficiency through diet management, and strategic feed selection. Without reductions in national herds, however, this approach will not result in net decreases of enteric methane. Historic and near-term projected trends show both a decreasing herd size and reduced methane emissions on a per unit product basis.

Technology Strategy

Technologies that would likely reduce methane emissions in addition to enhancing production efficiency include precision nutrition; and improvements in grazing management, feed efficiency, and livestock production efficiency. Research includes but is not limited to investigating between-animal differences to determine if traits for reduced methane production can be inherited, and dietary manipulation of grains, oils, and fats that reduce methane production. Key technologies include the following:

- Precision nutrition can minimize excess nutrients, particularly nitrogen, while meeting the nutritional needs of the ruminal microflora and those of the animal for growth, milk production, and digestion.

- Improved grazing management can increase forage yield and digestibility.

- Using ionophores to improve feed efficiency can inhibit the formation of CH_4 by rumen bacteria.

- Improving livestock production efficiency with natural or synthetic hormone feed additives or implants can increase milk production and growth efficiency and reduce feed requirements.

Current Portfolio

The current Federal research portfolio focuses on improved feed and forage management and treatment practices to increase the digestibility and reduce residence digestion time in the rumen, best-management practices to increase animal reproduction efficiency, and use of growth promotants and other agents to improve animal efficiency. Enteric emissions reduction goals focus on improved production efficiencies for forage and feedstuffs; increased digestibility; means to reach these goals include genetically designed forages; manipulation of ruminal microbial processes to sequester hydrogen, making it unavailable to methanogens; and genetically designed bacteria that can compete with natural microbes.[16]

Future Research Directions

The current portfolio supports the main components of the technology development strategy and addresses the highest priority current investment opportunities in this technology area. For the future, CCTP seeks to consider a full array of promising technology

[15] For additional information on the AgSTAR Program, see http://www.epa.gov/agstar/.

[16] See Section 4.2.3 (CCTP 2005): http://www.climatetechnology.gov/library/2005/tech-options/tor2005-423.pdf.

options. The current portfolio supports the main components of the technology development strategy and addresses the highest priority current investment opportunities in this technology area. For the future, CCTP seeks to consider a full array of promising technology options. From diverse sources, suggestions for future research have come to CCTP's attention. Some of these, and others, are currently being explored and under consideration for the future R&D portfolio.

In general, an improved understanding of the interaction and interrelationship among methane, carbon dioxide, and nitrous oxide emissions in agricultural environments is needed. This should involve a systems approach across gases and agricultural systems to synergize related technologies. Possible research activities include:

♦ Further research is needed with precision agriculture technologies such as genetic engineering of plants to enhance digestibility of feeds, reduce fertilizer requirements, and provide appropriate nutrients to enhance beneficial microbial competitiveness; development of livestock with increased productivity and dietary energy use efficiency that can be productive in various environments and use reduced feed resources; and development of models that represent accurate understanding of animal nutrient needs.

♦ Longer-term research is needed to improve understanding of specific rumen microbial processes to support development of methods for making desirable engineered microbes competitive with natural rumen microbes and development of vaccinations that can reduce methane production in the rumen.

It is estimated that an increase in production efficiency of approximately 25 percent could be realized if maximum implementation were to occur. A large potential exists as well in developing countries, where the livestock population is expected to increase significantly over the next few decades and where production efficiency is currently low (i.e., high methane per unit product).

Methane Emissions from Rice Fields

Another significant source of global anthropogenic methane is rice production. Rice is the dietary staple of a large proportion of the world's population. It is generally grown in flooded paddy fields, where methane is generated by the anaerobic decomposition of organic matter in the soil. Traditional wet cultivation emits an estimated 642 Tg CO_2 equivalent of methane (EPA 2004). Emissions from this source have leveled off in the past two decades.

Although water management, fertilizer selection, cultivar selection, and nutrient management are potential options for limiting methane emissions from rice fields, further research and development is needed to determine their cost-effectiveness and feasibility. Currently, there is no research ongoing in this area.

A number of opportunities for future research exist in this area, some of which include plant genetics, water management, and nutrient management. In general, the greatest challenges for mitigating methane emissions from rice fields arise from uncertainties in effecting changes in cultivation management, which affects rice yields; and developing feasible management practices that reduce methane emissions without increasing nitrogen losses and reducing yields. In addition, reduction of methane emissions

U.S. and Global Emissions of High-GWP Gases (2000 Emissions in Tg CO₂ Equivalent)

SOURCE	U.S. EMISSIONS	% OF TOTAL U.S. NON-CO₂ GHG EMISSIONS	GLOBAL EMISSIONS	% OF GLOBAL NON-CO₂ GHG EMISSIONS
Substitutes for Ozone-Depleting Substances	75	7	126	1
Industrial Use of High-GWP Gases	64	6	242	3
Total	139	13	368	4
Sources: EPA 2005, EPA 2004.				

Table 7-5. U.S. and Global Emissions of High-GWP Gases (2000 Emissions in Tg CO₂ Equivalent)

could be difficult to implement because, in many cases, the necessary actions could involve significant changes in agricultural practices (e.g., shifting to different water management regimes). In principle, application of known techniques could reduce methane emissions by 30 to 40 percent by the year 2020. Achieving these large emission reductions would, however, require finding suitable incentives and delivery mechanisms to induce changes in current practices.

7.3 Emissions of High Global-Warming Potential Gases

In 2000, high-GWP gases represented 13 percent of total U.S. non-CO_2 GHG emissions and 4 percent of global non-CO_2 emissions (Table 7-5). There are two different types of emission sources in this category, and each has different R&D priorities. As discussed below, emissions of high-GWP gases used as substitutes for ozone-depleting substances (ODSs) that are being phased out under the Montreal Protocol are currently increasing. High-GWP gases are also used or emitted by several other industries, and in many cases these emissions can be readily managed or eliminated. Table 7-5 shows emissions of substitutes for ODSs and high-GWP gases (Tg CO_2 equivalent).

Substitutes for Ozone Depleting Substances

High-GWP gases used as substitutes for ODSs are a growing emissions source in the United States and globally. These high-GWP gases are being used as replacements for chemicals (like CFCs) that deplete the stratospheric ozone layer (Box 7-2). ODSs, which are also GHGs, are being phased out under the Montreal Protocol and, thus, are not counted in national inventories. To address ozone depletion, the refrigeration, air conditioning, fire suppression, foam blowing, solvent cleaning, and other industries are in the midst of the ODS phaseout.

Potential Role of Technology

For many industries, the ODS phaseout is accomplished by switching to alternative chemicals. For most industries, the most popular and highest performing alternatives are chemicals like HFCs,

which do not deplete the ozone layer but are potent GHGs. At the same time, the phaseout is providing industries with an opportunity to improve processes and practices related to chemical use, management, and disposal in ways that reduce the emissions of HFCs and PFCs, where those chemicals are used as alternatives. As the ODS phaseout continues, opportunities exist to find better life-cycle climate performance alternatives and/or continue reducing emissions.

Technology Strategy

To reduce emissions of GHGs used as ODS substitutes, focus might be given to the following: (1) finding alternative gases with lower or no GWP to perform, safely and efficiently, the same function currently served by the HFCs and PFCs; (2) exploring technologies that can reduce the use of these chemicals and/or the rate at which they are emitted; and (3) supporting responsible handling practices and principles that reduce unintended and unnecessary emissions.

Current Portfolio

The Federal R&D portfolio is focused on the two largest sources of hydrofluorocarbon emissions. These emissions arise from the supermarket refrigeration and motor vehicle air conditioning sectors.

◆ **Motor Vehicle Air Conditioning: Hydrofluorocarbon Emissions** – The motor vehicle industry phased out the use of CFC-12 (with a GWP of about 10,000) in new car air conditioners between 1992 and 1994, and since then has used exclusively HFC-134a (with a GWP of 1300). R&D is underway to commercialize even lower-GWP refrigerants, mainly CO_2 (GWP=1) and HFC-152a (GWP=120). Due to the high-pressure and toxic effects of CO_2, and the flammability of HFC-152a, additional safety engineering and risk mitigation technologies are being developed. Furthermore, research and testing are needed to maintain or improve the energy efficiency (and hence gas usage and CO_2 emissions) of the new air conditioners. In the United States, direct refrigerant GWP emissions can be reduced by more than 95 percent and indirect fuel use emissions reduced by 30 percent or more, for a total reduction of total vehicle fuel emissions (in vehicles with air conditioning) by up to 2 percent.

◆ **Supermarket Refrigeration: Hydrofluorocarbon Emissions** – Supermarkets are phasing out the use of ozone-depleting refrigerants and substituting HFCs, which are potent GHGs. Technologies under development include distributed refrigeration, which reduces the need for excessive refrigerant piping (and hence emissions), and secondary-loop refrigeration, which segregates refrigerant-containing equipment to a separate, centralized location while using a benign fluid to transfer heat from the food display cases. The RDD&D goals for reducing HFC emissions from supermarket refrigeration include improving costs and energy-use performance of these new technologies and educating store designers and builders regarding new technologies and how these technologies can be integrated into new or retrofitted stores at a net savings.[17]

◆ The Significant New Alternatives Program (SNAP) has continued its progress in phasing down the use of global warming, ODSs like CFCs and hydrochlorofluorocarbons (HCFCs). SNAP has worked closely with industry to research, identify and implement climate and ozone friendly alternatives, supporting a smooth transition to these new technologies. In addition, SNAP has initiated programs with different industry sectors to monitor and minimize emissions of global warming gases like HFCs and perfluorocarbons (PFCs) used as substitutes to ozone-depleting chemicals.

Future Research Directions

The current portfolio supports the main components of the technology development strategy and addresses the highest priority current investment opportunities in this technology area. For the future, CCTP seeks to consider a full array of promising technology options. The current portfolio supports the main components of the technology development strategy and addresses the highest priority current investment opportunities in this technology area. For the future, CCTP seeks to consider a full array of promising technology options. From diverse sources, suggestions for future research have come to CCTP's attention. Some of these, and others, are currently being explored and under consideration for the future R&D portfolio.

◆ Continuation of the responsible-use practices developed to control emissions of ODSs has had and will continue to have a substantial effect on HFC and PFC emissions. Research indicates that approximately 80 percent of previous ODS uses have been replaced through conservation methods

Figure 7-6. Astron Remote Plasma Source (for NF3 CVD chamber cleaning), an important technology for reducing PFC emissions from semiconductor manufacturing.
Courtesy: EPA

and use of non-fluorocarbon technologies. Continued emphasis on this success is needed, for example, by using equipment and technologies to reduce emissions during service and maintenance.

◆ Long-term research could focus on technologies that hold the most potential for reducing or eliminating total GHG emissions, including associated energy production emissions, and are practical for their applications. Key areas for consideration over the long term are the investigation of new technologies and processes to replace current uses of ODSs and avoid or reduce emissions of high-GWP gases. This includes research to find alternative refrigerant/AC working fluids that are not high-GWP gases. Another approach is research on solid state refrigeration and AC systems that not only eliminates the working fluid, but reduces the overall energy use.

◆ A focused RD&D program to develop and deploy safe, high-performing, cost-effective climate protection technologies could result in U.S. emission reductions of 50 percent or more by 2020. However, due to the long lifetimes of many of the products that use these gases, efforts need to be taken in the near term to realize the stock turnover necessary to achieve these reductions in a cost-effective manner.

---

[17] See Section 4.3.6 (CCTP 2005): http://www.climatetechnology.gov/library/2005/tech-options/tor2005-436.pdf.

Industrial Use of High-GWP Gases

High-GWP synthetic gases are generally used in applications where they are critical to highly complex manufacturing processes and provide safety and system reliability, such as in semiconductor manufacturing, electric power transmission and distribution, and magnesium production and casting. High-GWP gases are also emitted as byproducts from the manufacture of refrigerants (HCFC-22) and from the production of primary aluminum.

Potential Role of Technology

Incremental improvements to current technology have been made through the initiation of voluntary public-private industry partnerships. EPA's partnerships with industries, including the U.S. primary aluminum producers, HCFC-22 manufacturing, electric utility industry, magnesium producers, and semiconductor industry, are identifying new technologies and process improvements that not only reduce emissions of high-GWP gases but also improve production efficiency, thereby saving money. With continued support, production technologies are expected to further improve, allowing these industrial sectors to cost effectively reduce and possibly eliminate emissions of high-GWP gases.

Technology Strategy

High-GWP gas-emitting industries are implementing an RDD&D strategy focused on pollution prevention. The industries have established long-term goals of reducing, and in some cases eliminating high-GWP emissions, and are pursuing these goals by investigating and implementing source reduction, alternative process chemicals, high-GWP gas capture and reuse, and abatement.

While the U.S. sources of high-GWP emissions are well defined, they are also very diverse, and thus a customized approach for each industry is required. New and enhanced R&D will accelerate and expand options to stabilize and reduce emissions. Opportunities exist for both near- and long-term RD&D on technologies, including alternative chemicals for plasma etching for semiconductors and magnesium melt protection, as well as continued demonstration of advanced plasma abatement devices for the semiconductor industry.

Current Portfolio

The current Federal portfolio for reducing industrial emissions of high-GWP gases focuses on five areas:

◆ **Research on the Semiconductor Industry: Abatement Technologies** – Abatement of high-GWP gases from the exhaust gas stream in semiconductor processing facilities may be achieved by two mechanisms: (1) thermal destruction and (2) plasma destruction. The RDD&D goals for the thermal-destruction mechanism target lowering high-GWP emissions from waste streams by more than 99 percent, while minimizing (1) NO_x emissions to levels at or below emissions standards, (2) water use and burdens on industrial wastewater-treatment systems, (3) fabrication floor space, (4) unscheduled outages, and (5) maintenance costs. Plasma-destruction mechanism goals focus on the application of plasma technology (Figure 7-6) to develop a cost-effective POU abatement device that lowers exhaust stream concentrations of high-GWP gases by two to three orders of magnitude from etchers and plasma-enhanced chemical vapor deposition chambers; and transforms those gases into molecules that can be readily removed from air emissions using known scrubbing technologies.[18]

◆ **Research on the Semiconductor Industry: Substitutes for High-GWP Gases** – One method of reducing high-GWP gas emissions from the semiconductor industry is to use an alternative chemical or production process. Identifying and replacing high-GWP gases with more environmentally friendly substitutes for chemical vapor deposition clean and dielectric etch processes is a preferred option when viewed from the perspective of EPA's pollution prevention framework. The goal of reducing high-GWP gases in the semiconductor industry is to identify the chemical and physical mechanisms that govern chemical vapor deposition chamber cleaning and etching with perfluorocarbons and non-perfluorocarbons as well as govern process performance so that emissions of high-GWP gases may be significantly reduced without either adversely affecting process productivity or increasing health and safety hazards.[19]

◆ **Semiconductors and Magnesium: Recovery and Recycle** – Three recovery-and-recycle technologies are being investigated and evaluated:

---

[18] See Section 4.3.1 (CCTP 2005): http://www.climatetechnology.gov/library/2005/tech-options/tor2005-431.pdf.

membrane separation, cryogenic capture, and pressure swing absorption. The goal in this area is to develop and demonstrate a cost-effective, universally applicable recovery-and-recycle technology (all fabrication facilities and all high-GWP gases) that can yield "virgin"-grade high-GWP gases for semiconductor fabrication or magnesium plant reuse or sufficiently pure high-GWP gases for further use or purification elsewhere.[20]

♦ **Aluminum Industry: Perfluorocarbon Emissions** – Current efforts to reduce perfluorocarbon emissions from primary aluminum production focus on using more efficient smelting processes to reduce the frequency and duration of anode effects, which create the PFC. Another concept, now in the research and development phase, involves replacing the carbon anode with an inert anode. Doing so would completely eliminate process-related PFC emissions. The goal to reduce PFC emissions in the aluminum industry is to develop a commercially viable inert anode technology design by 2007, with commercialization expected by 2010–2015. If successful, the nonconsumable, inert anode technology would have clear advantages over conventional carbon anode technology, including energy efficiency increases, operating cost reductions, elimination of PFC emissions, and productivity gains.[21]

♦ **Research for Electric Power Systems and Magnesium: Substitutes for SF_6** – The challenge is to identify substitutes to SF_6 with low or no global-warming potential that satisfy the magnesium industry's melt protection requirements and meet the electric power industry's high-voltage insulating needs (Figure 7-7).[22]

Future Research Directions

The current portfolio supports the main components of the technology development strategy and addresses the highest priority current investment opportunities in this technology area. For the future, CCTP seeks to consider a full array of promising technology options. The current portfolio supports the main components of the technology development strategy and addresses the highest priority current investment

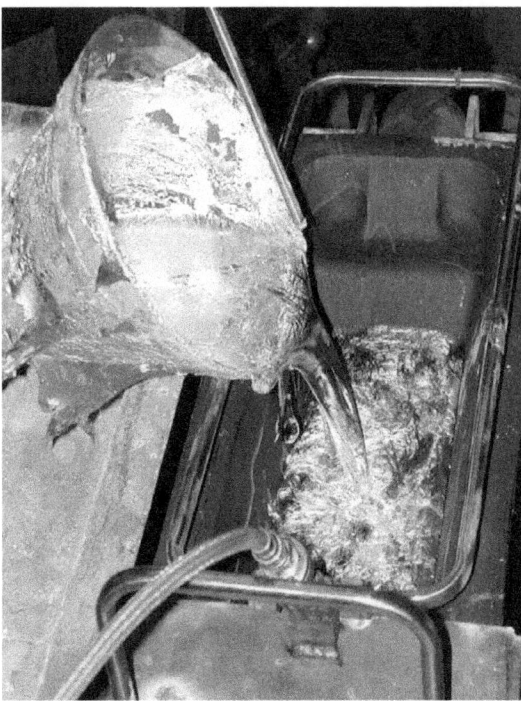

Figure 7-7. Improvements to production technologies, such as alternative cover gases, can cost-effectively reduce, and possibly eliminate, emissions of high global warming potential gases, in this case sulfur hexafluoride (SF_6).

Credit: 3M™ Performance Materials Division

opportunities in this technology area. For the future, CCTP seeks to consider a full array of promising technology options. From diverse sources, suggestions for future research have come to CCTP's attention. Some of these, and others, are currently being explored and under consideration for the future R&D portfolio.

Long-term research might focus on technologies that hold the most potential for reducing or eliminating total GHG emissions, including associated energy production emissions, and that are practical for their applications. Many of these research efforts may prove to be high risk due to unknown commercial viability, and thus are unlikely to be pursued by the industry without significant government funding. Possible research activities include:

♦ Research to identify environmentally friendly alternative cover gases to replace SF_6 for

---

[19] See Section 4.3.2 (CCTP 2005): http://www.climatetechnology.gov/library/2005/tech-options/tor2005-432.pdf.

[20] See Section 4.3.3 (CCTP 2005): http://www.climatetechnology.gov/library/2005/tech-options/tor2005-433.pdf.

[21] See Section 4.3.4 (CCTP 2005): http://www.climatetechnology.gov/library/2005/tech-options/tor2005-434.pdf.

[22] See Section 4.3.5 (CCTP 2005): http://www.climatetechnology.gov/library/2005/tech-options/tor2005-435.pdf.

magnesium melt protection. Another possible option is to develop and deploy manufacturing processes that eliminate the need for a cover gas such as injection molding of thixotropic metal alloys (i.e., semi-solid metal casting).

◆ Research focused on improved process controls and computer-based operator-training tools to further reduce PFC emissions from aluminum smelting.

◆ Research on the potential to use pure fluorine to replace SF_6 and PFCs in chemical vapor deposition (CVD) chamber cleaning and plasma etching processes in numerous electronics manufacturing processes. Although fluorine is not a GHG, it is a toxic and corrosive substance and designing systems to use it would be a challenge.

◆ Research could focus on finding alternatives to the use of SF_6 in high-voltage electric transformers, switchgear and circuit breakers, and in airborne military radar systems. Alternatives include new electric equipment designs that do not require GHG dielectric gases and solid state technologies and materials that do not require dielectric gases.

Significant opportunities exist to reduce emissions. A focused RD&D program to develop safe, high-performing, cost-effective climate protection technologies could result in emission reductions of 40 percent or more over the near term and a dramatic reduction and, in some cases, elimination of emissions by key industries within a few decades.

7.4 Nitrous Oxide Emissions from Combustion and Industrial Sources

Stationary and mobile source combustion and the production of various industrial acids account for about eight percent of non-CO_2 emissions in the United States and four percent globally (EPA 2005, 2004). U.S. emissions of N_2O associated with industrial acid production declined significantly after 1996 due to voluntary industry action and could remain relatively stable. Although generally not accounted for in N_2O emission inventories, significant emissions of NO_x from combustion sources are chemically transformed in the atmosphere and are eventually deposited as nitrogen compounds, which subsequently result in emissions of N_2O in a manner similar to emissions from fertilizer application (Figure 7-8). In 2000, the U.S. N_2O emissions from combustion and industry accounted for nearly 10 percent of total non-CO_2 GHG emissions, with the combustion sources accounting for over 70 percent of these (EPA 2005). Table 7-6 shows N_2O emissions from combustion and industrial sources. R&D priorities differ between N_2O combustion and industrial sources. The priorities for reducing N_2O emissions for each of the sources are discussed below.

U.S. and Global N_2O Emissions from Combustion and Industrial Sources
(2000 Emissions in Tg CO_2 Equivalent)

SOURCE	U.S. EMISSIONS	% OF TOTAL U.S. NON-CO_2 GHG EMISSIONS	GLOBAL EMISSIONS	% OF GLOBAL NON-CO_2 GHG EMISSIONS
Combustion	68	6	230	2
Industrial Sources	26	2	160	2
Total	93	9	390	4
Sources: EPA 2005, 2004.				

Table 7-6. U.S. and Global N_2O Emissions from Combustion and Industrial Sources (2000 Emissions in Tg CO_2 Equivalent)

Combustion

Combustion of fossil fuels by mobile and stationary sources is the largest non-agricultural contributor to N_2O emissions. Nitrous oxide can be formed under certain conditions during the combustion process and during treatment of exhaust or stack gases by catalytic converters. Since N_2O emissions do not contribute significantly to ozone formation or other public health problems, N_2O has not been regulated as an air pollutant and has historically not been a focus of emission control research.

Potential Role of Technology

A better understanding is needed of how and when N_2O forms and how N_2O emissions can best be prevented and reduced. For both stationary and mobile combustion sources, N_2O emissions appear to vary greatly with different technologies and under different operating conditions, and the phenomena involved are poorly understood. For stationary sources, catalytic NO_X reduction technologies can reduce N_2O emissions. Other NO_X control technologies either have no impact or can increase N_2O.

Technology Strategy

A key to identifying the most promising approaches and technologies for reducing N_2O emissions is understanding how N_2O is formed during combustion and under what circumstances catalytic technologies contribute to N_2O emissions. The main research thrust in the near term is to improve scientific understanding of these basic questions.

Current Portfolio

The current Federal research portfolio on N_2O emissions from combustion is focused on better understanding the formation and magnitude of N_2O emissions from fuel combustion and catalytic-converter operation; evaluating the climate-forcing potential of atmospheric nitrogen deposition, especially from combustion; and developing emission models to assess the potential climate benefits from changes in emissions from nitrogen oxide. The goal in this area is to determine linkages of NO_X emissions from transportation combustion and catalytic-converter operation to climate-change impacts due to nitrogen deposition and develop enhanced modeling capabilities.[23]

In addition, Federal research on advanced engine/combustion technologies and alternative fuel vehicles will contribute to a reduction in N_2O emissions. Research in these areas is described in the Transportation section of Chapter 4 (Reducing Emissions from Energy End-Use and Infrastructure).

Future Research Directions

The current portfolio supports the main components of the technology development strategy and addresses the highest priority current investment opportunities in this technology area. For the future, CCTP seeks to consider a full array of promising technology options. The current portfolio supports the main components of the technology development strategy and addresses the highest priority current investment opportunities in this technology area. For the future, CCTP seeks to consider a full array of promising technology options. From diverse sources, suggestions for future research have come to CCTP's attention. Some of these, and others, are currently being explored and under consideration for the future R&D portfolio.

Limited but recent additional collection of nitrous oxide test data have provided statistically reliable emissions estimates for most gasoline-powered passenger cars and light duty trucks. It will be important to develop vehicle- and engine-testing programs to generate nitrous oxide emissions data for a variety of vehicles and engines equipped with a range of current and advanced emission-control technologies and operated over a range of real-world operating conditions, particularly for diesel engines. In addition, future research could determine the effect of catalyst formulation including noble metal loadings and compositions for alternative catalysts that result in less nitrous oxide formation. Also, an intensified research effort is needed to assess the role of airborne nitrogen compounds emitted from combustion and deposited onto the ground, and how they interact with soil-generated nitrous oxide emissions.

The development of new combustion technologies and catalyst formulations that reduce or eliminate nitrous oxide emissions will require new Federal efforts to facilitate joint public-private RD&D activities that can effectively address the reduction of nitrous oxide emissions from combustion and industrial sources. This could include research that would form the basis for identification of new technologies in the future. Some areas for near-term study are outlined below:

---

[23] See Section 4.4.2 (CCTP 2005): http://www.climatetechnology.gov/library/2005/tech-options/tor2005-442.pdf.

Figure 7-8. Nitrogen oxides from combustion sources are chemically transformed in the troposphere, resulting in the formation of nitrogen compounds that are deposited on the ground. These compounds, in turn, give rise to emissions of nitrous oxide, a GHG.

Courtesy EPA

◆ Characterizing nitrous oxide from diesel and advanced technology engines through collaborative research between the EPA National Vehicle and Fuels Emission Laboratory (NVFEL), state air agencies and manufacturers of vehicles/engines. This research may include a variety of vehicles and engines equipped with a range of current and advanced emission control technologies and operated over a range of real-world operating conditions.

◆ Characterizing nitrous oxide from heavy-duty diesel vehicles that meet future (2007/2010) emission standards. Research is now being started in this area. As these vehicles will most likely use catalytic after-treatment, they may be an additional source of nitrous oxide that previously had not existed. Research on how to minimize these emissions is also needed. Emissions of nitrous oxide from combustion sources could be significantly reduced with improved catalyst technologies and other advances.

Industrial Sources

Nitric acid is an inorganic compound used primarily to make synthetic commercial fertilizer. As a raw material, it also is used for the production of adipic acid and explosives, for metal etching, and in the processing of ferrous metals. Facilities making adipic acid used to be high emitters of nitrous oxide, but

now that adipic acid plants in the United States have implemented nitrous oxide abatement technologies, nitric acid production is the largest industrial source of nitrous oxide emissions.

Potential Role of Technology

The nitric acid industry currently controls NO_X emissions using both non-selective catalytic reduction (NSCR) and selective catalytic reduction (SCR) technologies. NSCR is very effective at controlling nitrous oxide while SCR can actually increase nitrous oxide emissions. NSCR units, however, are generally not preferred in modern plants because of high energy costs and associated high gas temperatures. A catalyst to reduce nitrous oxide emissions from SCR plant is being developed in the Netherlands, and a manufacturer of nitric acid is testing a catalyst for use in the ammonia burners in nitric acid plants. Both research groups claim to be capable of reducing nitrous oxide emissions by up to 90 percent and their technology can be easily installed on existing plants. These technologies could be available for commercial application by 2010. Another manufacturer has developed an integrated destruction process; however, this process is only considered suitable for use on new plants because of the high capital costs and long operational down times needed to retrofit existing plants.

Technology Strategy

Additional research is needed to develop new catalysts that reduce nitrous oxide with greater efficiency, and to improve NSCR technology to make it a preferable alternative to SCR and other control options.

Current Portfolio

The current Federal portfolio focuses on developing catalysts that reduce nitrous oxide to elemental nitrogen with greater efficiency and promoting the use of NSCR over other NO_X control options such as SCR and extended absorption. The goal in this area is to focus on development of catalysts that reduce nitrous oxide to elemental nitrogen with greater efficiency and to promote the use of NSCR over other NO_X control options such as SCR and extended absorption.[24]

Future Research Directions

The current portfolio supports the main components of the technology development strategy and addresses the highest priority current investment opportunities in this technology area. For the future, CCTP seeks

---

[24] See Section 4.4.1 (CCTP 2005): http://www.climatetechnology.gov/library/2005/tech-options/tor2005-441.pdf.

to consider a full array of promising technology options. The current portfolio supports the main components of the technology development strategy and addresses the highest priority current investment opportunities in this technology area. For the future, CCTP seeks to consider a full array of promising technology options. From diverse sources, suggestions for future research have come to CCTP's attention. Some of these, and others, are currently being explored and under consideration for the future R&D portfolio.

The use of a catalyst that can reduce a higher percentage of nitrous oxide emissions might be a promising avenue for future research. Current technology is primarily implemented to reduce NO_X emissions, not to reduce nitrous oxide. In the longer term, in order to achieve further reductions in nitrous oxide emissions from nitric acid production, an advanced NSCR technology that is not energy intensive will likely need to be developed and implemented at most nitric acid production facilities.

7.5 Emissions of Tropospheric Ozone Precursors and Black Carbon

Understanding of the role of black carbon (BC) and tropospheric ozone in climate change is still evolving. Large uncertainties remain with regard to emission levels, atmospheric concentrations, net climatic effects, and mitigation potential. However, research to date indicates that these substances influence the global radiation budget, particularly at regional scales. Complicating our understanding is that BC, which tends to have a warming effect, is co-emitted with organic carbon (OC), which tends to have a cooling effect on climate, much like sulfate aerosols.

Mitigation options for BC and tropospheric ozone can already be identified in various sectors. However, for particular emission sources it is often difficult to precisely quantify the emission implications of different mitigation scenarios for these substances, and even more difficult to quantify the climatic implications of such scenarios. Activities to reduce tropospheric ozone precursors and BC will have large public health and local air quality benefits, in addition to their role in mitigating climate change. In fact, it is expected that even in the absence of climate-change-driven mitigation actions, reductions in tropospheric ozone and BC will be achieved as local and regional air quality concerns are addressed, in the United States and many other countries.

Potential Role of Technology

Ozone and particulate matter (PM), of which BC is a component, have been key targets of air pollution control efforts in the United States for many years. National, State, and local regulations have aimed at reducing the significant human health and environmental impacts from high levels of tropospheric ozone and particulate matter. Emission control programs directed toward reducing ozone have focused on the primary precursors that contribute to formation of 1-hour peak ozone concentrations in and near urban centers, such as i.e., emissions of NO_X and volatile organic compounds (VOC).

Programs aimed at reducing PM have led to significant advances in emission control technologies in the transportation, power generation, and industrial sectors, which have and will continue to reduce emissions of BC in the United States. Power plants and other large combustion sources use control technologies such as high-efficiency electrostatic precipitators, fabric filters, and scrubbers to reduce particulate matter, including BC. Regulatory efforts for other stationary sources have addressed biomass burning and include new source performance standards for residential wood heaters and limits on open and agricultural burning.

Technology Strategy

The approach to address the most significant sources of tropospheric ozone precursors and BC involve the following abatement technology areas:

◆ **Transportation control technologies:** PM emissions smaller than 2.5 microns (PM 2.5) from on- and off-road diesel vehicles (the largest source of BC emissions in the United States) are being targeted by stricter vehicle emission standards, where per-vehicle PM emissions are expected to be reduced by 90 percent over the next decade. Total national mobile source PM 2.5 emissions are expected, by 2020, to decline by 53 percent compared to 1996 levels and by 24 percent compared to projected 2020 baseline levels.

◆ **Temperature reduction in cities:** Heat islands form as cities replace natural vegetation with pavement for roads, buildings, and other structures. There are several measures available to reduce the urban heat island effect that can decrease ambient air temperatures, energy use for cooling purposes, GHG emissions, and the chemical formation of smog (ozone and precursors). (See Urban Heat Island Technologies in the Buildings subsection of Chapter 4.)

◆ **Biomass burning:** Important sources of BC aerosols in the United States include combustion of not only fossil fuels but also biomass. Available options to reduce open biomass burning include changing the frequency and conditions of prescribed burning and reducing open waste burning. However, open biomass burning emits greater amounts of OC relative to BC, meaning that, from a strictly climate-carbonaceous aerosol perspective, reducing these emissions could lead to net warming.

Current Portfolio

The current Federal portfolio focuses on the representative technologies listed below. Transportation goals are focused on developing cost-effective NO_X and PM (black carbon) engine and vehicle controls, especially for diesel engines, hybrid-diesel, and gasoline drive trains for medium- and heavy-duty vehicles (Figure 7-9). Goals for temperature reduction in cities are focused on understand and quantifying the impacts that heat island reduction measures have on local meteorology, energy use, GHG emissions, and air quality. Basic research goals are focused on better understanding of the joint role of BC and OC in climate change, including establishing linkages between air pollution and climate change by enhancing modeling capabilities; designing integrated emissions control strategies to benefit climate, regional and local air quality simultaneously.[25]

◆ Transportation control technologies include advanced tailpipe NO_X controls (including NO_X adsorbers), particulate matter filters (traps) for diesel engines (including catalyzed traps capable of passive regeneration), and hybrid and fuel cell vehicles.

◆ Representative technologies for temperature reduction in cities include the following:

• Strategically planted shade trees.

• Reflective roofs: There are over 200 ENERGY STAR® roof products, including coatings and single-ply materials, tiles, shingles and membranes. Energy savings with reflective roofs range as high as 32 percent during periods of peak electricity demand (and average 15 percent for the summer season).

• Reflective paving materials: There are several reflective pavement applications being developed, including new pavement and resurfacing applications, asphalt, concrete, and other material types.

◆ Alternatives to biomass burning include prescribed burning programs (which are directed at minimizing wildfires) and regulation or banning of open burning (such as in land clearing).

Future Research Directions

The current portfolio supports the main components of the technology development strategy and addresses the highest priority current investment opportunities in this technology area. For the future, CCTP seeks to consider a full array of promising technology options. The current portfolio supports the main components of the technology development strategy and addresses the highest priority current investment opportunities in this technology area. For the future, CCTP seeks to consider a full array of promising technology options. From diverse sources, suggestions for future research have come to CCTP's attention. Some of these, and others, are currently being explored and under consideration for the future R&D portfolio.

Basic research is needed to both better understand the role of black and organic carbon and tropospheric ozone precursors in climate change, and to achieve emission reductions in the near and long terms. Much of this research is a focus of the Administration's Climate Change Science Program. Some of the areas where basic research is needed include the following:

◆ The study of the roles of tropospheric ozone and BC and OC in global warming has begun only relatively recently. Although there are strong indications that these pollutants are important actors in climate change, much more research is needed to address the complex optical, chemical, and meteorological factors involved. For BC, this new research would be aimed at establishing more clearly how these pollutants affect solar radiation and cloud formation. For BC and tropospheric ozone, new research could focus on how

---

[25] See Section 4.5.1 (CCTP 2005): http://www.climatetechnology.gov/library/2005/tech-options/tor2005-451.pdf.

atmospheric concentrations vary with geography, time, and the presence of other compounds in the atmosphere.

◆ Greater understanding of the use of different definitions of and measurement protocols for BC (and its differentiation from elemental carbon and organic carbon), and the implications of such differences for climate assessments, is also needed. Much of this work is underway.

◆ Advanced, real-time measurement techniques for fine PM and carbonaceous soot are needed. It is difficult to measure the composition, number, volume, and mass densities of nanometer-size particles at combustion sources and in the atmosphere.

◆ Quantification of the synergies and potential tradeoffs among GHGs, BC, OC, tropospheric ozone, and other criteria air pollutants for different mitigation options, whether these options are targeted for climate, air quality, or both issues.

◆ Regarding BC emissions from open biomass burning, potential mitigation options include wildfire suppression and altering prescribed burning practices. However, it remains difficult to quantify emission reduction benefits due to large uncertainties in the time dynamics of wildfires and uncertainties in emissions factors resulting from different kinds of fires. Furthermore, the climate benefits are difficult to quantify because greater amounts of OC relative to BC are emitted from biomass burning. Further research into this area could support practices that reduce both BC and OC emissions for health and regional haze concerns, while at the same time understanding the net climatic effects. This type of effort could also enhance carbon sequestration on forestlands.

◆ A thorough study of life-cycle GHG and particulate matter emissions is needed to resolve questions of the overall climate impacts of vehicle emissions (including CO_2 and organic carbon particles) of vehicles operating on gasoline as compared to diesel fuel (taking into account the future schedule of diesel vehicle PM standards).

◆ Jet fuel additives could be found that minimize emission of carbonaceous particles (e.g., black carbon/soot) from aircraft engines during take-off, landing, and cruising.

Figure 7-9. Research is needed to better understand the role of particulate matter (e.g., black carbon) emissions from combustion in climate change mitigation.

Courtesy: EPA

◆ Computational models of soot formation are needed to enable inexpensive design of combustion devices and their optimum operational conditions.

Research and development of alternative, non-carbon based fuels could lead to significant reductions in emissions of tropospheric ozone precursors and BC in the longer term. Additional longer-term research needs include the following:

◆ Efforts to develop technologies to reduce NO_X emissions from on-road heavy-duty diesel engines are moving beyond engine-based technologies to exhaust after-treatment technologies.

◆ For both NO_X and particulate control technologies for diesel engines, designs capable of being retrofitted onto engines in the existing fleet could significantly accelerate the heath and climate benefits of these technologies by reducing the time that is otherwise required for engines to be retired and replaced by new models.

Improved understanding is necessary to translate these measures into quantifiable reductions in ozone precursors, BC, OC, and the associated climate effects.

7.6 Summary

This chapter reviews various forms of advanced technology, their potential for reducing emissions of non-CO_2 GHGs, and the R&D strategies intended to accelerate the development of these technologies. Although uncertainties exist about both the level at which GHG concentrations might need to be stabilized and the nature of the technologies that may come to the fore, the long-term potential of advanced technologies to reduce emissions of non-CO_2 GHGs is estimated to be significant, both in reducing emissions (as shown in the figure at the beginning of this chapter) and in reducing the costs for achieving those reductions, as suggested by Figure 3-21. Further, the advances in technology development needed to realize this potential, as modeled in the associated analyses, animate the R&D goals for each technology area focused on reducing emissions of non-CO_2 GHGs.

As one illustration among many hypothetical cases analyzed,[26] GHG emissions were constrained to a high level over the course of the 21st century in such a way that a stabilized GHG concentration levels could ultimately be attained. The lowest-cost arrays of advanced technology to reduce emissions of non-CO_2 GHGs, when compared to a reference case, resulted in reduced or avoided emissions of about 150 Gt of carbon equivalent over the 100-year planning horizon. This amounted to roughly 25 percent of all GHG emissions reduced, avoided, captured and stored, or otherwise withdrawn and sequestered needed to attain this level. Similarly, the costs for achieving such emissions reductions were reduced by roughly a factor of 3. See Chapter 3 for other cases and other scenarios.

As described in this chapter, CCTP's technology development strategy supports achievements in this range. The overall strategy is summarized schematically in Figure 7-10. Advanced technologies are seen entering the marketplace in the near-, mid-, and long-terms, where the long-term is sustained indefinitely. Such a progression, if successfully realized worldwide, would be consistent with attaining the potential for reducing emissions of non-CO_2 GHGs portrayed at the beginning of this chapter.

The timing and pace of technology adoption are uncertain and must be guided by science and supported by appropriate policies (see Approach 7, Chapters 2 and 10). In the case of the illustration above, the first GtC per year (1GtC/year) of reduced or avoided emissions, as compared to an unconstrained reference case, would need to be in place and operating, roughly, around 2050. For this to happen, a number of new or advanced technologies to reduce emissions of non-CO_2 GHGs would need to penetrate the market at significant scale before this date. Other cases would suggest faster or slower rates of deployment. See Chapter 3 for other cases and other scenarios.

Throughout Chapter 7, the discussions of the current activities in each area support the main components of this approach to technology development. The activities outlined in the current portfolio sections address the highest-priority investment opportunities for this point in time. Beyond these activities, the chapter identifies promising directions for future research, identified in part by the technical working group and assessments and inputs from non-Federal experts. CCTP remains open to a full array of promising technology options as current work is completed and changes in the overall portfolio are considered.

---

[26] In Chapter 3, various advanced technology scenarios were analyzed for cases where global emissions of GHGs were hypothetically constrained. Over the course of the 21st century, growth in emissions was assumed to slow, then stop, and eventually reverse in order to ultimately stabilize GHG concentrations in the Earth's atmosphere at levels ranging from 450 to 750 ppm. In each case, technologies competed within the emissions-constrained market, and the results were compared in terms of energy (or other metric), emissions, and costs.

Technologies for Goal #4: Reduce Emissions of Other Gases

	NEAR-TERM	MID-TERM	LONG-TERM
Methane from Energy & Waste	• Bioreactor Landfill Technology • Methane to Markets • New Drilling Techniques for Recovery of Coal bed Methane • Leak Detection, Measurement, and Mitigation Technologies for Oil & Natural Gas Systems	• Advanced Landfill Gas Utilization (e.g., Fuel Cells, Microturbines), Cover, and Collection Technologies • Ventilation Air Methane Technology • Advanced End-Use Technologies to Use Methane at Remote Well Sites	• Integrated Waste Management System with Automated Sorting, Processing & Recycle • Automated Coal Mining to Eliminate Methane Emissions • Smart Pipes and Self-Repairing Pipelines
Methane & N_2O from Agriculture	• Anaerobic Digesters that Produce Heat and Electricity • Precision Agriculture • Improved Livestock Production Efficiency	• Better Understand Relationship among CH_4, CO_2, N_2O, N_2 & C in Agriculture • Soil Microbial Processes • Prescription Release of Nutrients and Chemicals for Crops • Genetically Designed Forages and Bacteria to Improve Digestion Efficiency	• Zero-Emission Agriculture
High GWP Gases	• Advanced Refrigeration Technologies (Distributed and Secondary-Loop) • Advanced Abatement, Recovery, and Recycling Technologies • Advanced Aluminum Smelting Processes to Reduce Anode Effect	• Alternative Refrigeration Fluids (Non-GHG) • Substitutes for SF_6 in High-Voltage Applications and Magnesium Production • Inert Anode to Eliminate PFC Emissions in Aluminum Production	• Solid-State Refrigeration/AC Systems • New Equipment and Process Designs that do not Require High-GWP Gases
N_2O from Combustion	• Catalytic Reduction of N_2O in Nitric Oxide Plants • Better Understand N_2O Emissions from Vehicles	• Catalysts That Reduce N_2O to Elemental Nitrogen in Diesel Engines • Understand Role of N Compounds from Combustion with Soils and N_2O	• Advanced Vehicles and Non-Carbon Based Fuels
Ozone Precursors & Black Carbon	• Particulate Matter Control Technologies for Vehicles • Reflective Roofs to Reduce Heat Island Effects • Better Understand Effects of Ozone Precursors & Black Carbon	• Model Linkages Between Air Pollution and Climate Change • Jet Fuel Additives to Minimize Black Carbon and Soot	

Figure 7-10. Technologies for Goal #4: Reduce Emissions of Other Gases
(Note: Technologies shown are representations of larger suites. With some overlap, "near-term" envisions significant technology adoption by 10–20 years from present, "mid-term" in a following period of 20–40 years, and "long-term" in a following period of 40–60 years. See also List of Acronyms and Abbreviations.)

7.7 References

DeAngelo, B., F. de la Chesnaye, R. Beach, A. Sommer, and B. Murray. Forthcoming. Methane and nitrous oxide mitigation in agriculture. *Energy Journal.*

Delhotal, K. Casey, F. C. de la Chesnaye, A. Gardiner, J. Bates, and A. Sankovski. 2006. Mitigation of methane and nitrous oxide emissions from waste, energy and industry. *Energy Journal.*

Intergovernmental Panel on Climate Change (IPCC). 2001. *Climate change 2001: the scientific basis.* Cambridge, UK: Cambridge University Press.

Placet, M., K.K. Humphreys, and N.M. Mahasenan. 2004. *Climate change technology scenarios: energy, emissions and economic implications.* Richland, WA: Pacific Northwest National Laboratory. http://www.pnl.gov/energy/climatetechnology.stm

Schaefer, D., D. Godwin, and J. Harnish, 2006. Estimating Future Emissions and Potential Reductions of HFCs, PFCs, and SF_6. *Energy Journal* (forthcoming).

U.S. Climate Change Technology Program (CCTP). 2003. *Technology options for the near and long term.* DOE/PI-0002. Washington, DC: U.S. Department of Energy. http://www.climatetechnology.gov/library/2003/tech-options/index.htm

Update is at http://www.climatetechnology.gov/library/2005/tech-options/index.htm

U.S. Environmental Protection Agency (EPA). 1999. *U.S. methane emissions 1990 – 2020: inventories, projections, and opportunities for reductions.* #430-R-99-013. Washington, DC: U.S. Environmental Protection Agency. http://www.epa.gov/methane/reports/methaneintro.pdf

U.S. Environmental Protection Agency (EPA). 2003a. *Assessment of the worldwide market potential for oxidizing coal mine ventilation air.* #430-R-03-002. Washington, DC: U.S. Environmental Protection Agency. http://www.epa.gov/coalbed/pdf/ventilation_air_methane.pdf

U.S. Environmental Protection Agency (EPA). 2003b. Current status of farm-scale digesters. *AgSTAR Digest.* #430-F-02-028. Washington, DC: U.S. Environmental Protection Agency. http://www.epa.gov/agstar/pdf/handbook/appendixa.pdf

U.S. Environmental Protection Agency (EPA). 2004. *Global emissions of non-CO_2 greenhouse gases, 1990 – 2020.* Washington, DC: Environmental Protection Agency.

U.S. Environmental Protection Agency (EPA). 2005. *Inventory of U.S. greenhouse gas emissions and sinks: 1990 – 2003.* #430-R-05-003. Washington, DC: U.S. Environmental Protection Agency. http://yosemite.epa.gov/oar/globalwarming.nsf/UniqueKeyLookup/RAMR69V4ZS/$File/05_complete_report.pdf.

Enhancing Capabilities to Measure and Monitor Greenhouse Gases

The sources of greenhouse gas (GHG) emissions are varied and complex, as are the potential mitigation strategies afforded by advanced climate change technologies presented through this *Plan*. Measurement and monitoring (M&M) systems will be needed to complement these technologies to assess their efficacy and sustainability and to guide future research and enhancements. Contributing M&M systems cover a wide array of GHG sensors, instrumentation, measurement platforms, monitoring and inventorying systems, and associated analytical tools, including databases, models, and inference methods.

Development and application of such systems can provide accurate characterizations of GHG emissions from both existing and advanced technologies, enable increased understanding of performance, guide further research, reduce costs, and improve effectiveness. Research on and development of these systems is required to increase their capabilities and facilitate and accelerate their adoption (Figure 8-1).

Observations using M&M technologies can be used to establish informational baselines necessary for analytical comparisons, and to measure carbon storage and GHG fluxes across a range of scales, from individual locations to large geographic regions. If such baselines are established, the effectiveness of implemented GHG-reduction technologies can be assessed against a background of prior or existing conditions and other natural indicators. Many of the M&M technologies and the systems they can enable benefit from the ongoing R&D under the aegis of the Climate Change Science Program (CCSP), and from other Earth observation activities that are underway. All such M&M systems constitute an important component of a comprehensive Climate Change Technology Program (CCTP) R&D portfolio and could be improved through further development as outlined below.

On February 16, 2005, 55 countries endorsed a 10-year plan to develop and implement the Global Earth Observation System of Systems (GEOSS) for the purpose of achieving comprehensive, coordinated, and sustained observations of the Earth system. The U.S.

Figure 8-1. *Earth observation activities benefit both CCSP and CCTP research. For example, satellites such as CALIPSO (pictured above) take measurements of natural processes using advanced techniques such as Light Detection and Ranging (LIDAR). These measurements can be used in atmospheric research and, for CCTP, to help scientists estimte emissions reductions in energy end-use, energy supply, sequestration, and non-CO$_2$ gases.*
Credit: NASA

contribution to GEOSS is the Integrated Earth Observation System (IEOS). IEOS will meet U.S. needs for high-quality, global, sustained information on the state of the Earth as a basis for policy and decision-making in every sector of society. A strategic plan for IEOS[1] was developed by the United States Group on Earth Observation (USGEO), a Subcommittee reporting to the National Science and Technology Council's Committee on Environment and Natural Resources; the plan was released in April 2005. Both the GEOSS and the IEOS are focused on societal benefits, including climate variability and change, weather forecasting, energy resources, water resources, land resources, and ocean resources—all of which are relevant to CCSP and CCTP.

---

[1] Accessible at http://iwgeo.ssc.nasa.gov

8.1 Potential Role of Technology

M&M systems are important to addressing uncertainties associated with cycling of GHGs through the land, atmosphere, and oceans, as well as in measuring and monitoring GHG-related performance of various existing and advanced climate change technologies. R&D in this area offers the potential to:

♦ Characterize emissions, inventories, concentrations, and cross-boundary fluxes of carbon dioxide (CO_2) and other greenhouse compounds, including the size and variability of the fluxes.

♦ Characterize the efficacy and durability of particular mitigation technologies or other actions, and verify and validate claims for results.

♦ Measure (directly or indirectly through proxy measurements) anthropogenic changes in sources and sinks of GHGs and relate them to causes, to better understand the role of various technologies and strategies for mitigation.

♦ Identify opportunities and plans for guiding research investments in GHG M&M methods, technologies, and strategies.

♦ Explore relationships among changes in GHG emissions, fluxes, and inventories due to changes in surrounding environments.

♦ Optimize the efficiency, reliability, and quality of M&M that maximizes support for understanding and decision-making while minimizing the transaction costs of mitigation activities.

Ideally, an integrated observation system strategy would be employed to measure and monitor the sources and sinks of all gases that have an impact on climate change, using the most cost-effective mix of techniques ranging from local *in situ* sensors to global

Measurement and Monitoring Technologies for Assessing the Efficacy, Durability, and Environmental Effects of Emission Reduction and Stabilization Technologies

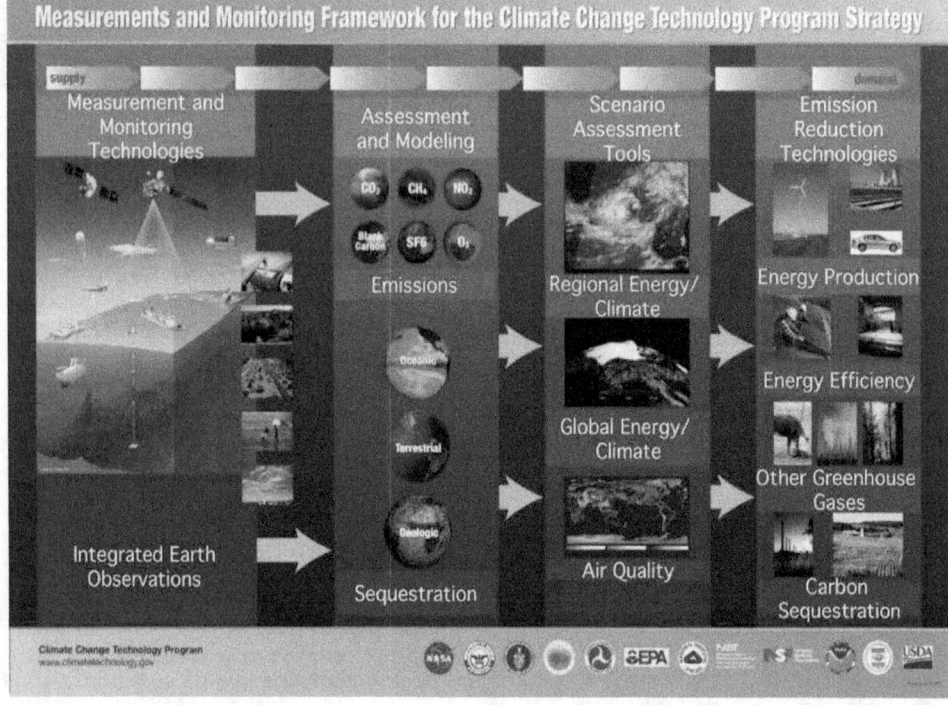

Figure 8-2. Measurement and Monitoring Technologies for Assessing the Efficacy, Durability, and Environmental Effects of Emission Reduction and Stabilization Technologies

remote-sensing satellites. This would involve technologies aimed at a spectrum of applications, including CO_2 from energy-related activities (such as end use, infrastructure, energy supply, and CO_2 capture and storage) and GHGs other than CO_2 (including CH_4, N_2O, fluorocarbons, ozone, and other GHG-related substances, such as black carbon [BC] aerosol). An integrating system architecture serves as a guide for many of the step-by-step development activities required in these areas. It establishes a framework for R&D that places M&M technologies in context with the Integrated Earth Observation System (IEOS) and other CCTP technologies (Figure 8-2). The integrated systems approach provides feedback through which demand-side actors (both in the public and private sectors) contribute to benchmarking results against expectations.

Such a framework facilitates coordinated progress evolving over time toward increasingly effective solutions and common interfaces of the gathered data and assessment systems. An integrating architecture would function within the context of, and in coordination with, other Federal programs (e.g., CCSP and the U.S. Group on Earth Observations) and international programs (e.g., the World Meteorological Organization and the Intergovernmental Panel on Climate Change) that provide or use complementary M&M capabilities across a hierarchy of temporal and spatial scales. It could, therefore, take advantage of the synergy between observations to measure and monitor GHG mitigation strategies and the research on observation systems for the CCSP, as well as the operational observations systems for weather forecasting, as described more fully in the CCTP report, *Technology Options for the Near and Long Term* (CCTP 2003).

In the near term, opportunities for advancing GHG measuring and monitoring systems present themselves as integral elements of the CCTP R&D programs and initiatives. Efforts must focus on the significant emission sources and sinks and on M&M of carbon sequestration and storage.

Technology can be developed to address knowledge gaps in GHG emissions and to improve inventories. In some cases, it is not necessary or cost-effective to measure emissions directly. In such cases, emissions can be measured indirectly by measuring other parameters as proxies, such as feedstock, fuel, or energy flows (referred to as "parametric" or "accounting-based" estimates); or by measuring

changes in carbon stocks. Under CCTP, there is a benefit to undertaking research to test, validate, quantify uncertainties, and certify such uses of proxy measurements.

The long-term approach is to evaluate data needs and pursue the development of an integrated and overarching system architecture that focuses on the most critical and supplementary data needs. Common databases would provide measurements for models that could estimate additions to and removals from various GHG inventories, forecast the long-term fates of various GHGs, and integrate results into relevant decision support tools and global-scale monitoring systems. This approach would include protocols for calibrated and interoperable (easily exchanged) data products, emissions accounting methods development, and coordination of basic science research in collaboration with CCSP. Tools would be validated by experimentation to benchmark protocols (to quantify the improvements that the tools provide), so that they would be recognized and accepted by the community-of-practice for emissions-related processes.

The M&M technologies that are emphasized in the following sections are based on their capacity to address one or more of the following criteria:

- M&M technology that supports the successful implementation and validation of a technological option that mitigates a substantial quantity of U.S. GHG emissions, on the order of a gigaton of carbon equivalent or more, over the course of a decade.

- M&M technology capacity to reduce a key uncertainty associated with a mitigation option.

- M&M technology sufficiently differentiated from, or adequately integrated with, comparable research efforts in the CCSP, IEOS, or other operational Earth observation systems.

- M&M technology helping to assure that a proposed advanced climate change technology does not threaten either human health or the environment.

8.2 Energy Production and Efficiency Technologies

M&M systems provide the capability to evaluate the efficacy of efforts to reduce GHG emissions through the use of (1) low-emission fossil-based power systems; (2) potentially GHG-neutral energy supply technologies, such as biomass energy systems (see Chapter 6) and other renewable energy technologies, including geothermal energy; and (3) technologies to more efficiently carry and/or transmit energy to the point of use. In this section, the M&M R&D portfolio for energy production and efficiency technologies is presented. Each of these technology sections includes a sub-section describing the current portfolio. The technology descriptions include a link to an updated version of the CCTP report, *Technology Options for the Near and Long Term*.[2]

Technology Strategy

M&M technologies can enhance and provide direct and indirect emissions measurements at point and mobile sources of GHG emissions. "Point sources" range from electric generation plants to industrial facilities. The term "mobile sources" typically refers to vehicles. Table 8.1 summarizes the nature of point and mobile sources and the potential roles for M&M technologies, which are broadly applicable across the range of emission sources and scales. The technology strategy emphasizes the potential role of M&M technologies in applications across a range of scales, from the individual vehicle to the larger power plant or industrial facility, as well as the balance between those M&M technologies needed in both the near- and long-terms. Development of software and tools that facilitate further integration of measurement data with emission modeling processes is a key dimension of the overall technology strategy. In the near term, the strategy focuses on technologies that measure multiple gases across spatial dimensions. In the long-term, the strategy focuses on development and evolution of a system of systems for remote, continuous, and global M&M that facilitates emissions accounting from the local to the global level.

Proposed R&D Portfolio for Measurement and Monitoring of Energy Production and Use Technologies

GHG EMISSION SOURCE	NATURE OF EMISSIONS AND SCALE	R&D PORTFOLIO OF MEASUREMENT AND MONITORING TECHNOLOGY
Power Generation	Large point sources	Component and system-level technologies to enable and demonstrate direct measurements, continuous emission monitoring, on-board diagnostics, remote sensing, data transmission and archiving, inventory-based reporting, and decision-support systems.
Industrial Facility	Many different processes, but mostly point sources	As above.
Transportation	Many mobile sources widely distributed	As above.

Table 8-1. Proposed R&D Portfolio for Measurement and Monitoring of Energy Production and Use Technologies

2 The full report is available at http://www.climatetechnology.gov/library/2005/tech-options/index.htm

Current Portfolio

R&D programs for M&M technologies spanning the Federal complex are focused on a number of areas, including the following:

◆ High-temperature sensors for NO_X and ozone, ammonia, and other gas emissions, with application in caustic industrial environments (e.g., steel mills, pulp and paper industries);

◆ Fast-response mass spectrometers and field-deployable isotope analysis systems;

◆ Continuous emissions monitors (CEMs) for measuring multiple gases at point sources (linked with energy use statistics at a facility); and

◆ Light Detection and Ranging (LIDAR) for remote monitoring of truck and aviation emissions.

The overall goals are to develop sensors and data transmission systems that allow quantification of emission reductions resulting from energy efficiency improvements.[3]

Future Research Directions

The current portfolio supports the main components of the technology development strategy and addresses the highest priority current investment opportunities in this technology area. For the future, CCTP seeks to consider a full array of promising technology options. From diverse sources, suggestions for future research have come to CCTP's attention. Some of these, and others, are currently being explored and under consideration for the future R&D portfolio.

◆ Improvements in performance, longevity, autonomy, spatial resolution of measurements, and data transmission of CEMs with the ability to measure multiple gases.

◆ More thorough process knowledge and life-cycle analysis for the estimation of changes in emission factors as a function of time and process.

◆ Satellite-based sensors for direct measurement of CO_2 and other gases or indicators, tracers, and isotopic ratios.

◆ Inexpensive, large-area systems for monitoring CO_2 leaks from energy production and end use.

◆ Low-cost, multiple wireless micro sensor networks to monitor migration, uptake, and distribution

patterns of CO_2 and other GHGs in soil and forests.

◆ Data protocols and analytical methods for producing and archiving specific types of data to enable interoperability and long-term maintenance of data records, data production models, and emission coefficients that are used in estimating emissions.

◆ Protocols and concurrent technologies for multiple assessments of the performance of end-uses of energy, including transportation, buildings, and industry. Assessments could ultimately provide real-time feedback to the end user.

◆ Direct measurements to replace proxies and estimates when these measurements are more cost-effective to optimize emissions from sources and improve understanding of the processes behind the formation of GHGs.

 ## CO_2 Capture and Sequestration

As discussed in Chapter 6, capture, storage, and sequestration of CO_2 can be accomplished by various approaches, including capture from point sources, accompanied by geologic or oceanic storage; and terrestrial sequestration. Advanced technologies can make significant contributions to measuring and monitoring GHG emissions that are captured, stored, and sequestered.

Innovations to assess the integrity of geologic structure, leakage from reservoirs, and accounting of sequestered GHGs are useful. Also useful are integrated carbon sequestration measurements of different components (geologic, oceanic, and terrestrial) across a range of scales and time, from the point of use at the present time to regional or larger scales over the future to provide a consistent net accounting of GHG inventories, emissions, and sinks. Advanced M&M technologies can provide histories of CO_2 concentration profiles near the sites of sequestration and track the potential release of CO_2 into the atmosphere. The development of software and tools that facilitate further integration of measurement data with emission modeling processes play an important, ongoing role in the M&M of sequestered CO_2 in conjunction with other technologies. Different M&M strategies associated

---

[3] For more details on the current R&D activities, see CCTP (2005): http://www.climatetechnology.gov/library/2005/tech-options/index.htm.

with the three alternative storage and sequestration approaches are described in the following sections.

Geologic Sequestration

M&M technologies are useful to assess the performance and efficacy of geologic storage systems. They will be critically important in assessing the integrity of geologic structures, transportation, and pipeline systems, the potential of leakage of sequestered GHGs in geologic structures, and in fully accounting for GHG emissions.

Technology Strategy

Realizing the possibilities of these technologies is the focus of a research portfolio that embraces a combination of M&M technologies for separation and capture, transportation, and geologic storage. In the near term, technologies can be improved to measure efficacy of separation and capture, and the

Proposed R&D Portfolio for Measurement and Monitoring Systems for Geologic Sequestration

SYSTEM CONCEPTS	R&D PORTFOLIO
Separation and Capture	• Monitors for CO_2 emissions using process knowledge • Sensors to monitor fugitive emissions around facilities
Transportation	• Leak detection systems from pipelines and other transportation • Pressure transducers • Remote detectors • Gaseous tracers enabling remote leakage detection
Geologic Storage	• Detectors for surface leakage • Indicators of leakage based on natural and induced tracers • Seismic/electromagnetic/ electrical resistivity/pressure monitoring networks

Table 8-2. Proposed R&D Portfolio for Measurement and Monitoring Systems for Geologic Sequestration

integrity of geologic formations for long-term storage. Within the constraints of available resources, a balanced portfolio addresses the objectives shown in Table 8-2.

Current Portfolio

Recent progress has been made in developing M&M technologies for geologic carbon sequestration. Many technologies for monitoring and measuring exist today. However, they may need to be modified to meet the requirements of CO_2 storage. The goals are to develop the ability to assess the continuing integrity of subsurface reservoirs using integrated system of sensors, indicators, and models; improve leak detection from separation and capture pipeline systems; apply remote sensors to fugitive emissions from reservoirs and capture facilities; improve, develop, and implement tracer addition and monitoring programs; evaluate microbial mechanisms for monitoring and mitigating diffuse GHG leakage from geologic formations; and more.[4]

Both surface and subsurface measurement systems for CO_2 leak detection and reservoir integrity estimates have been employed at sites currently storing CO_2. Large M&M efforts have taken place at Weyburn, Alberta, and at Sleipner in the North Sea. Within the measurement systems employed at these sites, seismic imaging using temporal analyses of 3-dimensional (3D) seismic structures (called 4D seismic analyses) have been commonly employed to characterize the reservoir, determine changes in reservoir structure and integrity, and to determine locations of CO_2 that have been pumped downhole. At the Sleipner site, for example, efforts to quantify the CO_2 have been undertaken through 4D seismic research. Other methods of subsurface reservoir analyses are cross-well seismic tomography, passive and active doublet analyses, microseismic analyses, and electromagenetic analyses.

Leak detection of CO_2 from storage reservoirs has been performed in the subsurface and surface regions. Within the subsurface, groundwater chemistry, precipitation of calcite, and subsurface CO_2 concentration measurements have been used to detect small gas emissions from reservoirs. At the ground surface, CO_2 flux changes, isotopes of CO_2 and other tracers, and vegetation changes have been monitored to detect surface leaks of CO_2 and identify the source.

Specific examples include four ongoing experiments: (1) Seismic methods are being used at the Sleipner test site to map the location of CO_2 storage;

---

4 For more information on the current R&D activities, see Section 5.3 (CCTP 2005): http://www.climatetechnology.gov/library/2005/tech-options/tor2005-53.pdf.

(2) Models, geophysical methods, and tracer indicators are being developed through the GEO-SEQ project (Box 8-1); (3) Detection of CO_2 emissions from natural reservoirs has been investigated by researchers at the Colorado School of Mines, University of Utah, and the Utah Geological Survey, including isotopic discrimination of biogenic CO_2 from magmatic, oceanographic, atmospheric, and natural gas sources; and (4) Fundamental research on high-resolution seismic and electromagnetic imaging and on geochemical reactivity of high partial-pressure CO_2 fluids is being conducted.

Future Research Directions

The current portfolio supports the main components of the technology development strategy and addresses the highest priority current investment opportunities in this technology area. For the future, CCTP seeks to consider a full array of promising technology options. From diverse sources, suggestions for future research have come to CCTP's attention. Some of these, and others, are currently being explored and under consideration for the future R&D portfolio.

◆ Tying the experimental research to the process models for geological storage systems, where fate and transport of the stored CO_2 are measured and verified with models. This contributes to verification of CO_2 storage in geologic structures in both the near- and long terms.

◆ The ability to assess the continuing integrity of subsurface reservoirs using an integrated system of sensors, indicators, and models. The heterogeneity of leakage pathways and probable changes over time make detection and quantification difficult.

◆ Indicators such as seismic, electromagnetic imaging, and tracers are needed for quantitative determination of CO_2 stored and specific locations of where the CO_2 is located underground.

◆ Improvements in leak detection from separation and capture and pipeline systems. Low leakage rates occurring at spatially separated locations make full detection difficult.

Terrestrial Sequestration

Sequestering carbon in terrestrial ecosystems (forests, pastures, grasslands, croplands, etc.) increases the total amount of carbon retained in biomass, soils, and wood products. Methods used to measure and

monitor terrestrial sequestration of carbon should address both the capture and retention of carbon in both above- and below-ground components of ecosystems. Determining measures of the desired levels of net sequestration will depend on evaluation of GHG emissions as a function of management practices and naturally occurring environmental factors (Post et al. 2004).

Technology Strategy

M&M systems employ an R&D portfolio that provides for integrated, hierarchical systems of ground-based and remote-sensing technologies of different system components over a range of scales. A system's utility is based on its applicability to a wide range of potential activities and a very diverse land base, an accuracy that satisfies reporting requirements of the 1605(b) voluntary reporting program (EIA 2004), and a cost of deployment such that M&M does not outweigh the value of the sequestered carbon. A balanced portfolio should address (1) remote sensing and related technology for land-cover and land-cover change analysis, biomass and net-productivity measurements, vegetation structure, etc.; (2) low-cost, portable, rapid analysis systems for *in situ* soil carbon measurements; (3) flux measurement systems; (4) advanced biometrics from carbon inventories; and (5) carbon and nutrient sink/source tracing and movement, including using isotope markers; and (6) analysis systems that relate management practices (e.g., life-cycle wood products, changes in agriculture rotations, energy use in ecosystem management, and others) to net changes in emissions and sinks over time (e.g., changes in agriculture rotations, energy use in ecosystem management, and others).

Current Portfolio

Current research activities associated with terrestrial sequestration are found across a number of Federal agencies. The goals of the current activities are to provide an integrated hierarchical system of ground-

based and remote sensing for carbon pools and CO_2 and other GHG flux measurements; reduce uncertainty on regional-to-country scale inventories of carbon stocks; develop low-cost, portable, rapid analysis systems for *in situ* soil carbon measurements; and develop standard estimates that relate management practices to net changes in emissions/sinks over time.[5]

The current portfolio includes the following:

- The Environmental Protection Agency (EPA), with assistance from the U.S. Department of Agriculture's (USDA) Forest Service, prepares national inventories of emissions and sequestration from managed lands. These inventories capture changes in the characteristics and activities related to land uses, and are subject to ongoing improvements and verification procedures.

- The USDA Forest Service Forest Inventory and Analysis Program and the Natural Resources Conservation Service's National Resources Inventory provide baseline information to assess the management, structure, and condition of U.S. forests, croplands, pastures, and grasslands. This information is then converted to State, regional, and national carbon inventories. Hierarchical, integrated monitoring systems are being designed in pilot studies such as the Delaware River Basin interagency research initiative.

- Prototype soil carbon analysis systems have been developed and are undergoing preliminary field testing.

BOX 8-2

AGRIFLUX

The Agriflux network is being developed by the USDA to measure the effects of environmental conditions and agricultural management decisions on carbon exchange between the land and the atmosphere. The network now comprises more than 125 sites in North and South America. Studies will identify crop management practices to optimize crop yield, crop quality, and carbon sequestration and other environmental conditions. Research will lead to new ways for prediction and early detection of drought in agricultural systems based on weekly and monthly climate forecasts.

BOX 8-3

AMERIFLUX

Flux towers such as the one pictured above are taking long-term measurements of CO_2 and water vapor fluxes in over 250 sites throughout the world, including the United States. Data gathered from these measurement sites are important to understanding interactions between the atmospheric and terrestrial systems. The network (http://public.ornl.gov/ameriflux/) is part of an international scientific program of flux measurement networks (e.g., FLUXNET-Canada, CarboEurope, and AsiaFlux) that seeks to better understand the role of the terrestrial biosphere carbon cycle. See http://www.fluxnet.ornl.gov/fluxnet/index.cfm for a global listing of flux towers.

Courtesy of DOE

---

5 For a detailed discussion on technologies and current research activities, see
Section 5.4 http://www.climatetechnology.gov/library/2005/tech-options/tor2005-54.pdf,
Section 3.2.3.1 http://www.climatetechnology.gov/library/2005/tech-options/tor2005-3231.pdf, and
Section 3.2.3.2 http://www.climatetechnology.gov/library/2005/tech-options/tor2005-3232.pdf (CCTP 2005).

♦ Methods are being developed for the use of Synthetic Aperture Radar in estimating forest bole volume at landscape scale.

♦ Satellite and low-altitude remote sensing systems have been developed that can quantify agricultural land features at spatial resolution of approximately 0.5 square meters and measure indicators of the carbon sequestration capacity of land use.

♦ Prototype versions of web-based tools are being developed for estimating carbon budgets for regions (e.g. CASA/CQUEST, CENTURY)

♦ Multidisciplinary studies are providing increased accuracy of carbon sequestration estimates related to land management and full accounting of land/atmosphere carbon exchange.

♦ The Agriflux and AmeriFlux programs (Boxes 8-2 and 8-3) are being implemented to improve the understanding of carbon pools and fluxes in large-scale, long-term monitoring areas. The flux measurements provide quantitative data for calibrating/validating remote sensing and other estimates of carbon sequestration. Approaches for scaling these results to regional estimates are under development (DOE-ORNL 2003).

♦ Other aerospace research activities focusing on imaging and remote sensing methods include LIDAR and RADAR, used for 3D imaging of forest structure for the estimation of carbon content in standing forests.

♦ Isotopes are being used to assess sequestration potentials by monitoring fluxes and pools of carbon in natural ecosystems.

♦ Increased accuracy of carbon sequestration estimates is being accomplished for use in land management and full carbon accounting procedures.

♦ Ongoing tillage and land conservation practices offer test beds for ground-based and remote-sensing methods, as well as verification of rules-of-thumb for emission factors.

♦ Many of the DOE National Laboratories are conducting research on *in situ* and remote-sensing technologies and laser-based diagnostics, supported by a variety of Federal agencies. These diagnostics include microbial indicators, Laser Induced Breakdown Spectroscopy (LIBS), LIDAR, Fourier Transform Infrared (FTIR) Spectroscopy, and a variety of satellite Earth observation programs (Box 8-4).

Future Research Directions

The current portfolio supports the main components of the technology development strategy and addresses the highest priority current investment opportunities in this technology area. For the future, CCTP seeks to consider a full array of promising technology options. From diverse sources, suggestions for future

> **BOX 8-4**
> ## DIAGNOSTIC TECHNOLOGIES
>
> **Laser Induced Breakdown Spectroscopy (LIBS)** is a robust chemical analysis technique that has found application in a range of areas where rapid, remote, and semi-quantitative analysis of chemical composition is needed. The technique in its essential form is quite simple. Light is used to ionize a small portion of the analyte and the spectral emission (characteristic of the electronic energy levels) from the species in the resulting plasma is collected to determine the chemical constituents. Most often the light comes from a laser since high-photon fluxes can be obtained readily with this type of light source. By focusing the light from the laser to a small spot, highly localized chemical analysis can be performed.
>
> **Light Detection and Ranging (LIDAR)** uses the same principle as RADAR. The LIDAR instrument transmits light out to a target. The transmitted light interacts with and is changed by the target. Some of this light is reflected/scattered back to the instrument where it is analyzed. The change in the light properties enables some property of the target to be determined. The time for the light to travel out to the target and back to the LIDAR is used to determine the distance to the target.
>
> **Fourier Transform Infrared Spectroscopy (FTIR)** technology has the capability to measure more than 100 of the 189 Hazardous Air Pollutants (HAPs) listed in Title III of the Clean Air Act Amendments of 1990. FTIR has the capability of measuring multiple compounds simultaneously, thus providing an advantage over current measurement methods, which measure only one or several HAPs. FTIR provides a distinct cost advantage since it can be used to replace several traditional methods.

research have come to CCTP's attention. Some of these, and others, are currently being explored and under consideration for the future R&D portfolio.

◆ Further development of imaging and volume-measurement sensors for land use/land cover and biomass estimates.

◆ Development of low-cost, practical methods to measure net carbon gain by ecosystems, and life cycle analysis of wood products, at multiple scales of agriculture and forest carbon sequestration.

◆ Research on isotope markers to identify and distinguish between natural and human sources and determine movement of GHGs in geological, terrestrial, and oceanic systems.

◆ Identification of new measurement technology needs that support novel sequestration concepts such as enhanced mechanisms for CO_2 capture from free air, new sequestration products from genome sequencing, and modification of natural biogeochemical processes.

Oceanic Sequestration

Sequestering carbon in oceans generally refers to two techniques: direct injection of CO_2 to the deep ocean waters, and fertilization of surface waters with nutrients. For direct injection, CO_2 streams are separated, captured, and transported using processes similar to those for geologic sequestration, and injected below the main oceanic thermocline (depths of greater than 1,000 to 1,500 meters). Fertilization of the oceans with iron, a nutrient required by phytoplankton, is a potential strategy to accelerate the ocean's biological carbon pump and thereby enhance the draw down of CO_2 from the atmosphere. For a description of oceanic sequestration approaches, see Section 6.4 in Chapter 6.

Measuring and monitoring technologies associated with CO_2 injection are directed towards the performance of the quantities of CO_2 injected and dispersion of the concentrated CO_2 plume. M&M technologies associated with ocean fertilization are focused on the quantity of carbon exported deeper in the water column and the stability and endurance of the carbon sink. Carbon sequestration in oceans can be enhanced significantly, but this has yet to be demonstrated, and the environmental impact of such an approach has not been fully evaluated.

Technology Strategy

These technologies could be advanced through R&D in direct measurement and model analysis, as well as indirect indicators that can be used across spatial scales for obtaining process information and for ocean-wide observations. In the near term, possible advances include (1) measurement of comprehensive trace gas parameters (total CO_2, total alkalinity, partial pressure of CO_2, and pH) to monitor the CO_2 concentration in seawater; (2) development of indirect indicators of fertilization effectiveness using remote-sensing technology; and (3) development of CO_2 sensors that "track" the dissolved CO_2 plume from injection locations. In the long term, advances could include a system that monitors CO_2 in the oceans, temporally and spatially, using integrated M&M concepts, satellite-based sensors, and other analysis systems that can avoid costly ship time.

Current Portfolio

The goal of the current research in support of M&M technologies associated with ocean sequestration is to develop integrated concepts that include direct measurement, model analysis, and indirect indicators that can be used across scales; data transmission and analysis systems that avoid costly ship time; quantitative satellite-based sensors; and development of plume dispersion models for direct injection of CO_2. Research activities in support of M&M technologies associated with ocean sequestration have been underway for several years.[6]

For example, for more than 13 years, DOE and the National Oceanic and Atmospheric Administration (NOAA) sponsored the ocean CO_2 survey during the World Ocean Circulation Experiment, monitoring the carbon concentration in the Indian, Pacific, and Atlantic Oceans from oceanographic ships (Box 8-5).

Another research and development effort underway is to develop low-cost, discrete measurement sensors that can be used in conjunction with the conductivity, temperature, depth, and oxygen sensors to measure the ocean profile on oceanographic stations.

Future Research Directions

The current portfolio supports the main components of the technology development strategy and addresses the highest priority current investment opportunities in this technology area. For the future, CCTP seeks to consider a full array of promising technology options. From diverse sources, suggestions for future

---

6 See Section 5.5 (CCTP 2005): http://www.climatetechnology.gov/library/2005/tech-options/tor2005-55.pdf

BOX 8-5

WORLD OCEAN CIRCULATION EXPERIMENT

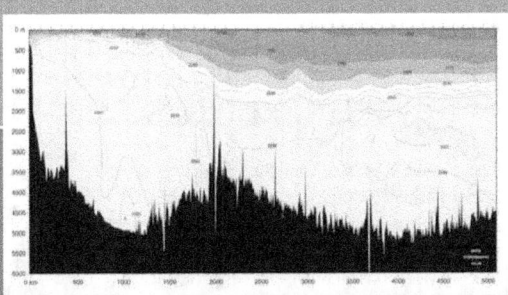

Courtesy WOCE Southern Ocean Atlas

The **World Ocean Circulation Experiment (WOCE)** was a component of the World Climate Research Program (WCRP) designed to investigate the ocean's role in decadal climate change. NSF, NASA, NOAA, the Office of Naval Research (ONR), and DOE supported U.S. participation in WOCE. Scientists from more than 30 countries collaborated during the WOCE field program to sample the ocean on a global scale with the aim of describing its large-scale circulation patterns, its effect on gas storage, and how it interacts with the atmosphere. As the data are collected and archived, they are being used to construct improved models of ocean circulation and the combined ocean-atmosphere system that should improve global climate forecasts.

In 2004, as its final activity, the WOCE program published a series of four atlases, concentrating respectively on the hydrograph of the Pacific, Indian, Atlantic, and Southern Oceans. The Southern Ocean is given a separate volume because of the importance of the circumpolar flow on the transport of heat, freshwater, and dissolved components. The volumes each have three main components: full-depth sections, horizontal maps of properties on density surfaces and depth levels, and property-property plots. The vertical sections feature potential temperature, salinity, potential density, neutral density, oxygen, nitrate, phosphate, silicate, CFC-11, 3He, tritium, 14C, 13C, total alkalinity and total carbon dioxide (see image above), against depth along the WOCE Hydrographic Program one-time lines.

research have come to CCTP's attention. Some of these, and others, are currently being explored and under consideration for the future R&D portfolio, including the following:

◆ Measurement of injected CO_2, and the tracking and dispersion of the concentrated CO_2 plume.

◆ Monitoring of the plume or pool to verify the trajectory and lack of contact with the mixed layer.

◆ Monitoring of the local fauna for adverse effects of enhanced acidity or alkalinity and/or pH changes.

With iron fertilization, it is not well understood whether the excess production stimulated thereby is exported out of the mixed layer, and on what time scale it remains out of contact with the atmosphere. To better understand this, the following R&D investments in measurement technologies would help:

◆ Measurement of the amount of CO_2 drawn down per unit of fertilization.

◆ Characterization of the fate and transport of organic carbon exported deeper in the water column and its longevity from using fertilization

technologies, including the spatial and temporal CO_2 concentration histories.

◆ Technologies that can provide accurate monitoring of local CO_2 concentrations and pH. Monitoring of fauna most likely will involve sampling bacterial populations using advanced biological techniques, but may also include macrofauna as appropriate.

◆ In addition to the specific measurements noted above, it will also be necessary to conduct ocean circulation studies and modeling support selection of injection and fertilization site and estimating storage timescale. As in deep ocean injection, the impact of fertilization on the ocean's biota and chemistry can be monitored carefully to determine the behavior and possible impacts (e.g., pH changes, fish behavior) to deep ocean systems, including the effects of nutrient fluxes on plankton biogeochemistry.

8.4

Other Greenhouse Gases

As discussed in Chapter 7, a wide variety of substances other than CO_2 contribute to the atmospheric burden of GHGs. Other GHGs include methane (CH_4), nitrous oxide (N_2O), chlorofluorocarbons (CFCs), perfluorocarbons (PFCs), sulfur hexaflourine (SF_6), hydrofluorocarbons (HFCs), tropospheric ozone precursors, and BC aerosols. These gases are emitted from both point sources (industrial plants) and diffuse sources (open pit coal mines, landfills, rice paddies, and others), and offer unique challenges for M&M of emissions due to their spatial and temporal variations. A robust R&D program should consider direct measurements of emissions and reporting methods that will become part of a larger integrated system. Moreover, the program should consider the needs for M&M both for point sources, and for the extensive and important diffuse sources, such as those associated with agriculture.

Technology Strategy

Advanced technologies can make important contributions to direct and indirect M&M approaches for point and diffused sources of emissions. Realizing the contributions of these technologies is the focus of an R&D portfolio that combines a number of areas, across a number of agencies, including NASA's A-Train (Figure 8-3).

In the near term, technical improvements to measurement equipment and sampling procedures can improve extended period sampling capabilities that would allow better spatial and temporal resolution of emissions estimates. Software development that allows further integration of measurement data with emission modeling processes can lead to improved estimates. In addition, instruments can be developed to measure from stand-off distances (tower measurements), and from airborne and space-borne sensors to address regional, continental, and global reductions of GHG emissions.

In the long term, development of inexpensive CEMs, satellite-based sensors, and improved accounting estimates of emissions offer promise. Integrating modeling techniques, including inverse modeling procedures that integrate bottom-up and top-down emissions data, regional data or global data are also desirable to identify data gaps or confirm source levels. To facilitate the delivery of cost-effective solutions, the strategy will couple academic and national laboratory R&D to benchmarking and transfer to industry for production and deployment.

Current Portfolio

A wide range of R&D programs currently exists in the area of M&M of emissions of other GHGs. The goals of these programs are to develop an integrated system that meshes observations (and estimations) from point sources, diffuse sources, regional sources, and national scales; inexpensive and easily-deployed sensors for a variety of applications, such as stack emissions, N_2O emissions across agricultural systems, CO_2 fluxes across forested regions, CO_2 and other

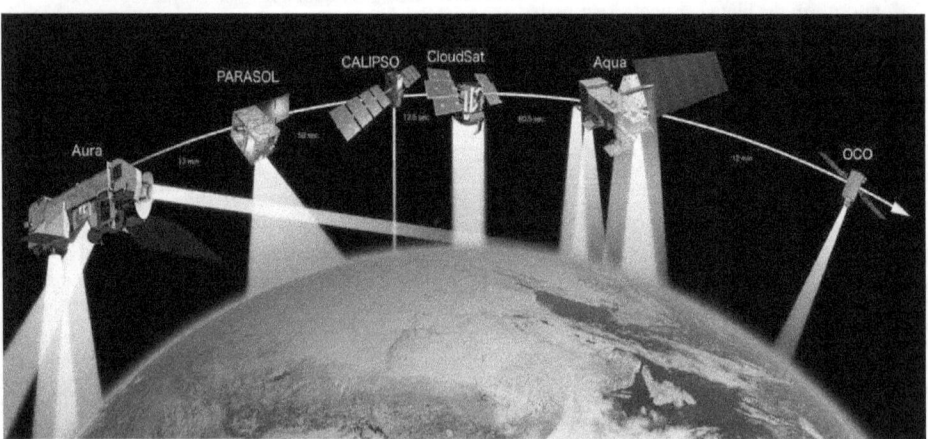

Figure 8-3. NASA's "A-Train" satellite constellation will consist of six satellites flying in formation around the globe. Each satellite will have unique measurement capabilities that greatly complement each other. Near simultaneous measurements of aerosols, clouds, temperature, relative humidity, radiative fluxes, and atmospheric constituents will be obtained over the globe during all seasons.

Courtesy: NASA

GHG emissions from transportation vehicles; accurate rules-of-thumb (reporting/accounting rules) for practices that reduce emissions or increase sinks; a high-resolution system that captures process-level details of sources and sinks (e.g., CO_2 or CO_2 isotopes) and a methodology to scale it up reliably; and data archiving and analysis system-to-integration observations and reporting information.[7]

The following is a summary of some of these programs:

◆ Annual national inventories prepared by EPA rely on both indirect modeling techniques and direct measurement data. These inventories capture changes in the characteristics and activities related to each source, and are subject to ongoing improvements and verification procedures. The indirect modeling procedures developed for these inventories are particularly important to capture emissions from diffuse area sources where individual measurements are not practical.

◆ Through the Advanced Global Atmospheric Gases Experiment (AGAGE) Network and other university-led measurement programs, NASA Earth science research includes measuring global distributions and temporal behavior of biogenic and anthropogenic gases important for both stratospheric ozone and climate. These include CFCs, HCFCs, HFCs, halons, N_2O, CH_4, hydrogen, and carbon monoxide. Measurements made at the sites in the NASA-sponsored AGAGE network, along with sites in cooperative international programs, are used in international assessments for updating global ozone-depletion and climate-forcing estimates and in NASA's triennial report to the Congress and the EPA on atmospheric abundances of chlorine and bromine chemicals.

◆ NOAA monitors the global atmospheric concentration of CH_4, N_2O, CFCs, HFCs, halons and SF_6, in addition to CO_2, through its network of observatories and global cooperative programs. Through these measurements, the global climate forcing by GHGs is updated annually.

◆ There are generally well-established measurement procedures for energy and industrial point sources, as well as for diffuse sources that are involved with voluntary programs of reduction (e.g., natural gas, coal mines) or are subject to monitoring through regulatory programs for other gases (e.g., landfills). Ongoing integration of these direct measurement results with indirect modeling

procedures is part of the national inventory process.

◆ Recent activities for sources such as agricultural soils, livestock, and manure waste focus on advanced modeling of emissions with verification and validation by direct measurements. Improvements to sampling and measurement techniques are a current priority for these sources.

◆ A number of measurement technologies have evolved to address the diffuse nature of many of the non-CO_2 sources. These include advanced chamber techniques for *in situ* sensors, FTIR, tracer gas, micrometeorological methods, and leak detection systems. The results of these measurements are being used to verify and feed back to emission factor development.

◆ BC and tropospheric ozone precursor emissions are an emerging area of importance. Although there is long history of monitoring particulate matter and ozone precursor emissions for criteria pollutant inventories, investigations into the particular sources, speciated forms, and fate of these gases and aerosols that are most applicable to climate forcing potential have become a priority research area.

◆ EPA is conducting analysis and research to improve GHG inventories and emissions estimation methods, implementing formalized quality control/quality assurance procedures and uncertainty estimation. This concentrated effort will improve all emission estimates for all source categories by identifying areas to target for improved or expanded M&M efforts.

◆ EPA and the aluminum industry have developed common protocols for the measurement of perfluorocarbon emissions from aluminum primary production facilities to ensure comparable global data.

Future Research Directions

The current portfolio supports the main components of the technology development strategy and addresses the highest priority current investment opportunities in this technology area. For the future, CCTP seeks to consider a full array of promising technology options. From diverse sources, suggestions for future research have come to CCTP's attention. Some of these, and others, are currently being explored and under consideration for the future R&D portfolio.

---

7 A detailed review of these R&D activities can be found in Section 5.6 (CCTP 2005): http://www.climatetechnology.gov/library/2005/tech-options/tor2005-56.pdf

These include:

◆ Further development of measurement, monitoring, and sampling techniques for agricultural sources, particularly in the area of N_2O from agricultural soils, and CH_4 and N_2O from manure waste. These techniques would address the temporal and spatial variation that is inherent to these emission sources.

◆ Development of high quality and current emission factors for BC, and, to some extent, tropospheric ozone precursors where there is limited measurement data available.

◆ CEMs that can measure multiple gases are well developed, but improvements in performance, longevity, autonomy, spatial resolution of measurements, and data transmission would improve measurement of multiple gases. CEMs have particular application to industrial and point sources; however, applying CEM technology to more diffuse sources is also an area for further research.

◆ Modeling activities that increase the accuracy of spatial and temporal estimates of CH_4 and N_2O from area-type sources such as wetlands, wastewater treatment plants, livestock, and agricultural soils. These are sources that are typically too numerous to measure and monitor on an individual basis, but can be addressed through indirect modeling techniques to account for global, national, and regional emissions. More sophisticated modeling practices could improve the accuracy of the estimates, particularly in terms of greater representation of changing conditions of operation.

◆ Space-based technologies for long-term monitoring of the global distribution and transport of BC aerosols and other aerosol types (Box 8-6). The NASA Orbiting Carbon Observatory (OCO) will also serve as a proof of concept for the measurements needed to derive surface sources and sinks of other GHGs, including CH_4, on regional scales. This measurement approach will have applications to future spaceborne measurements of GHGs. Planned collaborations with international partners—e.g., the Japan Aerospace Exploration Agency's GOSAT mission—will lead to a more complete suite of global GHG observations.

◆ Sophisticated modeling procedures that can fingerprint large-scale measurements to unique sources could help integrate continental and global measurements with regional and local emissions data.

BOX 8-6

CONCEPTS FOR GLOBAL CO₂ AND BC MEASUREMENTS

As part of its scientific research mission supporting the Climate Change Science Program, NASA conducts R&D of aerospace science and technology that is relevant to CCTP M&M needs. Several new measurement concepts have been developed by NASA. The Orbiting Carbon Observatory (OCO) concept involves space-based observations of atmospheric carbon dioxide (CO₂) and generates the knowledge needed.

An Aerosol Polarimetery Sensor (APS) for the NASA Glory space mission is being designed to provide improvements in monitoring of BC aerosols compared to the legacy satellite instruments that only measure the intensity of reflected sunlight.

Studies indicate that multi-angle spectro-polarimetric imager (MSPI) and a high spectral resolution LIDAR (HSRL) would have the capacity to provide column average estimates of aerosol optical depth, particle size distribution, single scattering albedo, size-resolved real refractive index, and particle shape to distinguish natural and anthropogenic aerosols and improve projections of future atmospheric CO₂.

◆ Collaborative research between EPA's National Vehicle and Fuels Emission Laboratory (NVFEL), manufacturers of vehicles/engines, emission control technology, and analytical equipment manufacturers on developing N_2O measurement techniques for emerging gasoline and diesel engines and their emission control systems. Measurement technology applies to both laboratory and field measurement.

Science questions driving future development of technologies for climate change M&M include:

◆ What effects do anthropogenic activities have on aerosol radiative forcing, at accuracies sufficient to establish climate sensitivity, i.e., < 1 W/m²?

◆ What are the separate impacts of anthropogenic and natural processes, including urban activities, fuel-use changes, emission controls, forest fires, and volcanoes, on trends in particulate pollution near the surface?

◆ What connections are there between cloud properties and aerosol amount and type?

8.5 Integrated Measurement and Monitoring System Architecture

The integrated system architecture established the context of a systems approach to delivering the information needed to plan, implement, and assess GHG reduction actions (Figure 8-4). This architecture provides a framework for assessing M&M technology developments in the context of their contribution to observation systems that support integrated system solutions for GHG reduction actions and helps in identifying more cost-effective solutions. It enables the benchmarking of planned improvements against current capabilities.

An integrated M&M capability has the ability to integrate across spatial and temporal scales and at many levels, ranging from carbon measurements in soils to emissions from vehicles, from large point sources to diffused area sources, from landfills to geographic regions. This capability is graphically depicted in Figure 8-5. The integrated system builds on existing and planned observing and monitoring

technologies of the CCSP and includes new technologies emerging from the CCTP R&D portfolio.

Advanced M&M technologies offer the potential to collect and merge global and regional data from sensors deployed on satellite and aircraft platforms with other data from ground networks, point-source sensors, and other *in situ* configurations. Wireless microsensor networks can be used to gather relevant data and send to compact, high-performance computing central ground stations that merge other data from aircraft and satellite platforms for analysis and decision-making. An integrated system provides the benefits of compatibility, efficiency, and reliability while minimizing the total cost of M&M.

Technology Strategy

The strategy for developing an integrated system is to focus on the most important measurement needs and apply the integrated concept design to ongoing technology opportunities as they arise. The near term focuses on development of observation systems at various scales. The longer term focuses on merging these spatial systems into an integrated approach employing IEOS. IEOS will enable and facilitate sharing, integration, and application of global, regional, and local data from satellites, ocean buoys, weather stations, and other surface and airborne Earth observing instruments (IEOS 2005). Although IEOS serves multiple purposes, one outcome will be the strengthening of U.S. capabilities to measure and monitor GHG emissions and fluxes. Development of software and tools to further

Integrating System Architectural Linking Measurement and Monitoring Observation Systems to Greenhouse Gas Reduction Actions

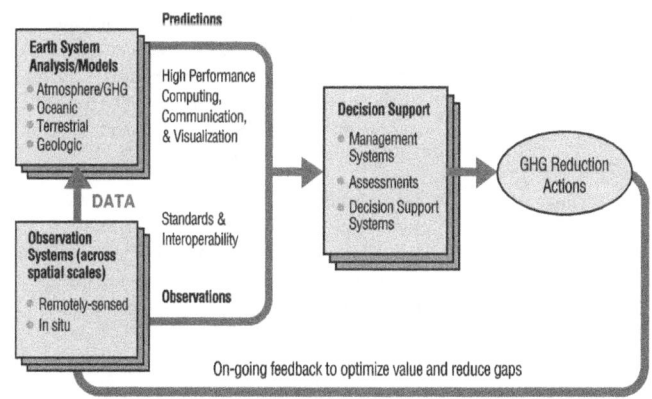

Figure 8-4. Integrating System Architectural Linking Measurement and Monitoring Observation Systems to Greenhouse Gas Reduction Actions

Courtesy NASA

Hierarchical Layers of Spatial Observation Technologies and Capabilities

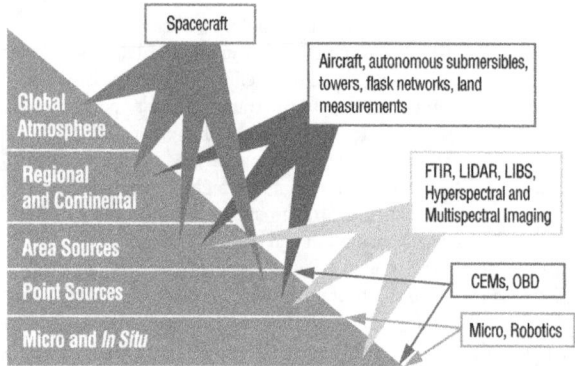

FTIR: Fourier Infared Spectrometer
LIDAR: Light Detection And Ranging
LIBS: Laser Induced Breakdown
   Spectoscopy
CEM: Continuous Emmission Monitors
OBD: On-Board Diagnostics Vehicles

Figure 8-5. Hierarchical Layers of
Spatial Observation Technologies
and Capabilities

Courtesy NASA

integrate measurement data with emission modeling processes will be an ongoing component of the technology strategy.

Current Portfolio

The current Federal R&D portfolio has been targeted at a number of developments, with the goal to develop an integrated system that meshes observations (and estimations) from point sources (e.g., power plant or geologic storage site), diffuse sources (e.g., from commercial and agricultural systems), regional sources (e.g., city/county), and national scales so that checks and balances up and down these scales can be accomplished. The system should be able to attribute emissions/sinks to both national level activities and individual/corporate activities and provide verification for reporting activities. The system must be inexpensive and use easily-deployed sensors for a variety of applications (stack emissions, N_2O emissions across agricultural systems, CO_2 fluxes across forested regions, CO_2 and other GHG emissions from transportation vehicles). In addition, the integrated system should have data archiving and analysis capability for system-to-integration observations and reporting information.[8]

Some examples of the current R&D activities include:

◆ **Global.** R&D programs enabled by NASA's Earth Observation System research satellites, NOAA's operational weather and climate satellites, and NOAA's distributed ground networks (including the Mauna Loa observatory) support improved understanding and measurements and monitoring capabilities relevant to CCSP and CCTP. The transition of NASA's research to NOAA operational use (referred to as "Research & Operations") enhances program planning and budget execution capabilities for the U.S. Earth Observation System.

◆ **Continental.** Recent research has tried to determine the net emissions for the North American continent using different approaches: inversion analysis based on CO_2 monitoring equipment as currently arrayed, remote sensing coupled with ecosystem modeling, and compilation of land inventory information. European researchers have embarked on a similar track by combining meteorological transport models with time-dependent emission inventories provided by member states of the European Union.

◆ **Regional.** Advanced technologies, such as satellites, are being developed to monitor and/or verify a country's anthropogenic and natural emissions. NOAA is building an atmospheric carbon monitoring system under the CCSP using small aircraft and tall communications towers that will be capable of determining emissions and uptake on a 1000-km scale (Box 8-7).

◆ **Local (micro or individual).** A number of techniques are currently used to directly or indirectly estimate emissions from individual sites and/or source sectors, such as mass-balance techniques, eddy-covariance methods (i.e.,

---

8   For a detailed analysis of the current research, see Section 5.1 (CCTP 2005): http://www.climatetechnology.gov/library/2005/tech-options/tor2005-51.pdf.

AmeriFlux sites, source identification using isotope signatures), application of emissions factors derived from experimentation, forestry survey methods, and CEMs in the utility sector.

Future Research Directions

The current portfolio supports the main components of the technology development strategy. Within constrained Federal resources, this portfolio addresses the highest priority current investment opportunities. For the future, CCTP remains open to and seeks to consider a full array of promising technology options. From diverse sources, including technical workshops, R&D program reviews, scientific advisory panels, and expert inputs, a number of such ideas have been brought to CCTP's attention.

◆ **An overarching measurement-and-monitoring system architecture that integrates a diverse set of models and data from local point sources.** The integrated management system would function at multiple scales (local-regional-global) and have the ability to integrate across spatial and temporal scales and from many sources, ranging from carbon measurements in soils to emissions from vehicles, from large point sources to diffused area sources, from landfills to geographic regions. Development would be facilitated via interagency planning and coordination.

◆ **Data fusion and integration technologies to support integration of information from numerous sources, such as satellite observations, real-time surface indicators, and reported emissions inventories.** Advances are needed in data handling and processing, and development of innovative sensors, platforms, advanced data protocols and mining algorithms, large storage systems (hardware and software), and computational models. Validation of data elements requires coordination with national and international standards-setting bodies to develop protocols for interoperability of datasets.

◆ **Platforms for all spatial scales and measurement layers, for example, from new types of global sensors on satellite platforms and from new airborne platforms (e.g., remotely operated or autonomous) facilitated by IEOS.** Monitoring of GHG emission sources and geologic sequestration would be supported by portable platforms for sensors and autonomous units that measure, analyze, and report emissions,

BOX 8-7

NOAA REGIONAL CARBON MONITORING

As part of the Climate Change Science Program (CCSP) and the North American Carbon Program (NACP), NOAA is building a Carbon Cycle Atmospheric Observing System mainly across the United States in order to reduce the uncertainty in the North American carbon sink. To measure carbon fluxes on a 1000-km scale over land, vertical profiling is necessary. From about 24 sites, small aircraft will, on a weekly basis, carry automatic flask sampling systems. These systems will collect 12 samples for analysis of carbon gases and isotopic carbon ratios at predetermined altitudes from the surface to about 8 km. In conjunction, tall communications towers (~ 500 m) will sample carbon and other GHGs continuously from about 12 U.S. sites. This technique will be capable of determining regional carbon sources and sinks and may have applications in the Climate Change Technology Program (CCTP) for monitoring the effectiveness of, for example, sequestration activities.

while ocean sequestration would be supported by autonomous submersible systems with appropriate sensors and reporting capabilities. An integrated system of sensors, indicators and models would be critical to platform development and use, as well as data collection and integration.

◆ **Capability for remote sensing of GHGs and aerosols from beyond low Earth orbit (geostationary L1).** Features would include multi spectral spectrometers, "stare" capability with high temporal resolution, spatial resolution on the order of a few kilometers, and ability to measure a variety of constituents.

◆ **Rapid prototyping and benchmarking of existing integrated system components (sensors, data handling, models, algorithms, decision support) and those evolving through R&D.** Laboratory capability will test and evaluate the efficacy of the solutions to systems integration.

◆ **Wide area networks (wireless mesh-communications with no towers or satellites** connected sensors) that provide robust communications. These networks would employ low-cost point GHG sensors that collect data at an appropriate frequency and spatial resolution.

◆ Decision support tools to incorporate data and information from M&M systems (e.g., change in emissions, regional or continental information, fate of sequestered gases), along with model sensitivities and model predictions generated by CCSP activities into interactive tools for decision makers. These tools would provide the basis for "what-if" scenario assessments of alternative emission reductions technologies (e.g., sequestration, emission control, differential technology implementation time schedules in key countries of the developing world). There is also a critical need for tools that can measure and monitor or simulate functionality in the design of climate change mitigation technologies.

8.6 Conclusions

Meeting the GHG measuring and monitoring challenge is possible with a thoughtful system design that includes near- and long-term technology advances. Figure 8-6 presents a set of representative M&M technologies that are featured in the technology strategies of this chapter and could arise over time from ongoing and future research investments. The resulting timeline illustrates the technology advances that, if realized, would produce continuing progress in GHG measuring and monitoring systems. Such systems are needed to support the design and implementation of strategies to ensure a future of near-net-zero GHG emissions.

Near-term opportunities for R&D include, but are not limited to (1) incorporating transportation M&M sensors into the onboard diagnostic and control systems of production vehicles; (2) preparing geologic sequestration M&M technologies for deployment with planned demonstration projects; (3) exploiting observations and measurements from current and planned Earth observing systems to measure atmospheric concentrations and profiles of GHGs

from planned satellites; (4) undertaking designs and deploying the foundation components for a national, multi-tiered monitoring system with optimized measuring, monitoring, and verification systems; (5) deploying sounding instruments, biological and chemical markers (either isotopic or fluorescence), and ocean sensors on a global basis to monitor changes in ocean chemistry; (6) maintaining *in situ* observing systems to characterize local-scale dynamics of the carbon cycle under changing climatic conditions; and (7) maintaining *in situ* observing systems to monitor the effectiveness and stability of CO_2 sequestration activities.

Through sustained R&D investments in monitoring and measurement capabilities, the United States can (1) enhance its ability to model emissions based on a dynamic combination of human activity patterns, source procedures, energy sources, and chemical processing; (2) develop process-based models that reproduce the atmospheric physical and chemical processes (including transport and transformation pathways) that lead to the observed vertical profiles of GHG concentrations due to surface emissions; (3) determine to what degree natural exchanges with the surface affect the net national emissions of GHGs; (4) develop a combination of space-borne, airborne, and surface-based scanning and remote-sensing technologies to produce 3D, real-time mapping of atmospheric GHG concentrations; (5) develop specific technologies for sensing of global methane "surface" emissions with resolution of 10 km; (6) develop remote-sensing methods to determine spatially resolved vertical GHG profiles, rather than column-averaged profiles; and (7) develop space-borne and airborne monitoring for soil moisture at resolutions suitable for M&M activities.

Technologies for Goal #5: Measure and Monitor Emissions

	NEAR-TERM	MID-TERM	LONG-TERM
Energy Production & Efficiency Technologies	• M&M Specifications and Performance Standards • Low-Cost Sensors and Communications • Samplings, Inventories, & Estimates	• Sensor Networks • Remote Sensing Prototype • Direct Measurement to Replace Proxies and Estimates	• Fully Operational Sensor and Satellite Networks that Feed the Integrated Architecture
Carbon Capture, Storage, & Sequestration	• M&M Specifications and Performance Standards • Low-Cost Sensors and Communications • Samplings, Inventories, & Estimates • Ability to Assess the Integrity of Geologic Reservoirs • Improved Leak Detection from Capture and Pipelines	• Sensor Networks • Remote Sensing Prototype	• Fully Operational Sensor and Satellite Networks that Feed the Integrated Architecture
Other GHGs	• M&M Specifications and Performance Standards • Low-Cost Sensors and Communications • Samplings, Inventories, & Estimates	• Sensor Networks • Remote Sensing Prototype • M&M Techniques for Agricultural Sources	• Fully Operational Sensor and Satellite Networks that Feed the Integrated Architecture
Integrated M&M Systems Architecture	• Identification of Metrics, Criteria, Sources, and Requirements for Measurements • Comprehensive Vision of Integrated Systems Architecture and Technology Needs	• Model and Data Specification • Large Scale, Secure Data Storage System • Data Visualization Tools • M&M Processes Incorporated into Design of Climate Change Technologies	• Fully Operational Integrated MM Systems Architecture (Sensors, Indicators, Data Visualization and Storage, Models)

Figure 8-6. Technologies for Goal #5: Measure and Monitor Emissions
(Note: Technologies shown are representations of larger suites. With some overlap, "near-term" envisions significant technology adoption by 10–20 years from present, "mid-term" in a following period of 20–40 years, and "long-term" in a following period of 40–60 years. See also List of Acronyms and Abbreviations.)

8.7 References

Energy Information Administration (EIA). 2004. *Voluntary reporting of greenhouse gases 2002.* DOE/EIA-0608(2002). Washington, DC: Energy Information Administration. January. http://www.eia.doe.gov/oiaf/1605/vrrpt/. For additional information on reporting requirements, see: EIA 1994. *General guidelines for the voluntary reporting of greenhouse gases under Section 1605(b) of the Energy Act Policy of 1992.* Washington, DC: Energy Information Administration. http://www.eia.doe.gov/oiaf/1605/1605b.html

Integrated Earth Observation System (IEOS). 2005. *Strategic plan for the U.S. integrated earth observation system.* http://usgeo.gov/docs/EOCStrategic_Plan.pdf

Post, W.M., R.C. Izaurralde, J.D. Jastrow, B.A. McCarl, J.E. Amonette, V.L. Bailey, P.M. Jardine, T.O. West, and J. Zhou. 2004. Enhancement of carbon sequestration in US soils. *Bioscience* 54:895-908.

U.S. Climate Change Technology Program (CCTP). 2003. *Technology options for the near and long term.* DOE/PI-0002. Washington, DC: U.S. Department of Energy. http://www.climatetechnology.gov/library/2003/tech-options/index.htm Update is at http://www.climatetechnology.gov/library/2005/tech-options/index.htm

U.S. Department of Energy (DOE), National Energy Technology Laboratory (NETL). 2004. *Geological sequestration of CO_2: the GEO-SEQ project.* http:/www.netl.doe.gov/publications/factsheets/project/Proj287.pdf

U.S. Department of Energy (DOE), Oak Ridge National Laboratory (ORNL). 2003. *Ameriflux.* http://public.ornl.gov/ameriflux/

Bolster Basic Science Contributions to Technology Development

The challenge of encouraging and sustaining economic growth, while simultaneously reducing greenhouse gas (GHG) emissions, calls for the development of an array of new and advanced technologies. Such an undertaking depends on scientific knowledge gained from basic research. Fundamental discoveries can reveal new properties and phenomena that can give rise to improved understanding of technical barriers and illuminate pathways toward innovative solutions. Fundamental discoveries can include breakthroughs in understanding biological functions, properties and phenomena of nano-materials and structures, improved computing architectures, applications and methods, progress in plasma and environmental sciences, and many more breakthrough developments now unfolding on the frontiers of active scientific and technical disciplines.

One of CCTP's core approaches (Approach #2) focuses on strengthening basic research in Federal laboratories, universities, and other research organizations. Basic research will give rise to knowledge and technical insights necessary to enable technical progress throughout CCTP's portfolio of applied research and development, explore novel approaches to new challenges, and bolster the underlying knowledge base for new discoveries (Figure 9-1).

In considering the roles for basic research and related organizational planning in advancing climate change technology development, CCTP characterizes opportunities for contributions as follows:

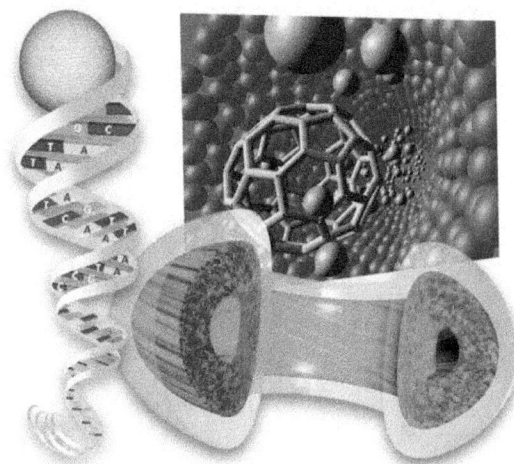

Figure 9-1. Fundamental science is critically important in the creation of new knowledge and improved understanding of technological innovation.

Courtesy: DOE, Office of Science

♦ **Fundamental Science:** Fundamental science is basic research that provides the underlying foundation of scientific knowledge that can lead to fundamental new discoveries. It is the systematic study of system properties and natural behavior that can lead to greater knowledge and understanding of the fundamental aspects of phenomena, processes, and observable facts, but without prior specification toward applications to design or develop specific processes or products. It includes scientific study and experimentation in the physical, biological, and environmental sciences and many interdisciplinary areas, such as computational sciences. Although not directly related to CCTP, it is a source of underlying knowledge that will enable future progress in CCTP.

♦ **Strategic Research:** Strategic research is basic research that is inspired by technical challenges in the applied research and development programs. It is research that may lead to fundamental discoveries (e.g., new properties, phenomena, or materials) or scientific understanding, but is primarily aimed at providing better understanding of underlying phenomena that are believed to be relevant to specific problems, challenges, or technical barriers impeding progress in technology development. For CCTP, strategic research are basic research endeavors relevant to R&D on energy supply; end-use; CO_2 capture, storage and sequestration; mitigation of emissions of non-CO_2 GHGs; and means for enhancing measurement and monitoring (M&M) of GHGs. This "strategic" research builds on knowledge gained from fundamental science and extends it to the technical challenges associated with technology R&D.

◆ **Exploratory Research:** Exploratory research is basic research, or early and exploratory study of application-inspired concepts, undertaken in the pursuit of high-risk, novel, emergent, integrative or enabling approaches, not elsewhere covered. Many such concepts may be pursued within existing applied research and development programs, but often truly novel concepts do not fit well within the established constructs of existing mission-directed or discipline-oriented programs. In addition, some early experimental research may be too risky or multi-disciplinary for a particular research program to justify or support unilaterally. Therefore, not all of the research on innovative concepts for climate-related technology is, or should be expected to be, aligned directly with an existing Federal R&D mission-related program. This *Plan* calls for exploratory research that could lead to new breakthroughs in technology development and thereby dramatically change the way energy is produced, transformed, and used in the global economy. Exploratory research of innovative and novel concepts, not elsewhere covered, is one way to uncover such "breakthrough technology," stimulate innovation across the research community, and enrich the overall R&D portfolio.

◆ **Integrated Planning:** Effective integration of fundamental science, strategic research, exploratory research, and applied technology research and development presents challenges to and opportunities for both the basic research and applied research communities. These challenges and opportunities can be effectively addressed through innovative, integrative planning processes, augmented by analysis and decision-support tools. These processes emphasize communication, cooperation and collaboration among the many associated communities. CCTP encourages and expects to build on the successful models and best practices in this area and plans to improve its analyses and tools.

This chapter discusses the potential research contributions to climate-related technology development of each of the above categories. Section 9.1, "Strategic Research," describes the basic, problem-inspired science underway, planned or under consideration that explores key technical challenges associated with CCTP's five strategic goals, as discussed in Chapters 4 through 8. Section 9.2, "Fundamental Science," describes the basic research that provides the underlying scientific foundation of knowledge needed to enable breakthrough technology. Section 9.3, "Exploratory Research," addresses research of high-risk, novel, emergent, integrative or enabling concepts, and others, important to climate change technology development, but not elsewhere covered. Finally, acknowledging that clarifying and communicating research needs of the applied technology research and development programs can help inform and guide basic research plans and programs, Section 9.4, "Toward Enhanced Integration in R&D Planning Processes," describes a generalized approach to integrate better basic research with the applied research and development programs related to climate change technology development.

9.1 Strategic Research

Scientific research enables both current and new generations of technologies that are needed to address the problem of GHG emissions. The outcomes expected from scientific research are time-variant:

◆ In the near term, a significant role of research is to overcome bottlenecks and barriers that presently limit or constrain the development and application of technologies that are progressing toward commercial status. Some of the barriers include a lack of suitable materials, advanced processing and manufacturing, the need for information on key processes, and the need for new instrumentation and methods (Figure 9-2). Research will contribute to studying the feasibility of new technologies, solving key materials and process issues, developing new instrumentation and methods, and reducing costs. For example, science-based analyses will help to assess the viability of carbon storage and sequestration over the next decade; to better understand the interactions between engineered systems and natural systems (e.g., in systems involving biotechnology); and to solve materials and chemistry problems in advanced energy systems, such as hydrogen production and fuel cells. The development of novel space-based monitoring systems could enhance GHG M&M strategies.

♦ In the mid-term, science will take nascent ideas and develop them to the point they can enter the technology cycle. For example, innovations achieved through the support of science programs may result in new nanomaterials and devices for energy transformation, the ability to capture bioenzymes in biomimetic membranes for various energy applications, advances in plasma science for the development of fusion energy, and identification of new materials and efficient processes for hydrogen production, storage, and conversion.

♦ In the long-term, the current wave of research "at the frontier" may open up entirely new fields involving genomics and the molecular basis of life, computational simulations, advanced analytical and synthetic technologies, and novel applications of nanoscience and nanotechnology. It is hard to predict discoveries that will open entirely new ways of producing and using energy or dramatically alter industrial processes. However, the history of science and technology contains many examples of such transformations.

Much of the research needed to address the challenges of climate change technology development requires cross-cutting strategic research approaches. These are discussed in the sections that follow, organized by the five CCTP strategic goals (Chapters 4 through 8):

♦ Reduce Emissions from Energy End-Use and Infrastructure

♦ Reduce Emissions from Energy Supply

♦ Capture and Sequester Carbon Dioxide

♦ Reduce Emissions of Non-CO_2 GHGs

♦ Enhance Capabilities to Measure and Monitor GHGs.

The section concludes with a description of cross-cutting strategic research areas that underpin the sixth CCTP goal: to strengthen the basic research foundations that enable climate technology advances.

Figure 9-2. Scientific research can help in overcoming barriers to the development of advanced process technology, such as the laser-assisted welding imaging system shown above.

Courtesy: DOE Office of Science

Research Supporting Emissions Reductions from Energy End-Use and Infrastructure

A broad array of research underpins emissions reductions from energy end-use and infrastructure, spanning the areas of transportation, buildings, industry, the electric grid, and infrastructure. The promising basic research directions for each of these areas are discussed below, with illustrative examples.

Transportation

Strategic research is needed to address major sources of CO_2 emissions from vehicles and other key transport modes. Research on reducing vehicle weight while maintaining strength and safety includes **materials science** that improves efficiency, economy, performance, environmental acceptability, and safety in transportation. Foci are ceramics and other durable high-temperature, wear-resistant materials and coatings; and strong and lightweight alloys, polymers, and composite materials for structural components. **Joining, welding, and corrosion sciences** will enable the application of new polymer composites and bimetallic alloys, which also require the development of low-energy techniques for materials processing.

The **nanosciences** can potentially contribute to many aspects of energy efficient vehicles, engines, and engine processes. Research can build on basic research in materials, chemistry, and computation to develop fundamentally new types of **materials with specific, tailored properties**, including innovative applications such as highly conductive nanofluids for lubrication and cooling.

Materials and membrane research for fuel cell stacks and advanced fuel cell concepts for vehicles will improve the efficiency of fuel cells along with their performance, durability, and cost. **Nanostructured catalysts** will reduce the need for noble metals and can be operated at lower temperatures while producing fewer side products.

Electrochemistry, materials, and catalyst research, including research at the nanoscale, may lead to innovations in onboard energy storage for electric hybrid and hydrogen-powered vehicles. For conventional and novel sources of power in mobile applications, energy conversion cycles can be made more efficient and **thermoelectric materials** can enable more beneficial use of waste heat.

Research in **thermo- and electro-chemistry and materials** for advanced sensors that are robust and inexpensive could improve vehicle fuel economy by predicting system failure and optimizing system parameters.

For both combustion and other transportation energy sources, research on the **energetics of chemical reactions** and the interactions of **chemistry at interfaces** may significantly improve or transform the efficiency of energy-producing reactions. The design and development of efficient, clean-burning designs can be accomplished more quickly and with a higher probability of success if combustion models are improved.

Research on intelligent transportation systems needs to include **complex systems science** for sustainable transportation as well as computational science and **improved mathematical algorithms and models** for improved traffic handling/management and for design and performance simulation.

Genomics, biochemistry, and other biological sciences will lead to more productive biomass feedstocks and more efficient conversion to biofuels. This strategic research is described in more detail in the following section on "Renewable Energy and Fuels."

Buildings

Three aspects of buildings that could significantly reduce CO_2 emissions would benefit from strategic research: the building envelope, building equipment, and integrated building design.

In improving energy efficiency in the building envelope, **materials science** will have a broad range of impacts, from a next generation of smart building insulation with phase change materials to transparent films for energy-efficient adaptive windows to new classes of lightweight structural materials. **Robotics,** along with the **joining and welding sciences,** will support the fabrication and construction of high-efficiency envelopes.

Building equipment will become more energy efficient through research in **plasma science** for arc lighting and **semiconductor alloys** for solid-state lighting, as well as light-emitting polymers. More efficient heating and cooling systems will be possible because of **combustion, materials, heat transfer, and engineering research**, and fundamentally new approaches to heating and cooling will result from research into **thermoacoustics and thermoelectrics**. Breakthroughs in **magnetism** will enable more efficient motors.

Research in whole-building integration will draw on the basic science research in **condensed matter physics** that enables improvements in smart transistors for energy-saving sensors and electronic devices to optimize space conditioning, new and improved self-powered smart windows through research in **constricted-plasma source thin film applications, electrochromics and dye-sensitized solar cells,** as well as **multilayer thin film materials and deposition processes** to control the interior environment, and smart filters for water systems based on **tailored pore sizes** and **pore chemistry**.

Industry

Strategic research is needed to address current and anticipated sources of emissions of CO_2 and other GHGs from energy conversions and process inefficiencies. Strategic research is also needed to facilitate improvements in energy efficiency and resource utilization.

Research on **advanced materials** with attributes such as the ability to operate in varied hostile environments, such as high temperatures and pressures and corrosive environments, can enable improved process efficiencies.

Advances in high-temperature **materials research** (Figure 9-3) will lead to increased energy efficiency in industrial processes; for instance, increased temperature will improve the efficiency of industrial boilers (super-critical steam cycles) and Integrated Gas Combined Cycle (IGCC) systems for recycle of byproduct streams in the paper and pulp industry. Other areas of materials science include ion implantation, thin films, carbon-based nanomaterials, ceramics, alloys, composites, and quasicrystals; welding, processing, and joining; and foundations for nanomechanics and nano-to-micro assembly.

Solid-state physics and related sciences will support advanced, energy-efficient computer chip concepts and manufacturing.

Because of the very wide diversity of industrial applications, environments, processes, and products, **strategic research in nearly all basic research disciplines** is needed for new and advanced industrial sensors. For example, superconducting quantum interference devices (SQUIDS) that can measure extremely weak signals via tiny variations in a magnetic field will provide feedback to systems and reduce energy use as situations change.

Research on **advanced separations, chemistry, and higher-selective catalysts** can increase resource recovery and utilization of industrial byproduct or waste material. **Advanced membranes and adsorption processes** can lead to improved industrial process efficiencies and costs.

Research into the **magneto-caloric effect** will lead to new, energy-efficient forms of industrial refrigeration.

Advances in **electronics research**, tailored to power electronics applications, will enable more efficient motor and drive systems with improved ability to vary motor speed to enable higher efficiencies in loads such as fans, pumps, and compressors.

Research on key **biotechnology platforms** and designs for biorefineries will enable chemical products to be derived from biomass rather than fossil fuels as described in the section below on "Renewable Energy and Fuels."

Electric Grid and Infrastructure

A balanced portfolio of strategic research addressing conductor technology, systems and controls, energy storage, and power electronics is needed to meet the need for secure and reliable power leading to reduced CO_2 emissions from electric generation.

Materials that improve the transmission and storage of electricity will achieve highly improved energy efficiency. **Solid-state physics and materials**

High-Temperature Materials Research

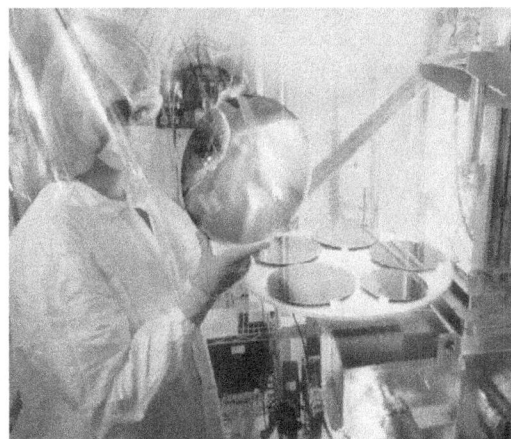

Figure 9-3. Advances in high-temperature materials research, facilitated by the use of synchrotron radiation, can make significant contributions towards improving the energy efficiencies of industrial equipment and electricity generation.

Courtesy: DOE, Office of Science

science will enable high-performance semiconductors and high-temperature superconductors for efficient, high-capacity transmission of electric power. **Superconductivity research** will also make possible innovative storage devices and efficient motors. The, 2006 workshop report, *Basic Research Needs for Superconductivity*, examined the prospects for superconducting grid technology and its potential for significantly increasing grid capacity, reliability, and efficiency to meet the growing demand for electricity over the next century (DOE-BES 2006).

Other **tailored materials research** can lead to highly conductive high-strength nanowires; superlattices; high-strength, lightweight composites and corrosion-resistant materials; nanostructured materials for semiconductors; and metallic glasses for vastly improved transformers and sensor implementation.

Basic Research Needs for Superconductivity (DOE-BES 2006)

Silicon carbides and thin-film diamond switching devices will improve performance and energy efficiency of power electronics and controls. **Sensors and adaptive controls** will enable optimization of the grid; development of responsive loads for peak shaving; and accommodation of distributed solar and wind supply on the grid.

Electrochemistry research, including electrolytes, electrode materials, thin films, and interfaces, will improve commercial batteries and other electric storage devices so important to integrating intermittent renewable resources into electric grid operations and for load leveling and optimized grid operations.

Computational science and computer/network science will improve real-time control of the utility transmission infrastructure and, thus, its energy efficiency.

Research Supporting Emissions Reductions from Energy Supply

Strategic research underpinning emissions reductions from energy supply targets low-emissions fossil-based power, hydrogen, renewable energy and fuels, nuclear fission, and fusion. Research in these areas includes the following:

Low-Emissions Fossil-Based Power

Strategic research is needed to achieve the principal fossil energy objective of a zero-emission, coal-based electricity generation plant that has the ability to co-produce low-cost hydrogen.

Since high temperatures result in lower GHG emissions, **combustion, materials research, and condensed matter physics (crystalline structure)** can contribute improved and new materials for high temperature, pressure, and corrosive environments. The result will be more efficient gasification processes for advanced coal plants and higher temperature turbine blades and heat exchangers to allow more efficient conversion of natural gas into electricity.

Research in **thermo- and electro-chemistry and materials** for advanced sensors will lead to improved monitoring and control of processes in fossil fuel combustion. Separation sciences will enable improved gas phase separations in coal liquefaction and reduced energy requirements for oxygen separations in oxycombustion options.

Opportunities for Catalysis in the 21st Century (BESAC 2002)

Computational sciences will advance simulation and design, especially for improved models and codes for fluid dynamics, turbulence, and heat transfer modeling.

Catalysis research employing **nanostructured materials** will find efficient pathways for the selective and efficient conversion of fossil fuels, including a catalyst for petroleum refining and chemical manufacturing and catalysis of carbon-hydrogen bonds. The *2002 Opportunities for Catalysis in the 21st Century* workshop report describes research directions for better understanding of how to design catalyst structures to control catalytic activity and selectivity (BESAC 2002).

Geosciences research for higher recovery rates of fossil fuels with lower societal impact will be needed to provide feedstocks for higher efficiency, new low-emission power plants.

Hydrogen

The development of energy-efficient and economically competitive technologies for H_2 delivery, storage, and production will require a broad portfolio of strategic research.

Research will focus on understanding the atomic and molecular processes that occur at the interface of hydrogen with materials in order to develop new materials suitable for use in a hydrogen economy. New research is needed for **tailored materials, membranes, and catalysts**, leading to fuel cell assemblies that perform at much higher levels, at much lower cost, and with much longer lifetimes.

In the hydrogen production area, a key focus is on **catalysts** and better understanding mechanisms for hydrogen production. **Biological enzyme catalysis, nanoassemblies** and bio-inspired materials and processes are areas of basic research related to hydrogen production from biomass. **Photoelectrochemisty and photocatalysis** research may lead to breakthroughs in solar production of hydrogen. Also, **thermodynamic modeling, novel materials research and membranes, and catalyst research** may support nuclear hydrogen production.

Hydrogen storage is a major challenge. Basic science research related to storage includes the study of **hydrogen storage-hydrides** and **tailored nanostructures** and the development of high-density reversible membranes. For instance, research on complex metal and chemical hydrides may support on-board recharging of fuel cell vehicles.

In the fuel cells area, **electrochemical energy conversion mechanisms** and materials research are important. In addition, there are identified needs for higher temperature **membranes and tailored nanostructures** that basic science research could support. The *2004 Basic Research Needs for the Hydrogen Economy* workshop report identifies fundamental research needs and opportunities in hydrogen production, storage, and use, with a focus on new, emerging, and scientifically challenging areas that have the potential to deliver significant impacts (BESAC 2004). An NSF (2004a) workshop on *Future Directions for Hydrogen Energy Research & Education* emphasized the promise that nanotechnology offers for hydrogen and fuel cell development, and the importance of developing a more interdisciplinary approach to future hydrogen research and development.

Renewable Energy and Fuels

Strategic research is needed to enable a transition from current reliance on fossil fuels to a portfolio that includes significant renewable energy sources, with a shift in infrastructure to allow for a more diverse mix of technologies.

Biochemistry, bioenergetics, genomics, and **biomimetics research** will lead to new forms of biofuels and capabilities for microbial conversion of feedstocks to fuels. This includes research on strategies for cellulose treatment, sugar transport, metabolism, regulation, and microbial systems designed to optimize the use of microbes that are known to break down different types of complex biomass to sugars and ferment those sugars to ethanol or other fuels. The 2006 workshop report, *Breaking the Biological Barriers to Cellulosic Ethanol: A Joint Research Agenda*, outlines a detailed research plan for developing new technologies to transform cellulosic ethanol—a renewable, cleaner-burning, and carbon-

Basic Research Needs for the Hydrogen Economy (BESAC 2004)

neutral alternative to gasoline—into an economically viable transportation fuel (DOE-BER-EERE 2006). The research may lead to scientific breakthroughs in the design of a single microbe for making ethanol from cellulose. **Nanoscale hybrid assemblies** will enable the photo-induced generation of fuels and chemicals. **Plant genomic research** and gene function studies will make possible increased crop yields, disease resistance, drought resistance, improved nutrient-use efficiency, tissue chemistry that enhances biofuel production and carbon sequestration. The NSF workshop report on *Catalysis for Biorenewables Conversion* (2004) identifies the need for a reinvigoration and redirection of U.S. catalysis research for the purpose of developing fuels and chemicals production from biorenewables.

Plant biology, metabolism, and enzymatic properties research will support the development of improved biomass fuel feedstocks and will enable the

Breaking the Biological Barriers to Cellulosic Ethanol: A Joint Research Agenda (DOE-BER-EERE 2006)

Genomics: GTL Roadmap (DOE-SC 2005a)

design of crops for bioconversion. For example, expression of thermostable cellulases directly in plant cell walls could facilitate enzymatic hydrolysis of cellulose. Research on key **biotechnology** platforms includes designs for biorefineries to produce biofuels, biopower, and commercial chemical products derived from biomass rather than fossil fuels. The *Roadmap for Biomass Technologies in the United States* prepared by the Biomass Research and Development Technical Advisory Committee (2002) itemizes a broad supporting R&D agenda that spans feedstock production, processing and conversion, and product uses and distribution. Complementing this agenda, the 2005 *Genomics: GTL Roadmap* (DOE-SC 2005a) describes an aggressive systems microbiology plan to accelerate the scientific discovery needed to support such developments.

Basic Research Needs for Solar Energy Utilization (DOE-SC 2005b)

Basic research in **photochemistry** and **photocatalysis** will provide foundations for future, alternative processes for light-energy conversion, thin-film, and nanosciences research for photovoltaics. Bio-inspired systems offer the promise of engineered systems that mimic photosynthesis at higher efficiencies and rates. The 2005 *Basic Research Needs for Solar Energy Utilization* (DOE-SC 2005b) workshop report examines the challenges and opportunities for the development of solar energy as a competitive energy source and to identify the technical barriers to large-scale implementation of solar energy and the basic research directions showing promise to overcome them.

Geosciences and **hydrology** research will support a broad range of siting issues related to hydro and geothermal power sources as well as assessing the availability of low-grade geothermal energy. Needed research includes mapping and monitoring geothermal reservoirs, predicting heat flows and reservoir dynamics, mapping the natural distribution of porosity and permeability in deep geologic media, and developing new methods to enhance or reduce permeability.

Research in **materials** and **composites** will lead to improved wind energy systems by enabling larger blades on wind turbine systems leading to lower unit costs for wind power and the economic use of wind turbines with low-speed wind resources. Research in **electro-chemistry, superconductivity, and solid-state physics** can aid advances in electric storage to help deal with the problem of intermittency that currently makes it difficult to integrate wind (and solar) energy into large-scale power dispatch systems. Research in **materials chemistry, electro-chemistry and solid-state physics** can lead to advances in development of power electronics, which will lead to power systems that can integrate with multi-level photovoltaics and inverters for solar and wind power systems and can convert DC power into 60-hz AC power.

Nuclear Fission Energy

Through strategic research addressing issues of safety, sustainability, cost-effectiveness, and proliferation resistance, advanced nuclear fission-reactor systems can play a vital role in diversifying the Nation's energy supply and reducing GHG emissions.

Heavy element chemistry, advanced **actinide and fission product separations** and extraction, and fuels research will support better process controls in nuclear fission cycles. The 2005 Workshop report on

The Path to Sustainable Nuclear Energy (DOE-BES-NP 2005)

The Path to Sustainable Nuclear Energy identifies new basic science that will be the foundation for advances in nuclear fuel-cycle technology in the near term, and for changing the nature of fuel cycles and of the nuclear energy industry in the long term (DOE-BES-NP 2005).

Fundamental research in **heat transfer** and **fluid flow** will lead to improved efficiency and containment.

Basic research will meet the **materials sciences** challenges of Gen IV reactor environments, with emphasis on the search for radiation-tolerant, ultra-strong alloy and composite materials. Increased temperatures enabled by high-temperature materials allow higher thermal-to-electricity conversion efficiencies. **Materials processing, welding, and joining** sciences will also play a critical role in reducing failure rates and ensuring system integrity, and research into **basic defect physics in materials**, equilibrium and radiation-modified **thermodynamics of alloys and ceramics** will improve reactor design. **Deformation and fracture studies** and analyses of **helium and hydrogen effects on materials** will contribute to safety and reliability of advanced nuclear energy systems, as will atomistic and 3D **dislocation dynamics studies**.

Geophysical research and **geological permeability engineering** will support nuclear siting and waste disposal in geologic repositories, potentially employing remote sensing technologies similar to the approaches used in siting renewable energy facilities.

Chemistry and **corrosion research** will improve design, operation, and predictability for performance.

Fusion Energy

Strategic research on magnetic confinement approaches, **materials science, plasma physics, and**

Computational Sciences

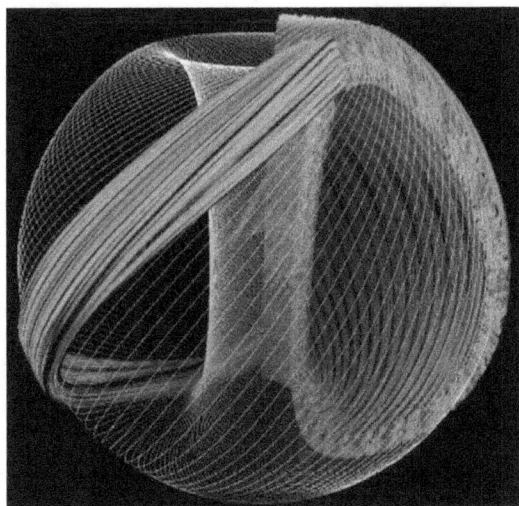

Figure 9-4. Computational sciences applied toward fusion energy research will aid in testing agreement between theory and experiment, and simulate experiments that cannot be conducted in the laboratory.
Courtesy: DOE, Office of Science

high energy density physics is needed to define and develop the most promising fusion concept.

Research in **burning plasmas** will validate the scientific and technological feasibility of fusion energy. Moreover, research aimed at a fundamental understanding of **plasma behavior** will provide a reliable predictive capability for fusion systems. Studies will identify the most promising approaches and configurations for confining hot plasmas for practical fusion energy systems.

Research in **materials tailored** for a fusion energy environment leading to components and technologies that will be necessary to make fusion energy a reality.

A broad underpinning of **computational sciences** (Figure 9-4) will advance fusion research, including computational modeling to test the agreement between theory and experiment, and simulating experiments that cannot readily be investigated in the laboratory.

Research Supporting CO_2 Capture and Sequestration

Research supporting carbon capture and sequestration underpins the development of technologies and strategies for CO_2 capture and sequestration that are described in Chapter 6.

Carbon Capture and Storage in Geologic Repositories

Realizing the possibilities for point source CO_2 capture requires a research portfolio covering numerous technology areas, including post-combustion capture, oxy-fuel combustion, and pre-combustion decarbonization to reduce costs and energy penalties. Carbon storage in geologic repositories will require comprehensive understanding of the economic, health, safety, and environmental implications of long-term, large-scale geologic storage.

Membranes and **chemistry** research will enable separating CO_2 in post-combustion stack gases, capturing it, and if needed, transforming it to another form of carbon that may be more useful, or more safely or permanently stored.

Geophysics, geochemistry, hydrology, and geological permeability engineering research of CO_2 repositories in geological formations will increase understanding of how CO_2 injected into such formations interacts with minerals and what the long-term fate of CO_2 would be after injection. This research will probe the factors that determine the residence time of carbon sequestered in soils, and ways in which the quantity and residence time of carbon sequestered in soils can be increased. Such research provides the scientific foundation for credible calculation of sequestration by terrestrial ecosystems.

Modeling, simulation, and assessment of geological repositories research are necessary to identify sites that have been or could be selected for use in storing CO_2 removed from industrial flue gases. Such research will help meet the need for a more definitive understanding of geologic storage potential.

Research is also needed on **microbial processes** that act to metabolize CO_2 in geologic structures.

Terrestrial and Ocean Sequestration

Realizing the potential to sequester carbon in terrestrial systems requires research on equipment, processes, management systems, and techniques that can enhance carbon stocks in soils, biomass, and wood products, while reducing CO_2 concentrations in the atmosphere.

Basic **biological** and **environmental research** on terrestrial carbon sequestration could enhance the natural carbon cycle—plants that store even more CO_2. For example, this could involve development of technologies for enhancing the ability of trees to sequester carbon by modifying their root systems. Another possibility is genomic research on black cottonwood to characterize key biochemical functions related to photosynthesis, tree growth, and carbon storage. **Environmental science** research can analyze how efforts to increase terrestrial carbon sequestration might influence other environmental processes, such as nutrient cycling, the emissions of other GHGs, and albedo effects on climate at all scales.

Genomic research will identify traits that would enable plant species to grow and persist in environments that are of marginal quality and, hence, may not be useful for purposes other than capturing carbon in plant biomass. Genomic research on microalgae and photosynthetic bacteria may identify traits that enable the organisms to efficiently capture and fix CO_2 separated from other industrial flue gases before it is released into the atmosphere. Research related to **modifying plants** and **soil micro-organisms** can provide the basis for capturing and retaining nitrogen and other essential plant nutrients and engineering the pathways for lipid synthesis to trap a larger fraction of photosynthate directly in hydrocarbon precursors.

Soil science research on the formation and transformation of soil organic matter will enable efficient application of technologies to enhance soil carbon sequestration, increase plant productivity, and reduce non-CO_2 GHGs (e.g., nitrous oxide [N_2O]) from soil.

Materials research may enhance carbon sequestration by substituting carbon-based products for steel, cement, and other commodities. Examples include carbon fiber from black liquor used in the manufacture of carbon composite lightweight materials and wood composites used in place of steel beams.

Research will explore ways of injecting CO_2 into the deep ocean, how long the injected CO_2 would remain isolated from the atmosphere, and what the potential **ecological** and **chemical** effects might be of injecting relatively pure streams of CO_2 into the deep ocean. Research on methods of enhancing the abiotic uptake of CO_2 by the ocean, and/or storing carbon in the ocean in forms other than acid-producing, easily degassible CO_2 will also be considered.

A roadmap of various technology development approaches to carbon sequestration is described in *Carbon Sequestration Research and Development* (DOE-SC-FE 1999).

Research Supporting Emissions Reductions of Non-CO$_2$ GHGs

Basic and applied research is also supported by Federal agencies to develop ways of reducing emissions of non-CO$_2$ GHGs. This includes research in the **physical sciences, biological, and environmental sciences, and in computational sciences**.

Work on **materials** and **chemistry** will lead to replacements for high global warming potential non-CO$_2$ GHGs, such as sulfur hexafluoride (SF$_6$) and perfluorocarbons that are used in industrial processes. For example, research in materials chemistry, electro-chemistry, and solid-state physics can lead to advances in development of power electronics needed to minimize SF$_6$ emissions from transformers by leak reduction, replacement of SF$_6$ with other dielectric material, and development of new power transmission equipment that does not require SF$_6$ insulation.

Research on **thin films and membranes** will isolate non-CO$_2$ GHGs in industrial flue gases and other waste streams; **combustion research** will reduce emissions of N$_2$O, ozone precursors, and soot; and **catalysis research** will reduce emissions of non-CO$_2$ GHGs.

Basic research in the **biological and environmental sciences**, including microbial processes in the rumen of farm animals, animal metabolism, and animal grazing will enable reductions in methane emissions by livestock. **Biological research** will increase understanding of soil microbes to reduce methane emissions from livestock feedlots.

Basic **biogeochemistry** coupled with **microbial ecology and soil science research** may enable reductions in N$_2$O emissions from soils.

Basic Research Supporting Enhanced Capabilities to Measure and Monitor GHGs

There is a continuing need to enhance capabilities to measure and monitor GHG emissions and concentrations across a range of scales and applications so that carbon management strategies can

Figure 9-5. Advances in material chemistry can support the development of technologies that reduce GHG emissions. Shown here, laser-based surface analysis using resonant ionization of sputtered atoms identifies and accurately measures trace impurities in solid materials.

Courtesy: DOE, Office of Science

be designed and implemented consistent with economic and environmental goals. Basic research in this area includes the following:

Various kinds of measurement for GHGs in the atmosphere are necessary. Observed vertical profiles of GHG concentrations are a result of surface emissions and atmospheric physical and chemical processes. Remote sensing methods will determine spatially resolved vertical GHG profiles rather than column-averaged profiles. Combined airborne and surface-based scanning techniques for remote sensing will yield 3D, real-time mapping of atmospheric GHG concentrations. Specific technologies for airborne remote sensing will measure methane surface emissions at a 10-km spatial resolution. Technologies for the long-term monitoring of global black carbon (BC) sources and transports, along with other aerosols, will enable solutions tailored to emission sources and their regional impacts.

Research in **materials chemistry, electro-chemistry and solid-state physics** can lead to advances in development of high-fidelity sensors needed for making precise and accurate measurements of GHGs in remote and hostile environments (Figure 9-5). Innovative technologies for non-invasive measurement of soil carbon will provide rapid methods for monitoring the effectiveness of carbon management approaches applied to terrestrial ecosystems and agricultural practices. **Microbial**

genomics research will seek to identify or develop eco-genomic sensors and sentinel organisms and communities for use in monitoring the effects of sequestering CO_2 in terrestrial soils and in the ocean.

Models will simulate and predict GHG emissions based on dynamic combinations of human activity patterns, energy technologies and energy demand, and industrial activities. **Environmental science and computational science** can develop models that can simulate and predict carbon flows resulting from, for example, specific carbon management policy actions that provide a consistent picture of the effectiveness of efforts to reduce GHG emissions.

Cross-cutting Strategic Research Areas

As the five previous sections illustrate, dissimilar technologies often require similar scientific advances to succeed. For example, degradation-resistant materials that perform well under high temperature conditions are essential to improving the energy efficiency of industrial processes; they are also needed to advance to a next generation of fossil and nuclear power plants, and are essential to realizing fusion energy. Similarly, plant and microbial genomics are central to advancing biofuels and biobased chemicals, improving terrestrial sequestration of carbon dioxide, and reducing methane emissions from landfills and livestock. Table 9-1 describes twenty cross-cutting strategic research areas and identifies the climate change technology goals that depend on their success.

This broad agenda of strategic research is inspired by the technical challenges of specific climate change technologies. If adequately funded, research will successfully convert many of today's emerging technologies into cost-competitive and attractive products and practices. However, to address the

century-scale problem of climate change, scientific breakthroughs will be needed to broaden the range of today's options. The next section (9.2) describes the role of fundamental research, which enriches the underlying foundation of scientific knowledge necessary for problem-solving. Attention then turns to the novel approaches, advanced, integrative, and enabling concepts that fall under the category of "exploratory research" (Section 9.3).

9.2 Fundamental Science

At the outset of the 21st century, science is in the midst of an information revolution that is bringing on the rapid development of many new and promising discoveries across a variety of fields. In addition, the rapidly developing global infrastructure for computing, communications, and information is expected to accelerate scientific processes through computational modeling and simulation and to reduce the time and cost of bringing new discoveries to the marketplace. These potential discoveries and infrastructure developments portend a rapid advancing of capabilities to further the development of CCTP technologies. Fundamental research is needed in the following areas, which are representative of the opportunities afforded and serve as a reminder of the importance of sustained leadership and continued support of the pursuit of fundamental scientific knowledge.

Physical Sciences

Many of the advances in lowering energy intensity stem from developments in the materials and chemical sciences, such as new magnetic materials; high strength, lightweight alloys and composites; novel electronic materials; and new catalysts, with a host of energy technology applications. Two remarkable explorations—observing and manipulating matter at the molecular scale, and understanding the behavior of large assemblies of interacting components—may accelerate the development of more efficient, affordable, and cleaner energy technologies. Nanoscale science research—the study of matter at the atomic scale—will enable structures, composed of just a few atoms and molecules, to be engineered into useful devices for desired characteristics such as super-lightweight and ultra-strong materials. The 2004 *National Nanotechnology Initiative Workshop* report describes many of these opportunities (DOE-BES-NSET 2005).

Nanoscience Research for Energy Needs
(DOE-BES-NSET 2005)

Cross-Cutting Strategic Research Areas

Fundamental Research Area	Strategic Research Area	Goal 6: Basic Research	Goal 1: Energy End Use				Goal 2: Energy Supply					Goal 3: Capture & Sequestration			Goal 4	Goal 5
			Transportation	Buildings	Industry	Grid	Fossil	Hydrogen	Renewable	Nuclear	Fusion	Capture	Geo-Storage	Terrestrial Sequestration	Non-CO$_2$ Gases	Measurement and Monitoring
Physical Sciences	Materials: High Temperature																
	Materials: Tailored Mechanical Chemical Properties																
	Materials: Tailored Electrical Magnetic Properties																
	Heat Transfer & Fluid Dynamics																
	Combustion																
	Chemistry (Electro, Thermo)																
	Chemistry (Photo, Radiation)																
	Membranes & Separations																
	Condensed Matter Physics																
	Nanosciences																
	Geosciences & Hydrology																
	Chemical Catalysis																
Biological Sciences	Bio-Catalysis																
	Plant and Microbial Genomics (Biotechnology)																
	Bio-Based & Bio Inspired Processing																
Environmental Sciences	Environmental Science																
	Atmospheric Science																
Advanced Scientific Computing	Computational Sciences (Models & Simulations)																
Fusion Sciences	Plasma Sciences																
Enabling Research	Strategic Research for Sensors & Instrumentation																

A strategic research area that is **central to advancing** the technology approach.

A strategic research area that is **expected to contribute significantly** to the technology approach.

A strategic research area that has the **potential to contribute significantly** to the technology approach.

A strategic research area that is **not expected to contribute** significantly to the technology approach.

*Table 9-1.
Cross-Cutting
Strategic
Research Areas*

Underpinning these basic research explorations are the powerful tools of science, including a suite of specialized nanoscience centers and the current generation synchrotron x-ray and neutron scattering sources, terascale computers, higher resolution electron microscopes, and other atomic probes. Fundamental research in the physical sciences includes research in material sciences, chemical sciences, and geosciences, all of which are described in more detail below.

◆ **Materials sciences** research helps in the development of energy generation, conversion, transmission, and use. Research currently being conducted by the U.S. Department of Energy (DOE) and relevant to climate-related technology involves fundamental research for the development of advanced materials for use in fuel cells, exploration of corrosion and high-temperature effects on materials with potential cross-cutting impacts in both energy generation and energy use technologies, investigations of

Nanoscale Materials Science

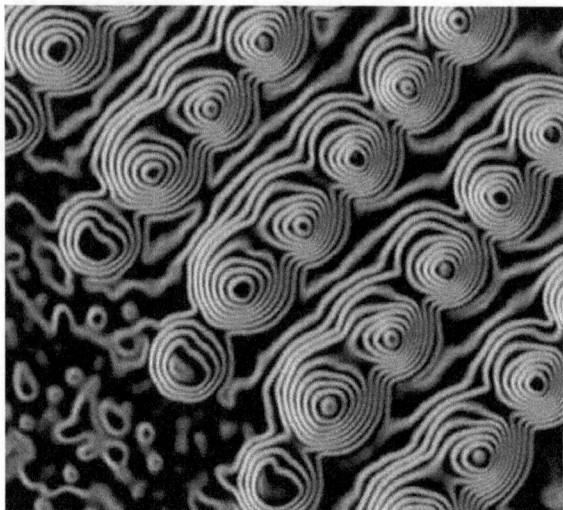

Figure 9-6. Nanoscale materials science contributes to the design of materials and devices at the atomic and molecular level to achieve entirely new functionality.

Courtesy: DOE, Office of Science

radiation-induced effects relevant to nuclear fission and fusion technologies, fundamental research in condensed matter physics and ceramics that might lead to high-temperature superconductors and solid-state materials, chemical and metal hydrides research related to hydrogen storage, and nanoscale materials science (Figure 9-6) and technology that offer the promise of designing materials and devices at the atomic and molecular level to achieve entirely new functionality.

♦ **Chemical sciences** research provides the fundamental understanding of the interactions of atoms, molecules, and ions with photons and electrons; the making and breaking of chemical bonds in gas phase, in solutions, at interfaces, and on surfaces; and the energy transfer processes within and between molecules. The fundamental understanding resulting from this research—an understanding of the chemistries associated with combustion, catalysis, photochemical energy conversion, electrical energy storage, electrochemical interfaces, and molecular specific separation from complex mixtures—could result in reductions in carbon dioxide emissions. Advances in chemical sciences will enable the development of hydrogen as an energy carrier; new alternative fuels; low-cost, highly-active, durable cathodes for low-temperature fuel cells; separations and capture of CO_2; catalysts for new industrial and energy processes; and better energy storage devices.

Being at the interface of materials and biology, chemical sciences will also be key to bio-inspired materials and processes.

♦ **Geosciences** research supports mineral-fluid interactions; rock, fluid, and fracture physical properties; and new methods and techniques for geosciences imaging from the atomic scale to the kilometer scale. The activity contributes to the solution of problems in multiple Federal agency mission areas, including development of the scientific basis for evaluating methods for sequestration of CO_2 in subsurface regions; for the discovery of new fossil resources, such as oil and gas, and methane hydrates; and for techniques to locate geothermal resources, to map and model geothermal reservoirs, and to predict heat flows and reservoir dynamics. The nanoscience capabilities of geochemistry are answering fundamental questions involving the electric double layer, which occurs at the interface of electrolytes.

Biological Sciences

The revolution in genomics research has the potential to provide entirely new ways of producing forms of energy, sequestering carbon, and generating materials that require less energy to produce. It includes research to investigate the underlying biological processes of plants and microorganisms, potentially leading to new processes and products for energy applications, thereby enabling the harnessing of natural processes for GHG mitigation. Research includes:

♦ **Genomic research on microbes** focusing on their ability to generate, harvest, store, and manipulate energy-supplying compounds in almost any form to carry out life's functions. Current genomic research is focused on sequencing microbes that either aid in carbon sequestration or produce fuels, particularly ethanol or hydrogen.

♦ **Genomic research on plants**—for example, on the genome of Poplar, a common tree species—is identifying genes that determine key biochemical functions which could improve the ability of these trees to sequester carbon in their root systems or produce biofuels more efficiently. Selecting, propagating, and modifying crops specifically for soil carbon sequestration (i.e., below-ground storage) would accelerate carbon fixation while minimizing terrestrial impacts because harvesting would not be necessary. Other crops could be modified to produce more lipids and other useful products. Fast-growing crops could be modified to improve bioconversion.

◆ Research on **biological catalytic reactions** aims to improve the understanding of reactions in photoconversion processes and advanced techniques for screening and discovering new catalysts and biomemitic approaches to materials synthesis. This research could provide insights into biochemical regulatory interventions that could improve the rate or efficiency of these processes.

◆ Research related to modified **plants and soil micro-organisms** can provide a basis for using and renewing marginal lands for bio-based energy feedstocks, incorporating stress-resistant plants and microbes, and developing advanced bioengineering approaches to capturing and retaining nitrogen and other essential plant nutrients. Sustainability could be ensured by engineering traits such as increased below-ground storage of photosynthate.

◆ **Biotechnology** has the potential to provide the basis for direct conversion of sunlight into hydrogen, lipids, sugars, and other fuel precursors. Work in this field can accelerate an understanding of fundamental aspects of microbial and plant production systems, including thermophilic, algal, and fermentative approaches.

◆ New **bio-based industrial processes** can be developed, involving combining biological functionality with nano-engineered structures to achieve new functionalities and phenomena. Incorporating biological molecular machines (such as elements of photosynthetic chromophores) into nanostructures has the potential to achieve the selectivity and efficacy of biological processes with the high intensity and throughput of engineered processes.

◆ **Advanced structural biocomposites research** could help replace high-energy-intensity material such as concrete and steel with renewable carbon-based natural products.

◆ Research on key **biotechnology platforms** includes designs for biorefineries to produce biofuels, biopower, and commercial chemical products derived from biomass rather than fossil fuels; fuel cells powered by bio-based fuels or bio-generated hydrogen; engineered systems to support processes such as direct photo-conversion utilizing bio-based processes of water, CO_2, and

nitrogen to produce useful fuels; and small modular biopower systems for incorporation of biological processes.

Environmental Sciences

Research in the environmental sciences is rapidly evolving with the development and application of new tools for measuring and monitoring environmental processes both *in situ* and remotely at scales never before possible. These new tools will provide data on the functioning of ecological systems, including the provision of goods and services such as sequestering carbon and how they are affected by environmental factors. Genomics research is and will continue to contribute to the advances in environmental sciences by providing understanding of the fundamental processes, structures, and mechanisms of complex living systems, including ecological systems. Examples of such fundamental research include the following:

◆ **Carbon sequestration research** could identify how efforts to increase terrestrial carbon sequestration might influence other environmental processes, such as nutrient cycling, the emissions of other GHGs, and local, regional, and global climate through impacts on heat balances and albedo (Figure 9-7);

◆ In **biological and ecological systems** there is a need to understand, quantify, predict, and manage biological and ecological processes affecting carbon allocation, storage, and capacity in terrestrial systems;

Free-Air CO₂ Enrichment Facility

Figure 9-7. *Carbon sequestration research, such as that being conducted at the Free-Air CO₂ Enrichment Facility, can help to identify how efforts to increase terrestrial sequestration might influence other environmental processes.*

Courtesy of DOE, Office of Science

*Advanced Computerized Materials Science
(DOE-SC-NEST 2004)*

◆ Research can be focused on the development of sensors that allow **measuring and monitoring** of environmental carbon flows. Computational models can be developed that can simulate and predict carbon flows resulting from, for example, specific carbon management policy actions and that provide a consistent picture of the effectiveness of efforts to reduce anthropogenic emissions; and

◆ Research can be conducted on **indoor air quality** and its inter-relationship with other buildings-related environmental factors, so as to understand the possible ramifications of increasing the energy efficiency of buildings.

Advanced Scientific Computation

Computational science is increasingly central to progress at the frontiers of almost every scientific discipline. The science of the future demands advances beyond the current computational capabilities. Accordingly, new advanced models, tools, and computing platforms are necessary to dramatically increase the effective computational capability available for scientific discovery in such areas as fusion, nanoscience, climate and environmental science, biology, and complex systems. With advances in computation, its role will become even more central to a broad range of future discoveries and subsequent innovations in climate change technologies. Examples of areas in which exploratory modeling and simulation research are being employed to assist in the development of advanced energy systems include the following:

◆ Modeling and simulation of advanced fusion energy systems to support ITER and the National Ignition Facility (NIF).

◆ Modeling of combustion for advanced diesel engines and other combustion systems; modeling of heat transfer in thermoelectric power systems.

◆ Modeling and simulation of nanoscale systems, including thin-film polymer-based photovoltaics. The computational effort required to simulate nanoscale systems far exceeds any computational efforts in materials and molecular science to date.

◆ Improved models of the aerodynamics of wind turbines and other fluid dynamics processes.

◆ 3D computational fluid dynamics modeling for next-generation nuclear reactors. This should include coupling of neutronics physics to heat transfer and fluid dynamics, as well as 3D models for critical heat flux or dry-out prediction. Significant validation will be required for such powerful models to be used in a regulatory environment.

◆ Further development of tools to model gasification accurately and reliably. This should include multiphase model development to improve computational efficiency for the unsteady flows found in gasification combustion systems. Ash, slagging, and fouling are some of the most important technical challenges facing gasification systems, and the relevant computational fluid dynamics models need to be developed and validated to better understand these chemical processes.

◆ Integrated assessment models of global climate change.

The 2004 *Advanced Computational Materials Science Workshop* report describes how an increased effort in modeling and simulation could help bridge the gap between the data that is needed to support the implementation of advanced nuclear technologies and the data that can be obtained in available experimental facilities (DOE-SC-NEST 2004).

Fusion Sciences

The majority of fusion energy sciences research is aligned, generally, with the goal of providing the knowledge base for environmentally and economically attractive energy sources (summarized in Section 9.1.2); the remainder of the basic research is fundamental in nature. This research includes general plasma sciences, the study of ionized gases as the underpinning scientific discipline for fusion research, through university-based experimental research, theory, plasma astrophysics, and plasma processing and other applications. See also Section 5.5.

9.3 Exploratory Research

Typically, applied R&D programs, as described in Chapters 4 through 8, focus on well-defined research projects and deployment activities designed to achieve results-oriented, specific metrics and meet deadlines. As described in Section 9.1, strategic research has a long-term, basic research focus, yet it is still oriented toward and inspired by the need to understand and contribute to solving problems associated with currently supported technology development thrusts. To meet the challenges associated with the CCTP goals, there is another need, that is, to augment existing applied R&D and strategic research programs with exploratory research. Such research would pursue novel, advanced or emergent, enabling and integrative concepts that do not fit well within the defined parameters of existing programs, and are not elsewhere covered.

Exploratory research would not duplicate, but complement, and potentially enrich, the existing R&D portfolio of climate-change-related strategic research and applied technology R&D. If the explored concepts proved meritorious, it would be expected that they would then become better positioned to be considered favorably in future plans among the existing Federal R&D programs, or form the basis for new R&D programs. This approach would stimulate innovative, novel, or cross-cutting technical approaches, not predisposed to one technology or another, and ensure that a full measure of the most promising technology options was explored.

CCTP plans to review agency experiences with exploratory research programs, including those of the Defense Advanced Research Projects Agency, and encourage the pursuit of exploratory approaches, as appropriate, within the Federal climate change technology portfolio. An exploratory research program would be expected to support research to explore novel "out-of-the-box" transformational technologies. Projects would be selected through widely advertised competitive solicitations. Awards would be made through merit-based grants, cooperative agreements, and contracts with both public and private entities, including businesses, Federal research and development centers, and

institutes of higher education. Multi-agency coordination might be required for integrative ideas that span technical disciplines and economic sectors. Exploratory research conducted under such a program could add to scientific and engineering knowledge and contribute to U.S. technological leadership, while addressing long-term challenges in global warming.

Some important generic areas for exploratory research include novel, advanced, integrative, and enabling concepts, as elaborated upon below. Exploratory research also includes the development of decision-support tools to assess and better understand the role, impacts, and potential limits of technology in meeting CCTP goals.

Novel Concepts

Novel concepts, by definition, are "atypical" ideas. They often do not have funding support within the boundaries of traditional research and development organizations or other means to demonstrate their potential applications and value. They may build on scientific disciplines outside the usual disciplines in that field or attempt to apply previously unexplored methods, and may offer approaches that compete with the more traditional approaches already being pursued. These novel approaches may lead to better ways to reduce GHG emissions, reduce GHG concentrations, or otherwise address the effects of climate change.

Novel concepts might include, for example, innovative ways to produce or convert energy (e.g., high-altitude wind kites, direct energy conversion, immiscible liquid/liquid heat exchangers). Novel concepts might be used to mitigate the effects of global warming in the stratosphere (e.g., geo-engineered solar insulation) or sequester carbon (e.g., enhancement of the natural carbon cycle, or microbial fixation of carbon in geologic formations). Another approach might be to combine the biosciences with fields such as nanotechnology, chemistry, computers, medicine, and others (e.g., Bio-X) to create novel solutions for technology challenges. This could lead to innovative concepts such as the use of "enzyme machines" or even new materials (e.g., bio-nano hybrids) that could replace traditional technology altogether (Figure 9-8). Spaceborne measurements of the Earth system from the Lagrange points or other non-traditional orbits are another novel concept that could advance our understanding of regional and global GHG distributions.

Advanced Concepts

Advanced concepts are high-risk, long-term ideas that are often too risky or unconventional for applied R&D programs to support, but are also often too purposeful or applied for basic research programs to support. These ideas draw upon conventional scientific disciplines and concepts but seek to take them beyond the realm of current capabilities. Over the long-term, the pursuit of advanced concepts in such fields as solar energy, biotechnology, ocean and tidal energy, and other fields could lead to dramatic changes in the way we produce and use energy.

For example, advanced concepts are emerging in the field of biotechnology and could be applied to the production of bioenergy as well as methods for sequestering carbon. Plant metabolic engineering could be used to improve the properties of plants for conversion to bioenergy, or to increase the storage of photosynthates as hydrocarbons that could be extracted as energy. Advances in plant genomics could be utilized to develop new crop species that maximize soil carbon storage in marginal lands, or to convert annual crops to perennial crops to facilitate carbon sequestration and provide more viable feedstocks for bioenergy. Further, the natural chemical reactivity of CO_2 could be exploited to remove CO_2 from the air or from waste streams, while forming stable, storable carbon compounds or useful products.

Advances in other areas might include solar fuels derived from carbon dioxide via artificial photosynthetic systems. Alternatively, solar-powered photo-catalyzed systems could be developed to produce liquid transportation fuels from hydrogen and carbon dioxide. Energy could potentially be derived from wave and tidal power conversion systems using slow wave motion, or from tidal dams producing energy based on ebb tides.

Integrative Concepts

Integrative concepts cut across traditional R&D program boundaries and combine systems, technologies, disciplines, and in some cases, sectors of the economy. For example, a net-zero GHG emission building could integrate energy for heating, cooling, and lighting with on-site power production for an electric vehicle. Developing such integrative concepts is an interdisciplinary, complex undertaking and would involve coordination across multiple

agencies or across existing R&D program or mission areas. A more concerted effort might be needed to explore these concepts and manage multi-mission R&D. The combing of multiple concepts into integrated, more efficiently functioning systems could, however, have potentially large implications for climate change and should be encouraged.

Integrative concepts might include, for example, the combination of coal power and aquifer sequestration, or biochemical and thermochemical conversion of biomass. Also, an integrated process that converts biomass wastes into hydrogen fuel and a char-based fertilizer (sequestering carbon), while scrubbing CO_2 and other flue gases, may have potential. Another example is engineered urban design, where land use is designed to reduce vehicle-mile requirements and allow co-location of activities with common needs for conserving energy, water, and other resources. The integration of transport, electricity, residential and commercial buildings, and industrial complexes within communities is a potential way to optimize the use of energy and reduce GHGs through co-location of energy sources and sinks. A related concept is the integration of plug-in hybrid electric vehicles with wind, solar, zero-energy buildings, and utility peak-saving, which could dramatically reduce GHGs from vehicles and optimize use of intermittent energy sources such as solar and wind.

Energy used to support the water infrastructure is another area where integration of systems could be beneficial. Technologies to minimize energy requirements for water use could include, for example, buildings designed to reduce use and conveyance of water; gray-water re-use; and integration of water storage and treatment with intermittent renewable energy supplies.

Enabling Concepts

Enabling technologies contribute indirectly to the reduction of GHG emissions by facilitating the development, deployment, and use of other important technologies that reduce GHG emissions. Enabling technologies often represent the scientific and engineering breakthroughs needed to move next-generation concepts forward along the development cycle. In addition, enabling technologies often cut across multiple disciplines and can lead to the diffusion of new concepts in multiple areas.

Enabling technologies may span or be applicable to many aspects of energy end-use, supply, and GHG mitigation and sequestration, from power generation to instrumentation to separations, new materials, and storage. For example, enabling technology is needed to support a next-generation electricity grid and supply a potential nationwide fleet of hybrid electric vehicles. Such research might encompass the development of electrochemical, kinetic, thermal, or electromagnetic storage systems. Enabling technology (e.g., electron transmission via nanotubes) is also needed to make low-resistance power transmission possible without the use of cryogenics. Similarly, wireless transmission of electrical energy— or power beaming—would enable large solar-based energy conversion systems to be located far distances from population centers, such as in the Earth's deserts, in low-Earth orbit, outer space, or even on the moon, and still supply large quantities of energy to where it would be needed on Earth. Two technologies that might be pursued under this category of exploratory research suitable for power beaming involve microwave or laser energy.

Research to develop breakthrough processing technology will be needed to take advantage of emerging fields with great promise, such as nanotechnology. Exploratory research might include the integration of nanotechnology with engineering and other disciplines, joining technologies for new nanomaterials, and advanced processing technologies for nanomaterials and nano-bioengineering systems. Advances in these areas could enable the greater use of nanomaterials in applications that could lead to reduction and/or mitigation of GHG emissions.

Integrated Planning and Decision-Support Tools

Decision-support tools include analytical, assessment, software, modeling, or other quantitative methods for better understanding and assessing the role of technology in long-term approaches to achieving stabilization of GHG concentrations in the atmosphere. While individual R&D programs sponsor the development of such tools, the tools developed are applicable mainly within each program's respective areas of responsibility or technologies. Broader analytical tools can provide intelligence for integrated planning and making research decisions that span disciplines, industries, and agencies.

Figure 9-8. Nanoscale research to develop breakthrough processing technology can lead to greater use of nanomaterials in advanced technologies that reduce or mitigate GHG emissions. One such example is shown above, where stable enzymes are embedded in a biomimetic nanomenbrane.

Credit: DOE/PNNL.

An important evaluation tool is the analysis of the net environmental benefits of various climate change technologies (i.e., bringing science to the decision process). Understanding the response of the environment and long-term impacts can influence how the research portfolio will be structured to achieve the maximum benefits. Lifecycle analysis is another important assessment tool. Guidelines and standards are needed to develop lifecycle analysis that examines carbon flux, land use, energy use, economics and other factors related to the adoption of technologies that could potentially impact climate change. Along with lifecycle analysis, ecosystem models are needed to integrate improved genetics and metabolic processes for land use, CO_2 fixation, and soil processes.

An integrated system architecture approach for GHG M&M across varying spatial and temporal scales would provide a framework for assessing and implementing GHG reduction strategies.

9.4 Toward Enhanced Integration in R&D Planning Processes

Effective integration of fundamental science, strategic research, exploratory research, and applied technology research and development presents challenges and opportunities for any mission-oriented research campaign. These challenges and opportunities can be addressed by CCTP through enhanced and integrative R&D planning processes that emphasize communication, cooperation, and collaboration among the affected scientific and technical research communities. Appropriate means for regularly assessing the success of basic research in supporting the technology development programs also need to be established, including criteria, to ensure that basic research is a productive, indeed, enabling component of the larger CCTP R&D portfolio.[1]

Technology development programs are often hindered by incomplete knowledge and lack of innovative solutions to technical bottlenecks. Information can be shared and potential pathways to solutions can be suggested by bringing together multidisciplinary research expertise and applied technology developers. Increased discussion among research personnel from various complementary fields and face-to-face exploration of ideas is a good way to foster innovative ideas and create synergies. The traditional structure of research, operating mainly within the narrower confines of specific disciplinary groups, will not be sufficient.

A model integrated planning process would include the following:

◆ Systematic exploration of various technology program issues, challenges and impediments to progress.

◆ Mechanisms to communicate technology program needs to the basic research community.

◆ Exploration of a wide range of potential research avenues to address the identified issues, challenges and impediments; this can effectively be accomplished through an R&D or technology roadmapping activity.

◆ Design of strategic research program areas to pursue the most promising avenues, including clear articulation of research goals.

◆ Solicitations of research proposals to address the identified areas.

◆ Funding of specific meritorious research projects, selected by a peer review process.

The first few steps in the model process described above could be accomplished using workshops and other multi-party planning mechanisms. Technical workshops can bring together the applied and basic research communities and focus on research strategies and barriers impeding development in a particular technology area. The resulting report can form the basis for a framework of high priority research needs, a solicitation for proposals, and awards.

For instance, in recognition of the growing challenges in the area of energy and related environmental concerns, the Department of Energy's Office of Basic Energy Sciences (DOE-BES) initiated a series of workshops in 2002 focusing on identification of the underlying basic research needs related to energy technologies. The first of these workshops, held in October 2002, undertook a broad assessment of basic research needs for energy technologies to ensure a reliable, economical, and environmentally sound energy supply for the future (BESAC 2003). More than 100 people from academia, industry, the national laboratories, and Federal agencies participated in this workshop.

More than a dozen such workshops have been held since 2002. Many apply directly to goals and

Basic Research Needs fto Assure a Secure Energy Future (BESAC 2003)

---

[1] Criteria for "success" in basic research are well established, as are the multiple modes for evaluating related programs and projects. Such evaluations occur continuously in both pre-award and post-award settings.

technical challenges of CCTP. A number of these are cited throughout this chapter. More are scheduled for the near future, including basic research needs for superconductivity (2006); solid-state lighting (2006); advanced nuclear energy systems (2006); and energy storage (2007).

CCTP seeks to encourage continued and broadened application, across all agencies, of best practices in integrated research planning. In its periodic reviews of the adequacy of the CCTP R&D portfolio, CCTP identified a number of topical areas for consideration, in addition to those already planned, for future basic research needs assessments in support of CCTP technology development. These areas include: architecture and control systems for the electric grid; thermoelectrics by application (e.g., refrigeration, power generation); "bio-x", combining nanosciences and genomics; plant genetic engineering; measuring and monitoring of climate change mitigation, with an international focus; sensors, controls, and communication technologies; batteries–power & energy (basic chemistry); heat transfer–material insulation, cryogenics, thermal conducting coolants; power electronics–conversion; and ocean sequestration of carbon dioxide.

Based on the experiences of past and successful workshops held by the Office of Science, Basic Energy Sciences, and Biological and Environmental Research, the following principles are identified to help guide future planning:

◆ **Make Merit-Based Decisions:** All decisions should be based on merit and need. Once this principle is compromised, the process degenerates quickly.

◆ **Share Ownership:** Long-term commitment and ownership by those in positions of authority and responsibility is a must for success.

◆ **Understand and Formalize Relationships:** Roles, responsibilities, rules of integration, allocation of resources, and terms of dissolution should be formalized at the start.

◆ **Measure Performance:** At the start, participants must agree on goals, objectives, operational elements, and methodology for measuring progress, outputs, and outcomes. (Avoid collaboration for collaboration's sake and integration for integration's sake.)

◆ **Ensure Commitment and Stability:** Team members must commit to work seamlessly, with the goal of a stable operation for the time necessary to achieve results.

◆ **Provide Flexibility:** Within general guidelines, flexibility ensures accountability and fosters innovation and experimentation. The process must allow for unanticipated results and empower people to act on their own.

◆ **Have a Customer Focus:** A clear understanding of who the customer is, what the customer wants, and the customer's complete involvement in all phases of the activity is critical to success.

Achieving the CCTP vision will likely require discoveries and innovations well beyond what today's science and technology can offer. Better integration of basic scientific research with applied technology development may be key to achievement of CCTP's other goals related to energy efficiency, energy supply, carbon capture and sequestration, M&M, and reducing emissions of non-CO_2 gases. Basic science research is likely to provide the underlying knowledge foundation on which new technologies are built.

The CCTP framework aims to strengthen the basic research enterprise so that it will be better prepared to find solutions and create new opportunities. The CCTP approach includes strengthening basic research in national laboratories, academia, and other research organizations by focusing efforts on key areas needed to develop insights or breakthroughs relevant to climate-related technology R&D. Importantly, in the process, these basic research activities will enable training and developing of the next-generation of scientists who will be needed in the future to provide continuity of such research to find solutions and create new opportunities.

9.5 References

Basic Energy Sciences Advisory Committee (BESAC). 2002. *Basic Energy Sciences Advisory Committee Subpanel workshop report: opportunities for catalysis in the twenty-first century.* http://www.sc.doe.gov/bes/besac/CatalysisReport.pdf

Basic Energy Sciences Advisory Committee (BESAC). 2003. *Basic research needs to assure a secure energy future.* http://www.sc.doe.gov/bes/besac/Basic_Research_Needs_To_Assure_A_Secure_Energy_Future_FEB2003.pdf

Basic Energy Sciences Advisory Committee (BESAC). 2004. *Basic research needs for the hydrogen economy: report on the basic energy sciences workshop on hydrogen production, storage, and use.* Argonne, IL: Argonne National Laboratory. http://www.sc.doe.gov/bes/hydrogen.pdf

Biomass Research and Development Technical Advisory Committee. 2002. *Roadmap for biomass technologies in the United States.* http://usms.nist.gov/roadmaps/object.cfm?ObjectID=61

DOE (See U.S. Department of Energy)

National Science Foundation (NSF). 2004a. *Future directions for hydrogen energy research & education.* www.getf.org/hydrogen/index.cfm

National Science Foundation (NSF). 2004b. *Catalysis for biorenewables conversion,* www.egr.msu.edu/apps/nsfworkshop

U.S. Department of Energy (DOE), Office of Basic Energy Sciences (BES). 2006. *Basic research needs for superconductivity.* Washington, D.C.: U.S. Department of Energy, Office of Science. http://www.sc.doe.gov/bes/reports/abstracts.html#SC

U.S. Department of Energy (DOE), Office of Basic Energy Sciences (BES) and the Nanoscale Science, Engineering, and Technology (NSET) Subcommittee of the National Science and Technology Council (NSTC). 2005. *Nanoscience research for energy needs.* http://www.sc.doe.gov/bes/reports/files/NREN_rpt.pdf

U.S. Department of Energy (DOE), Office of Basic Energy Sciences (BES) and Office of Nuclear Physics (NP). 2005. *The path to sustainable nuclear energy, basic and applied research opportunities for advanced fuel cycles.* Washington, D.C.: U.S. Department of Energy, Office of Basic Energy Sciences and Office of Nuclear Physics. http://www.sc.doe.gov/bes/reports/files/PSNE_rpt.pdf

U.S. Department of Energy (DOE), Office of Biological and Environmental Research (BER), and Office of Energy Efficiency and Renewable Energy (EERE). 2006. *Breaking the biological barriers to cellulosic ethanol: a joint research agenda.* Washington, D.C.: U.S. Department of Energy, Office of Science. http://www.doegenomestolife.org/biofuels/b2bworkshop.shtml

U.S. Department of Energy (DOE), Office of Science (SC). 2005a. *DOE genomics: GTL Roadmap.* Washington, D.C.: U.S. Department of Energy, Office of Science. http://doegenomestolife.org/roadmap/index.shtml

U.S. Department of Energy (DOE), Office of Science (SC). 2005b. *Basic research needs for solar energy utilization.* Washington, D.C.: U.S. Department of Energy, Office of Science. http://www.sc.doe.gov/bes/reports/files/SEU_rpt.pdf

U.S. Department of Energy (DOE), Office of Science (SC) and Office of Fossil Energy (FE). 1999. *Carbon sequestration research and development.* Washington, D.C.: U.S. Department of Energy, Office of Science and Office of Fossil Energy. http://www.fe.doe.gov/programs/sequestration/publications/1999_rdreport/index.html

U.S. Department of Energy (DOE), Office of Science (SC) and Office of Nuclear Energy, Science, and Technology (NEST). 2004. *Advanced computational materials science: application to fusion and Generation IV fission reactors.* Washington, DC. U.S. Department of Energy Office of Science and Office of Nuclear Energy, Science, and Technology. http://www.sc.doe.gov/bes/reports/files/ACMS_rpt.pdf

U.S. Climate Change Science Program (CCSP). 2003. *Strategic plan for the U.S. Climate Change Science Program.* http://www.climatescience.gov.

Summary and Next Steps

C limate change is a complex, long-term challenge. United States' climate change goals are consistent with and supportive of the United Nations Framework Convention on Climate Change (UNFCCC), which set an ultimate goal of stabilizing greenhouse gases (GHGs) in the atmosphere at a level that avoids dangerous human interference with the climate system. There is international recognition that climate change concerns cannot be

addressed in isolation from other pressing needs, such as economic development, energy security, and pollution reduction, especially in developing economies. Successfully addressing these complementary concerns will require the development and commercialization of advanced technologies, particularly, but not exclusively, those that have the potential to fundamentally alter the way we produce and use energy. It will also require a sustained, long-term commitment by all nations over many generations and a substantial degree of international cooperation.

It is within this broad context that the United States is pursuing a comprehensive strategy on climate change. This strategy includes implementing policies and measures to slow the growth of GHG emissions, investing in climate science to improve understanding of climate change and provide information to decision-makers, accelerating the development and adoption of energy and other technologies that reduce, avoid, or capture and sequester GHG emissions, and promoting international collaboration. In this way, the United States is working to ensure a bright and secure energy and economic future for the Nation and a healthy planet for future genertions (Figure 10-1).

Under the auspices of the Cabinet-level Committee on Climate Change Science and Technology Integration (CCCSTI), the Climate Change Technology Program (CCTP), led by the Department of Energy (DOE), represents the technology component of this strategy. Authorized by the Energy Policy Act of 2005, CCTP functions as a multi-agency planning and coordination entity. CCTP is charged with reviewing and prioritizing the Federal Government's climate-related technology programs. Its activities are guided by the CCCSTI and carried out by representatives of the participating R&D agencies. CCTP provides strategic direction for and coordinates an investment portfolio of climate-change

Figure 10-1. The United States is working to ensure a bright and secure energy and economic future for the Nation and a healthy planet for future generations.

Courtesy: NASA, Hasler Laboratory for Atmospheres Goddard Space Flight Center, Credit: Nelson Stockli

related technology research, development, demonstration and deployment (R&D) of nearly $3 billion in Fiscal Year 2006.

 ## Summary

The CCTP *Strategic Plan* is an important milestone on the road to accelerating the development and adoption of advanced climate change technologies. The *Plan* articulates a vision for the role for advanced technology in addressing climate change concerns, provides a supporting long-term planning context with insights from analysis, and establishes goals, approaches, and guiding principles for Federal R&D agencies to use in formulating climate change-related technology components of their respective R&D portfolios.

CCTP's strategic vision is to attain, on a global scale and in partnership with others, a technological capability that can provide abundant, clean, secure, and affordable energy and other services needed to encourage and sustain economic growth, while simultaneously achieving substantial GHG emissions reductions to mitigate the potential risks of climate change from increasing GHG concentrations. To give substance to this vision, CCTP's portfolio pursues six complementary strategic goals: (1) reduce emissions from energy use and infrastructure; (2) reduce emissions from energy supply; (3) capture and sequester CO_2; (4) reduce emissions of other GHGs; (5) measure and monitor emissions; and (6) bolster the contributions of basic science. The *Plan* outlines seven approaches to attain those goals, including next steps.

Long-Term Planning Under Uncertainty

CCTP operates within a planning environment characterized by uncertainty. First, the complex relationships among population growth; economic development; energy demand, mix, and intensity; resource availability; technology advancement; and many other societal variables make it difficult to estimate with confidence future global GHG emissions over CCTP's long-term planning horizon. This creates uncertainty about the scope and scale of

the technological challenge. Second, evolving climate science, as well as the uncertain nature of the (as yet undetermined) UNFCCC's stabilization objective, adds uncertainty about the appropriate pace of technology development. Finally, research and development itself is risky, such that the future readiness, cost, and performance characteristics of the many advanced technologies envisioned to facilitate GHG emissions reductions are unknown. This adds uncertainty about deployment and which, if any, technologies will ultimately emerge as successful.

Goals for Technology Development

Uncertainties notwithstanding, CCTP sets ambitious goals for technology development in each of its strategic areas. By addressing uncertainties systematically through analyses of options, or scenarios, under a range of varying assumptions, CCTP can illuminate its technological challenges in the form of bracketed insights regarding reductions or avoidances in GHG emissions, comparative costs, and the timing of significant technology adoption. These insights, in turn, can be used to guide R&D portfolio planning, as well as inform specific near-, mid- and long-term technological development objectives.

Regarding reductions or avoidances in GHG emissions, CCTP envisions significant contributions

Estimated Cumulative GHG Emissions Mitigation (GtC) from Accelerated Adoption of Advanced Technologies over the 21st Century, by Strategic Goal, Across a Range of Hypothesized GHG Emissions Constraints

CCTP STRATEGIC GOAL	VERY HIGH CONSTRAINT[1]	HIGH CONSTRAINT	MEDIUM CONSTRAINT	LOW CONSTRAINT
GOAL #1. Reduce Emissions from Energy End Use & Infrastructure	250 - 270	190 - 210	150 - 170	110 - 140
GOAL #2. Reduce Emissions from Energy Supply	180 - 330	110 - 210	80 - 140	30 - 80
GOAL #3. Capture and Sequester Carbon Dioxide	150 - 330	50 - 140	30 - 70	20 - 40
GOAL #4. Reduce Emissions of Non-CO_2 GHGs	160 - 170	140 - 150	120 - 130	90 - 100

Table 10-1. Estimated Cumulative GHG Emissions Mitigation (GtC) from Accelerated Adoption of Advanced Technologies over the 21st Century, by Strategic Goal, Across a Range of Hypothesized GHG Emissions Constraints[2].

[1] A "very high constraint" scenario is associated with low GHG emissions and low GHG concentrations. A "low constraint" scenario is associated with higher GHG emissions and concentrations.

[2] Source: Clarke et al. 2006, as described in Chapter 3. The 100-year cumulative values shown are consistent with Figure 3-19. Values have been rounded to the nearest 10 GtC.

from the accelerated adoption of advanced technologies in each of its four emissions-reduction strategic goals.[3] Across a wide range of atmospheric GHG concentration levels and hypothesized emissions constraints (from "low" to "very high"), CCTP explored the potential contributions of three contrasting technological futures over a 100-year planning horizon. Table 10-1 provides a representative sample of the range of GHG emissions reductions, accumulated over this period, that may be possible should the promise of advanced technologies associated with these goals be realized (see Chapter 3).

Across all four goals, as well as among the many technologies within each goal, the accelerated adoption of advanced technologies is shown to be potentially capable of contributing significantly to GHG emissions reductions and avoidances. The underlying assumptions for technology performance can help inform long-term technology R&D planning and goal-setting. The significance of the potentials across all four goals suggests the importance of pursuing a diversified R&D portfolio.

Potential to Reduce Mitigation Costs

CCTP analysis suggests that significant economic benefits could accrue, if advanced technologies with high emission mitigation potentials were to be successfully deployed. Figure 10-2 presents the results of a comparative analysis of the cumulative costs over the 21st century of GHG mitigation, with and without the accelerated adoption of advanced technologies, across a range of advanced technology scenarios and variously hypothesized GHG emissions constraints. The relative cost reductions are significant in all cases. As one would expect, the absolute cost reductions are more significant under the higher emissions constraints.[4] The results of modeling these hypothetical scenarios suggest the potential for advanced GHG-reducing technologies to reduce mitigation costs.

Considerations of Timing

With regard to considerations of timing and commercial readiness of the advanced technology options, insights were gained by analysis of the lowest

Comparative Analysis of Estimated Cumulative Costs Over the 21ˢᵗ Century of GHG Mitigation, With and Without Advanced Technology, Across a Range of Hypothesized GHG Emissions Constraints

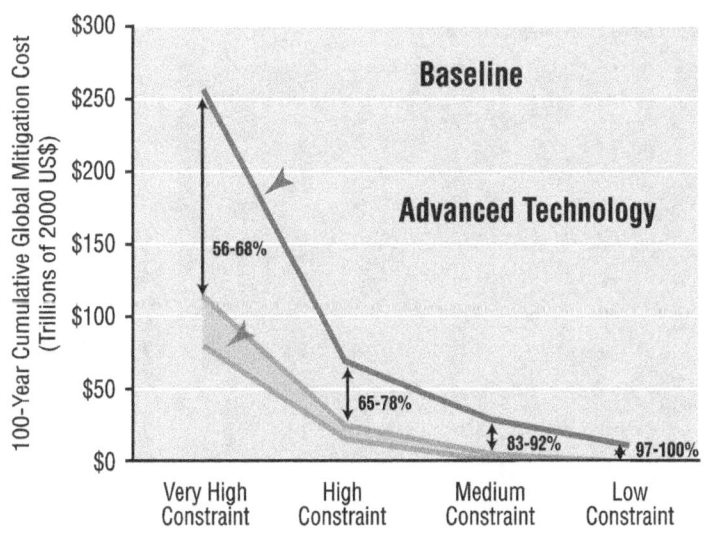

Figure 10-2. Comparative Analysis of Estimated Cumulative Costs Over the 21ˢᵗ Century of GHG Mitigation, With and Without Advanced Technology, Across a Range of Hypothesized GHG Emissions Constraints.

---

[3] Non-CO$_2$ GHGs include a diverse group of gases, such as methane, nitrous oxide, and chlorofluorocarbons. They are expected to contribute as much as 20 percent to total radiative forcing throughout the 21ˢᵗ century.

[4] In the associated analysis (Chapter 3), the cumulative costs over the 21ˢᵗ century are undiscounted. Accordingly, relative comparisons (percents) are likely to be more meaningful than those showing absolute costs. Variations among the advanced technology scenarios are shown by the shaded area in Figure 10-2.

cost solutions across a range of technology scenarios and hypothesized GHG concentration levels. Generally speaking, the lower the level of GHG concentrations hypothesized, the earlier the need for the commercial readiness of the advanced technologies.

Using the metric of when the first GtC per year of reduced or avoided GHG emissions could be achieved by the accelerated adoption of advanced technologies, across a range of assumptions, scenarios, and GHG emissions constraints, the analysis indicates that, in relative terms, advanced technologies for end-use energy efficiency contribute first, followed soon thereafter by technologies for mitigation of non-CO_2 GHGs, advanced energy supply technologies, and carbon capture, storage, and sequestration. The timing of such contributions for each CCTP strategic goal, however, may vary by decades depending on the GHG concentration level assumed, as shown in Table 10-2. Allowing for capital stock turnover and other inertia inherent in the global energy system and infrastructure, it is noted that for advanced technologies to achieve one GtC per year, they would need to be available in the commercial marketplace and gaining market share years before the periods indicated on Table 10-2.

Strategies for Technology Development and Deployment

In Chapters 4 through 7, the *Plan* lays out dozens of technology-specific strategies for research and development for key technologies including, where appropriate, strategies for technology deployment that show promise for realizing such gains. In Chapter 8, the *Plan* addresses the cross-cutting technology area for measuring and monitoring GHG emissions. For each technology area, the *Plan* examines the role that advanced technologies can play in contributing to each CCTP strategic goal, establishes strategic direction for R&D, highlights current R&D activities, and identifies promising directions for future research. A thematic timeline for technology development, evolution and adoption is provided for each of these five CCTP strategic goals.

Figure 10-3 presents an integration of these five timelines, for the near-, mid-, and long-terms. In general, advanced technologies appear in the figure at the point where commercialization is achieved and a favorable investment environment supports their adoption. Although many technologies are shown, only those that offer cost-effective options, compared to their competitors, would achieve significant market representation.

Estimated Timing of Advanced Technology Market Penetrations, as Indicated by the First GtC-Eq./Year of Incremental Emissions Mitigation[5], by Strategic Goal, Across a Range of Hypothesized GHG Emissions Constraints

CCTP STRATEGIC GOAL	VERY HIGH CONSTRAINT	HIGH CONSTRAINT	MEDIUM CONSTRAINT	LOW CONSTRAINT
GOAL #1. Reduce Emissions from Energy End Use & Infrastructure	2010 - 2020	2030 - 2040	2030 - 2050	2040 - 2060
GOAL #2. Reduce Emissions from Energy Supply	2020 -2040	2040 - 2060	2050 - 2070	2060 - 2100
GOAL #3. Capture and Sequester Carbon Dioxide	2020 - 2050	2040 or Later	2060 or Later	Beyond 2100
GOAL #4. Reduce Emissions of Non-CO_2 GHGs	2020 - 2030	2050 - 2060	2050 - 2060	2070 - 2080

Table 10-2. Estimated Timing of Advanced Technology Market Penetrations, as Indicated by the First GtC-Eq./Year of Incremental Emissions Mitigation[5], by Strategic Goal, Across a Range of Hypothesized GHG Emissions Constraints.

5 The years shown in the table represent the period, according to the analysis (Chapter 3), in which the first GtC (or GtC-eq.) of incremental emissions reduction (below an assumed Reference Case) is projected to occur due to the lowest-cost, accelerated adoption of each class of advanced technology in any one of the advanced technology scenarios. The Reference Case, without advanced technology, includes significant penetration of energy-efficient end-use technologies, nuclear, renewable, and biomass energy, terrestrial sequestration and non-CO_2 emission reductions. The table shows ranges of values because results vary among the advanced technology scenarios. "Energy Supply" means net-zero or very low-emissions energy supply technologies.

Climate Change Technology Development and Deployment for the 21st Century [6]

	NEAR-TERM	MID-TERM	LONG-TERM
GOAL #1 Energy End-Use & Infrastructure	• Hybrid & Plug-In Hybrid Electric Vehicles • Engineered Urban Designs • High-Performance Integrated Homes • High Efficiency Appliances • High Efficiency Boilers & Combustion Systems • High-Temperature Superconductivity Demonstrations	• Fuel Cell Vehicles and H_2 Fuels • Low Emission Aircraft • Solid-State Lighting • Ultra-Efficient HVACR • "Smart" Buildings • Transformational Technologies for Energy-Intensive Industries • Energy Storage for Load Leveling	• Widespread Use of Engineered Urban Designs & Regional Planning • Energy Managed Communities • Integration of Industrial Heat, Power, Process, and Techniques • Superconducting Transmission and Equipment
GOAL #2 Energy Supply	• IGCC Commercialization • Stationary H_2 Fuel Cells • Cost-Competitive Solar PV • Demonstrations of Cellulosic Ethanol • Distributed Electric Generation • Advanced Fission Reactor and Fuel Cycle Technology	• FutureGen Scale-Up • H_2 Co-Production from Coal/Biomass • Low Wind Speed Turbines • Advanced Biorefineries • Community-Scale Solar • Gen IV Nuclear Plants • Fusion Pilot Plant Demonstration	• Zero-Emission Fossil Energy • H_2 & Electric Economy • Widespread Renewable Energy • Bio-Inspired Energy & Fuels • Widespread Nuclear Power • Fusion Power Plants
GOAL #3 Capture, Storage & Sequestration	• CSLF & CSRP • Post Combustion Capture • Oxy-Fuel Combustion • Enhanced Hydrocarbon Recovery • Geologic Reservoir Characterization • Soils Conservation • Dilution of Direct Injected CO_2	• Geologic Storage Proven Safe • CO_2 Transport Infrastructure • Soils Uptake & Land Use • Ocean CO_2 Biological Impacts Addressed	• Track Record of Successful CO_2 Storage Experience • Large-Scale Sequestration • Carbon & CO_2 Based Products & Materials • Safe Long-Term Ocean Storage
GOAL #4 Other Gases	• Methane to Markets • Precision Agriculture • Advanced Refrigeration Technologies • PM Control Technologies for Vehicles	• Advanced Landfill Gas Utilization • Soil Microbial Processes • Substitutes for SF_6 • Catalysts That Reduce N_2O to Elemental Nitrogen in Diesel Engines	• Integrated Waste Management System with Automated Sorting, Processing & Recycle • Zero-Emission Agriculture • Solid-State Refrigeration/AC Systems
GOAL #5 Measure & Monitor	• Low-Cost Sensors and Communications	• Large Scale, Secure Data Storage System • Direct Measurement to Replace Proxies and Estimators	• Fully Operational Integrated MM Systems Architecture (Sensors, Indicators, Data Visualization and Storage, Models)

Figure 10-3. Climate Change Technology Development and Deployment for the 21st Century [6].

In the near-term (in less than 20 years), the CCTP strategy envisions commercial readiness, such that significant market entry can occur, by hybrid cars, high-efficiency buildings and industrial processes, selected technologies to capture, store and sequester CO_2, coal-based integrated gasification-combined cycle power plants, and methane capture and use technologies. In the mid-term (20 to 40 years later), the early technologies would be followed by significant market shares of hydrogen fuel cell vehicles, "smart" buildings, transformational technologies for energy intensive industries, improved

CO_2 capture, methane emission reductions, and advanced nuclear energy In the long-term, such technologies would be improved and extended more broadly and more advanced technologies would enter the marketplace. Ultimately, societies could see extensive adoption of low emissions infrastructure and communities, low emissions intelligent transport systems, wide-spread adoption of renewable energy and nuclear power, large scale adoption of zero-emission power plants with carbon sequestration, fusion power plants, and high levels of management of emissions of non-CO_2 GHGs.

---

[6] Note: Technologies shown are representations of larger suites. With some overlap, "near-term" envisions significant technology adoption by 10–20 years from present, "mid-term" in a following period of 20–40 years, and "long-term" in a following period of 40–60 years. See also List of Acronyms and Abbreviations.

Figure 10-4. CCTP's portfolio supports near-term deployment of civilian nuclear power, such as Westinghouse's AP1000. Nuclear power safely produces electricity with little or no GHG emissions.

Courtesy: Westinghouse Electric Company, LLC

Key Initiatives

Since early 2002, when CCTP was first organized, there have been a number of portfolio realignments. Foremost among these has been the identification of key technology initiatives that advance multiple technology goals, such as enhancing energy security, reducing air pollution, and promoting economic growth and productivity, while also addressing important thrusts of CCTP strategic goals. These key initiatives complement a core portfolio of technologies in energy efficiency, renewable energy, nuclear power (Figure 10-4), and highly efficient and clean use of coal (Figure 10-5). Although many of the key initiatives are not motivated exclusively by CCTP strategic goals, each is an important contributor to these goals. Collectively, they lend strategic focus and coherence to the many underlying research activities across programs within and among the agencies. They also give visibility to major technology thrusts and complement the regional and international partnerships. Key initiatives are mentioned throughout the *Plan* in their respective technology areas.[7] A current list of the key initiatives, with links to current programmatic information, may be found at the CCTP Web site.[8]

Contributions of Basic Science

Regarding CCTP's Goal #6 to bolster basic science contributions to technology development, the *Plan* recognizes the critical importance of new knowledge in enabling and accelerating progress in the applied areas. Discoveries in biology, nanosciences, computational modeling and simulation, physical processes, and environmental sciences could result in important breakthroughs for both emissions-related technologies and in measuring and monitoring capabilities. Three strategic thrusts to be pursued by CCTP in basic science include: (1) conducting basic research in areas inspired by the technical challenges in applied climate change R&D; (2) carrying out an exploratory research on innovative concepts and enabling technologies that have great potential for breakthroughs; and (3) improving R&D planning and integrative R&D processes.

Emerging Priorities

The CCTP working groups plan to review regularly the adequacy of the CCTP portfolio to attain CCTP's strategic goals and, where needed, make recommendations to prioritize and strengthen the portfolio, as outlined in Chapter 2. Each goal encompasses many different technologies with varying degrees of GHG reduction potential, likelihood of technological and commercial success, and private sector interest and incentive to invest. CCTP must consider these factors, along with criteria presented in Section 2.4, when recommending priorities for Federal investment.

In this way the CCTP portfolio is strengthened on an ongoing basis. This has resulted in the realignment of the portfolio in certain areas, evidenced by changes in investment levels over time. Climate change strategy also has provided compelling rationales for entirely new or revamped programs. These are described in Chapters 4 through 9 in the current portfolio sections.

Within the overall CCTP portfolio, certain activities are identified as priorities associated with President's National Climate Change Technology Initiative (NCCTI). NCCTI priorities are defined as discrete research, development, demonstration, or deployment activities that address technological challenges that, if solved, could advance technologies with the potential

---

7   See also CCTP Vision and Framework for Strategy and Planning (2005).

8   For CCTP-related Key Initiatives, see: http://www.climatetechnology.gov.

to dramatically reduce, avoid, or sequester greenhouse gas emissions. The NCCTI priorities appearing in the Fiscal Year 2007 budget request include, as ordered by strategic goal, the following:

1. Transportation fuel cell systems; solid state lighting; and "Climate Leaders" (Strategic Goal 1).

2. Low wind speed technology, cellulosic biomass (Figure 10-6), biochemical platform R&D, hydrogen storage, nuclear hydrogen initiative, integrated gasification combined cycle (IGCC), and advanced fuel cycle/advanced burner reactor (Strategic Goal 2).

3. Sequestration (Strategic Goal 3).

4. Methane partnership initiatives (Strategic Goal 4).

Also included among the NCCTI priorities is CCTP program support. Details on these activities can be found in Appendix B. Current and updated programmatic information in this regard may be found at the CCTP Web site.[9]

Advances in climate change science under the Climate Change Science Program (CCSP) are expected to improve our knowledge of climate change so that its potential impacts and uncertainties about the causes and effects of climate change will be better understood. This will help to inform the potential risks and benefits of various courses of action. Similarly, advances in climate change technology under the CCTP are expected to bring forth an expanded array of advanced technology options that can meet a range of societal needs, including reducing GHG emissions, with better performance and lower costs. Improved options will enable and facilitate actions that may be called for and informed by science.

Finally, should widespread adoption of advanced climate change technologies be pursued, as guided by science, it would likely need to be supported by appropriate technology policy, potentially including market-based incentives. As Federal efforts to

Figure 10-5. CCTP's portfolio supports the development and demonstration of revolutionary coal-based technologies, such as FutureGen, that is expected to produce electricity, hydrogen, fuels, and chemicals with nearly zero emissions of GHGs.

Courtesy: DOE/FE

advance technology go forward, broadened participation by the private sector in these efforts will be increasingly important to accelerate both the innovation and the commercialization of advanced technologies.

Such participation can be encouraged by appropriate and supporting technology policy. This is evidenced today, in part, by a number of market-based incentives already in place, by others proposed by the Administration,[10] and by still others, as may be appropriate, to GHG-reducing technology investments, soon to be implemented in accord with the provisions of Title XIII of the Energy Policy Act of 2005.[11] Title XVI of the Energy Policy Act of 2005 sets the stage for future development of policies that would facilitate new technology adoption. CCTP will support CCCSTI in implementing the provisions of Title XVI. Additionally, CCTP will explore a number of technology policy options, as listed in its next steps below.

 **Next Steps**

CCTP's next steps focus on a number of broad thrusts. First, the CCTP will continue its coordinating role and provide support to CCCSTI and its IWG. These activities are expected to include multi-agency planning and analysis, portfolio reviews,

9 For NCCTI priorities, see: http://www.climatetechnology.gov.

10 Federal Climate Change Expenditures Report to Congress, April 2006. http://www.whitehouse.gov/omb/legislative/fy07_climate_change.pdf

11 Financial incentives in Title XIII for technologies related to climate change goals are scored at more than $11 billion over 10 years.

213

Figure 10-6. CCTP's portfolio supports research to transform switchgrass (shown), other agricultural and forest products, cellulosic plant matter, and associated detritus (e.g., residential waste) into bio-based fuels, which are low in net emissions of GHGs.

Courtesy: USDA NRCS

interagency communications and information exchanges, technical assessments of research needs, and formulating and presenting recommendations. The CCTP will contribute its expertise and provide support to the CCCSTI and IWG as they address issues of climate change science and technology, weigh policies and priorities on related science and technology matters, make informed decisions, and make recommendations to the President and the agencies. Second, the CCTP will continue to work with and support the participating agencies in developing plans and carrying out activities needed to advance CCTP's vision, mission, and strategic goals. Third, CCTP anticipates opportunities to work with technical representatives of other countries to further international cooperation in related planning and program coordination, particularly on joint and multi-lateral technology initiatives.

CCTP intends to pursue specific activities to ensure progress toward achieving CCTP's strategic goals. These activities, organized by approach, are outlined below. CCTP does not expect to pursue all activities immediately, or simultaneously.

Strengthen Climate Change Technology R&D

◆ Continue to review, realign, reprioritize, and expand, where appropriate, Federal support for climate change technology research, development, demonstration, and deployment.

◆ Periodically assess the adequacy of the multi-agency portfolio with respect to its ability to achieve or make technical progress toward CCTP's strategic goals, identify gaps and opportunities, and make recommendations supported by analysis.

◆ In key technology areas, perform long-term assessments of technology potentials, including market considerations and potentially limiting factors.

◆ Improve methods, tools, and decision-making processes for climate technology planning and management, and R&D planning and assessment, including tools that allow portfolio planning to address risks through hedging strategies.

Strengthen Basic Research Contributions

◆ Establish or improve within each of the participating Federal R&D agencies a process for the integration with, and application of, basic research to help overcome barriers impeding technical progress on climate change technology development (Figure 10-7).

◆ Develop means for expanding participation in climate change technology R&D, including relevant basic research, at universities and other non-Federal research institutions.

◆ Review agencies' experiences with basic and exploratory research programs aimed at novel, advanced, integrative or enabling concepts not covered elsewhere as a means of stimulating innovation within the research community and enriching the technology R&D portfolio.

Enhance Opportunities for Partnerships

◆ Review status and encourage further formation of public-private partnerships as a common mode of conducting R&D portfolio planning and program execution.

◆ Encourage formation of non-R&D partnerships.

◆ Establish means for enabling and encouraging the secure sharing of potentially sensitive partner information regarding GHG emissions and related performance of GHG-intensity reducing technologies.

Increase International Cooperation

◆ Expand international participation in key climate change technology R&D activities and build on the many cooperative international initiatives already underway.

◆ Assist the Department of State and CCSP in the coordination of U.S. input and support of Working Group III on Mitigation of the IPCC's periodic Assessment Reports and other technology-related IPCC Special Reports, as means of stimulating international efforts to develop advanced technologies.

◆ Support continued efforts to negotiate, execute, and support bilateral agreements[12] that encourage international cooperation on climate change science and technology research. Pursue opportunities for outreach and communication to build relationships and encourage other similar initiatives by other countries.

◆ Pursue additional means to enhance the effective use of existing international organizations, such as the Organization of Economic Cooperation and Development, International Energy Agency, Intergovernmental Panel on Climate Change,

Figure 10-7. CCTP's portfolio supports programs to promote efficiency in energy use, such as advanced lighting concepts including Los Alamos National Laboratory's (LANL) research on white light from inorganic, multi-color light-emitting diodes (LEDs).

Courtesy: DOE/LANL

Group of Eight (G8),[13] Global Environmental Observing System of Systems, and others, to shape and encourage expanded R&D on climate change technology development worldwide.

◆ Develop globally integrated approaches, such as the Asia-Pacific Partnership for Clean Development and Climate (Figure 10-8),[14] to foster capacity building in developing countries, encourage cooperative planning and joint ventures and, enable the development, transfer, and deployment of advanced climate change technology.

---

[12] The current bilateral agreements are: Australia, Brazil, Canada, China, Central America, Germany, the EU, India, Italy, Japan, Mexico, New Zealand, Republic of Korea, the Russian Federation, and South Africa. The countries included in the Central American agreement, apart from the United States, are: Belize, Costa Rica, El Salvador, Guatemala, Honduras, Nicaragua, and Panama.

[13] The countries are, in alphabetical order, Canada, France, Germany, Italy, Japan, Russia, United Kingdom, and United States. The G8 meetings often include the European Commission.

[14] The Asia Pacific Partnership for Clean Development and Climate was announced in July 2005. Six countries are participating, namely: Australia, China, India, Japan, South Korea, and the United States.

Figure 10-8. Asia-Pacific Partnership on Clean Development and Climate; Second Policy and Implementation Committee and Task Force Meeting, Berkeley, California - April 18–21, 2006, http://www.asiapacificpartnership.org.
Courtesy: Asia Pacific Partnership on Clean Development and Climate

Support Cutting-Edge Technology Demonstrations

◆ As part of the agencies' regular planning and budgeting processes, consider additional cutting-edge technology demonstrations relevant to CCTP strategic goals.

Ensure a Viable Technology Workforce of the Future

◆ Explore the establishment of graduate fellowships for promising candidates who seek a career in climate-change-related technology R&D.

◆ Explore possibilities of expanding internships related to climate change technology development in Federal agencies, national and other laboratories, and other Federally Funded Research and Development Centers (FFRDCs).

◆ Explore possibilities for establishing CCTP-sponsored educational curricula in K–12 programs related to climate change and advanced technology options.

Provide Supporting Technology Policy

◆ Evaluate technology policy options for stimulating private sector investment in CCTP-related research, development, and experimentation activities.

◆ Evaluate technology policy options for stimulating private investment in and adoption of advanced climate change technology and other GHG-intensity reducing practices.

◆ Support, as needed, policy-related activities undertaken by CCCSTI in furtherance of the Energy Policy Act of 2005.

◆ Evaluate various technology policy options for stimulating land-use and land management practices that promote carbon sequestration and GHG emission reductions.

In carrying out these activities, and in accord with its management structure (Chapter 2), CCTP will be advised by the CCTP Steering Group, assisted by its multi-agency CCTP Working Groups, informed by inputs from varied sources, and supported by CCTP staff and resources. Results will be conveyed to the CCCSTI via the IWG. The CCTP also plans to issue periodically reports on its current activities, future plans, and research progress.

10.3 Closing

The United States, in partnership with others, is now embarked on a near- and long-term global challenge, guided by science and facilitated by advanced technology, to address concerns about climate change and increasing concentrations of GHGs. This CCTP *Strategic Plan* is a first step toward guiding Federal investments in R&D to accelerate technologies that will address these concerns. The *Plan* will be updated periodically, as needed.

Appendix A

Federal Research, Development, Demonstration, and Deployment Investment Portfolio for Fiscal Years 2005 and 2006, with Budget Request Information for Fiscal Year 2007, U.S. Climate Change Technology Program

In order for the U.S. Climate Change Technology Program (CCTP) to carry out its mission, it is necessary to assess on a periodic and continuing basis the adequacy of Federal investments in the CCTP-relevant research portfolio and make recommendations. A first step in this regard is to establish and maintain a current inventory, or baseline, of all the Federal research, development, demonstration and deployment (R&D) activities among the participating agencies relevant to the vision, mission, and goals of the CCTP. This baseline, and subsequent years of data, can be used to identify and track trends and other changes in the portfolio. It also serves as an index or guide to relevant Federal R&D investments and programs.

In 2003, the CCTP, Office of Management and Budget (OMB) and other agencies agreed on a set of classification criteria to identify R&D activities that would be included as part of the CCTP. These criteria are provided on page A-2.

The baseline information for the Federal R&D budget shown in this Appendix are for Budget Authority as Enacted for Fiscal Years 2005 and 2006, and for the Administration's Budget Request for Fiscal Year 2007. For each year, respectively, the participating Federal agencies submitted budget data for R&D activities that meet the CCTP/OMB criteria. Table A-1 is a summary table for all participating agencies. This process is updated annually. Current versions of Table A-1 may be found at the CCTP Web site.[1]

This baseline activity and resulting portfolio contribute to and are consistent with the Congressional requirement that the President report annually on Federal climate change expenditures. The multi-agency R&D baseline for CCTP constitutes the technology component of OMB's Federal Climate Change Expenditures Report to Congress.[2]

Climate Change Technology Program Classification Criteria

Research, development, and deployment activities[3] classified as part of the Climate Change Technology Program (CCTP) must be activities funded via discretionary accounts that are relevant to providing opportunities for:

◆ Current and future reductions in or avoidances of emissions of greenhouse gases (GHGs),[4]

◆ Greenhouse gas capture and/or long-term storage, including biological uptake and storage;

◆ Conversion of GHGs to beneficial use in ways that avoid emissions to the atmosphere;

◆ Monitoring and/or measurement of GHG emissions, inventories and fluxes in a variety of settings;

◆ Technologies that improve or displace other GHG emitting technologies, such that the result would be reduced GHG emissions compared to technologies they displace;

◆ Technologies that could enable or facilitate the development, deployment and use of other GHG-emissions reduction technologies;

◆ Technologies that alter, substitute for, or otherwise replace processes, materials, and/or feedstocks, resulting in lower net emission of GHGs;

◆ Technologies that mitigate the effects of climate change, enhance adaptation or resilience to climate change impacts, or potentially counterbalance the likelihood of human-induced climate change;

---

[1] See http://www.climatetechnology.gov.

[2] Fiscal Year 2007 "Federal Climate Change Expenditures Report to Congress," April 2006. This report is an account of Federal spending for climate change programs and activities, both domestic and international. The report is provided annually to Congress.

[3] In this context, "research, development, demonstration, and deployment activities" is defined as: applied research; technology development and demonstration, including prototypes, scale-ups, and full-scale plants; technical activities in support of research objectives, including instrumentation, observation and monitoring equipment and systems; research and other activities undertaken in support of technology deployment, including research on codes and standards, safety, regulation, and on understanding factors affecting commercialization and deployment; supporting basic research addressing technical barriers to progress; activities associated with program direction; and related activities such as voluntary partnerships, technical assistance/capacity building, and technology demonstration programs that directly reduce greenhouse gas emissions in the near-and long-term.

[4] GHGs are gases in the Earth's atmosphere that vary in concentration and may contribute to long-term climate change. The most important GHG that arises from human activities is carbon dioxide (CO_2), resulting mainly from the oxidation of carbon-containing fuels, materials or feedstocks; cement manufacture; or other chemical or industrial processes. Other GHGs include methane from landfills, mining, agricultural production, and natural gas systems; nitrous oxide (N_2O) from industrial and agricultural activities; fluorine-containing halogenated substances (e.g., HFCs, PFCs); sulfur hexafluoride (SF_6); and other GHGs from industrial sources. Gases falling under the purview of the Montreal Protocol are excluded from this definition of GHGs.

- Basic research activities undertaken explicitly to address a technical barrier to progress of one of the above climate change technologies; and

- GHG emission reductions resulting from clear improvements in management practices or purchasing decisions.

Climate Change Technology Program Classification Example Activities

Specific examples of climate change technology activities include, but are not limited to:

- Electricity production technologies and associated fuel cycles with significantly reduced, little, or no net GHG emissions;

- High-quality fuels or other high-energy density and transportable carriers of energy with significantly reduced, little, or no net GHG emissions;

- Feedstocks, resources or material inputs to economic activities, which may be produced through processes or complete resource cycles with significantly reduced, little or no net GHG emissions;

- Improved processes and infrastructure for using GHG-free fuels, power, materials, and feedstocks;

- CO_2 capture, permanent storage (sometimes referred to as sequestration), and biological uptake;

- Technologies that reduce, control or eliminate emissions of non-CO_2 GHGs;

- Advances in sciences of remote sensing and other monitoring, measurement and verification technologies, including data systems and inference methods;

- Technologies that substantially reduce GHG-intensity, and therefore limit GHG emissions;

- Voluntary government/industry programs designed to directly reduce GHG emissions; and

- Programs that result in energy efficiency improvements through grants or direct technical assistance.

Note: Programs and activities presented for consideration can include Congressionally mandated "earmarks," but earmarked activities must be relevant to one or more of the CCTP criteria, and descriptions and funding levels must be clearly called out as such in the information provided. Programs and activities funded by mandatory authorizations should not be included.

CCTP Participating Agencies, Budgets and Requests

In the following budget table, data are provided on CCTP-related activities, per the criteria above, for Fiscal Years 2005 and 2006, and for the President's Budget Request for Fiscal Year 2007, across all CCTP participating agencies. In each FY, budget data includes activities for CCTP-related research, development and demonstration (R&D).

Table A-1 CCTP Participating Agency – FY 2005 to FY 2007 Budgets and Requests
Categorization of RDD&D Funding To Climate Change Technology
(Funding, $ Millions) [5,6]

DEPARTMENT AND ACCOUNT(S)	FY 2005 ENACTED	FY 2006 ENACTED	FY 2007 REQUEST
Department of Agriculture			
Natural Resources Conservation Service (NRCS) – Biomass R&D (Section 9008 Farm Bill)	13.0	12.0	12.0
– NRCS Carbon Cycle	0.5	0.5	0.5
Forest Service R&D – inventories of carbon biomass	0.0	0.5	0.5
Agricultural Research Service – Bioenergy Research	2.4	2.4	2.4
Cooperative State Research, Education and Extension Service (CRSEES) – Biofuels/Biomass Research; formula funds, National Research Initiative	4.7	4.7	3.4
Forest Service – Biofuels/Biomass, Forest and Rangeland Research	2.4	2.4	2.8
Rural Business Service – Renewable Energy Program and Value Added Prducer Grants	24.8	25.3	12.7
Subtotal – USDA	**48.2**	**47.8**	**34.2**
Department of Commerce - ITA			
International Trade Administration (ITA) - Asia Pacific Partnership	0.0	0.0	2.0
Subtotal – DOC/ITA	**0.0**	**0.0**	**2.0**
Department of Commerce - NIST			
National Institute of Standards and Technology (NIST) Scientific and Technological Research and Services	7.7	7.2	7.2
Industrial Technical Services – Advanced Technology Program	18.1	10.3	0.0
Subtotal – DOC/NIST	**25.8**	**17.4**	**7.2**
Department of Defense			
Army	27.0	36.5	5.5
Navy	18.1	23.4	6.9
Air Force	1.0	0.0	0.0
Defense Advanced Research Projects Agency (DARPA)	11.0	7.1	3.0
R&D, Office of Secretary of Defense	2.0	3.6	0.0
Subtotal – DOD	**59.1**	**70.6**	**15.4**
Department of Energy			
Energy Efficiency and Renewable Energy (EERE)	1,234.3	1,174.0	1,176.3
Fossil Energy	373.8	404.5	419.1
Nuclear Energy	291.4	332.5	463.3
Science	385.5	422.6	551.4
Electricity Delivery and Energy Reliability	57.4	73.0	100.3
Climate Change Technology Program[7]	0.0	0.0	1.0
Subtotal – DOE	**2,342.4**	**2,406.5**	**2,711.4**

5 This table is consistent with the Fiscal Year 2007 "Federal Climate Change Expenditures Report to Congress" prepared by the Office of Management and Budget (OMB), http://www.whitehouse.gov/omb/, and published in April 2006. Minor differences, if any, are due to arithmetic corrections after the OMB report was finalized and due to differences in rounding.

6 All agency data are current, as of April 2006. Totals may not add due to rounding.

7 In Fiscal Year 2005, $1.5M was enacted for CCTP Program Direction within DOE's EERE Program Direction account.

DEPARTMENT AND ACCOUNT(S)	FY 2005 ENACTED	FY 2006 ENACTED	FY 2007 REQUEST
Department of Interior			
US Geological Survey – Surveys, Investigations and Research - Geology Discipline, Energy Program	2.4	0.0	0.0
Subtotal – DOI	2.4	0.0	0.0
Department of Transportation			
Office of the Secretary for Technology – Transportation, Policy, R&D	0.8	0.0	0.0
National Highway Traffic Safety Admin	0.0	0.9	0.9
Research and Innovative Technology Admin	0.5	0.5	0.5
Subtotal – DOT	1.3	1.4	1.4
Environmental Protection Agency			
Environmental Programs and Management	90.5	90.0	91.9
Science and Technology	19.0	18.6	12.5
Subtotal – EPA	109.5	108.6	104.4
National Aeronautics and Space Administration[8]			
Exploration, Science & Aeronautics	207.8	104.4	85.8
Subtotal – NASA	207.8	104.4	85.8
National Science Foundation			
Research and Related Activities	10.6	17.7	18.6
Subtotal – NSF	10.6	17.7	18.6
Total for CCTP	2,807.1	2,774.4	2,980.4

ACTIVITIES ASSOCIATED WITH CCTP[9]

USAID Activities Associated with CCTP

	FY 2005	FY 2006	FY 2007
Energy Technology Development	80.1	92.0	57.3
Carbon Capture and Sequestration Measures	87.3	80.3	71.7
Subtotal – USAID	167.5	172.2	129.0
Department of State Activities Associated with CCTP			
Asia Pacific Partnership	0.0	0.0	30.0
Methane to Markets	0.8	6.0	6.0
Subtotal - STATE	0.8	6.0	36.0
Total CCTP and Associated Activities	2,975.3	2,952.6	3,145.4

[8] For Fiscal Year 2006 and Fiscal Year 2007, NASA is realigning its Aeronautics Research and is no longer pursuing previously reported activities in certain vehicle systems areas.

[9] STATE and USAID activities are not included in the totals for CCTP, as they are associated expenditures promoting deployment and adoption of climate change technologies abroad. They are shown here for completeness to the extent that such activities are consistent with the criteria for inclusion in CCTP.

Appendix B

National Climate Change Technology Initiative Priorities
Investment Portfolio for Fiscal Years 2005 and 2006, with Budget Request Information for Fiscal Year 2007, U.S. Climate Change Technology Program

CCTP continues to prioritize its multi-agency portfolio of Federally funded climate change technology R&D, consistent with the goals and objectives of the President's National Climate Change Technology Initiative (NCCTI). NCCTI priorities are defined as discrete research, development, demonstration, or deployment activities that address technological challenges, which, if solved, could advance technologies with the potential to dramatically reduce, avoid, or sequester greenhouse gas emissions. Activities are identified from within the larger CCTP portfolio as shown on Table B-1, are for Budget Authority as Enacted for Fiscal Years 2005 and 2006, and for the Administration's Budget Request for Fiscal Year 2007 by agency.

Table B-1 National Climate Change Technology Initiative Priorities
FY 2005 to FY 2007 Budgets and Requests
(Funding, $ Millions)[1]

AGENCY/ PROGRAM/ ACTIVITY	FY 2005 ACTUAL BUDGET AUTHORITY	FY 2006 ENACTED BUDGET AUTHORITY	FY 2007 PROPOSED BUDGET AUTHORITY	NCCTI PRIORITY ACTIVITIES DESCRIPTION
Department of Energy				
Energy Efficiency and Renewable Energy				
Hydrogen Storage	22.4	26.6	34.6	Addresses key challenge to advancing a hydrogen-based transportation system, which could substitute for oil and dramatically reduce GHG emissions. A major technological breakthrough is needed to be able to store enough hydrogen on board a fuel cell vehicle to provide a driving range comparable to today's vehicles.
Low Wind Speed Technology	9.9	5.0	19.1	Currently, wind power is only cost competitive in areas of high-wind speeds, which are relatively sparse and not near major load centers. Improving technologies to make wind power competitive in low-wind speed areas could expand this GHG-free power producer and displace (or reduce future need for) coal- and gas-fired electricity generation. Includes R&D on deepwater off-shore systems.
Solid State Lighting	13.8	19.3	19.3	Such lighting has the potential to double the efficiency of conventional lighting. Deployment could reduce GHG emissions and slow the growth of future base load electricity generation capacity, which will largely use coal.

[1] This table is consistent with the FY 2007 "Federal Climate Change Expenditures Report to Congress" prepared by the Office of Management and Budget http://www.whitehouse.gov/omb/ and published in April 2006. Minor differences are due to rounding.

AGENCY/ PROGRAM/ ACTIVITY	FY 2005 ACTUAL BUDGET AUTHORITY	FY 2006 ENACTED BUDGET AUTHORITY	FY 2007 PROPOSED BUDGET AUTHORITY	NCCTI PRIORITY ACTIVITIES DESCRIPTION

Department of Energy

Energy Efficiency and Renewable Energy

AGENCY/ PROGRAM/ ACTIVITY	FY 2005 ACTUAL BUDGET AUTHORITY	FY 2006 ENACTED BUDGET AUTHORITY	FY 2007 PROPOSED BUDGET AUTHORITY	NCCTI PRIORITY ACTIVITIES DESCRIPTION
Cellulosic Biomass (Biochemical Platform R&D)	11.1	10.4	32.8	The research focuses on converting complex cellulosic carbohydrates of biomass into simple sugars. Ultimately, this could lead to use of "waste" biomass to produce power, chemicals, and fuel, such as ethanol. Cellulosic biofuels can displaces fossil fuel products and have the potential to be nearly "carbon neutral" by cyclically capturing and releasing carbon dioxide, the main GHG, to the atmosphere.
Transportation Fuel Cell Systems	7.5	1.1	7.5	This activity works to incorporate fuel cells into vehicles—converting hydrogen into electricity and water vapor—directly displacing the burning of fossil fuels in vehicles.
EERE Sub-total	64.7	62.4	113.3	

Nuclear Energy

AGENCY/ PROGRAM/ ACTIVITY	FY 2005 ACTUAL BUDGET AUTHORITY	FY 2006 ENACTED BUDGET AUTHORITY	FY 2007 PROPOSED BUDGET AUTHORITY	NCCTI PRIORITY ACTIVITIES DESCRIPTION
Nuclear Hydrogen Initiative	8.7	24.8	18.7	This program aims to develop technologies that will apply heat available from advanced nuclear energy systems, in combination with power production, to produce hydrogen at a cost that is competitive with other alternative transportation fuels. Although it is but one of many hydrogen production methods, nuclear energy provides an emissions-free way to produce large amounts of hydrogen.
Advanced Fuel Cycle/Advanced Burner Reactor	0.0	5.0	25.0	Advances in nuclear fuel recycling can make nuclear power, which emits no GHG emissions, more attractive. The Advanced Burner Reactor (ABR) is a component of a multifaceted research program aimed at recycling spent nuclear fuel; reducing waste; promoting non-proliferation; and enabling the expansion of nuclear power—a GHG-free energy source. With ABR technology, the only waste to be placed in a repository is of a less challenging content, absent long-lived radioactive isotopes and other transuranics. One Yucca Mountain size repository would be able to accommodate the waste from many reactor-years of operation—a content that would fill as many as 21 equal repositories taking all that spent fuel directly.
NE Sub-total	8.7	29.8	43.7	

AGENCY/ PROGRAM/ ACTIVITY	FY 2005 ACTUAL BUDGET AUTHORITY	FY 2006 ENACTED BUDGET AUTHORITY	FY 2007 PROPOSED BUDGET AUTHORITY	NCCTI PRIORITY ACTIVITIES DESCRIPTION
Department of Energy				
Fossil Energy				
Sequestration	44.3	66.3	78.2	The continued use of fossil fuels, particularly coal, to generate electricity may be important to maintain both a diversified fuel mix and ensure adequate energy supplies at a reasonable price. A successful carbon sequestration research and development effort could allow the continued use of economical fossil fuels, while also limiting GHG emissions to the atmosphere.
Integrated Gasification Combined Cycle (IGCC)	44.6	55.9	55.6	Instead of burning coal, IGCC technology gasifies coal in such a way so as to enable the more efficient conversion of coal and other carbon-based feed-stocks into electricity and other useful products, providing the potential for over 50 percent reduction in CO_2 emissions, compared to today's more conventional combustion technologies. It also facilitates capture and sequestration processes.
FE Subtotal	89.0	122.2	133.8	
Climate Change Technology Program Direction	–[2]	0.0	1.0	The CCTP is the multi-agency planning and coordination activity, led by DOE, that carries out the President's climate change technology initiative and implements relevant climate change provisions of the Energy Policy Act of 2005. CCTP provides strategic direction, planning, analysis and multi-agency coordination for the participating Federal R&D agencies.
Total – DOE	162.4	214.4	291.8	
Environmental Protection Agency				
Methane Partnership Initiatives	9.0	10.0	13.0	Includes EPA's domestic partnership programs with industry, as well as the international Methane to Markets Partnership. These programs encourage development and deployment of technologies to reduce methane emissions and make a substantial contribution to achievement of the President's GHG-intensity reduction goal.
Climate Leaders	2.0	2.0	2.0	Climate Leaders is a set of flagship voluntary industry-government partnerships that encourage private entities to develop and implement long-term, comprehensive climate strategies, and set GHG emission reduction goals.
Total – EPA	11.0	12.0	15.0	
TOTAL – NCCTI[3]	173.4	226.4	306.8	

[2] In FY 2005, $1.5M was enacted for CCTP Program Direction within DOE's EERE Program Direction account.

[3] Totals may not add due to rounding. All Agency data are as of April 2006.

www.ingramcontent.com/pod-product-compliance
Lightning Source LLC
Chambersburg PA
CBHW080636180526
45168CB00008B/3190

* 9 7 8 1 5 0 8 4 9 9 9 2 3 *